THE ART OF SHOOTING GREAT ORDNANCE

A History of the Development, Manufacture and Use of Artillery, 1494–1628

Jonathan Davies

Helion & Company

Dedicated to
Thomas Davies and Timothy Jones

Who have devoted almost as much time to completing our falconet Elinor
as I have to this book

Helion & Company Limited
Unit 8 Amherst Business Centre
Budbrooke Road
Warwick
CV34 5WE
England
Tel. 01926 499 619
Email: info@helion. co. uk
Website: www. helion. co. uk
Twitter: @helionbooks
Visit our blog http://blog. helion. co. uk/

Published by Helion & Company 2022
Designed and typeset by Serena Jones
Cover designed by Paul Hewitt, Battlefield Design (www. battlefield-design. co. uk)

ISBN 978-1-915070-28-9

British Library Cataloguing-in-Publication Data.
A catalogue record for this book is available from the British Library.

For details of other military history titles published by Helion & Company Limited, contact the above address, or visit our website: http://www. helion. co. uk

We always welcome receiving book proposals from prospective authors.

Contents

Preface iv
Acknowledgements vi
Introduction vii

1. Medieval Firepower 11
2. The Gunner's Art 23
3. Artillery in Action 58
4. Designing the Gun 105
5. Guns from Forge and Foundry 143
6. Gun Carriages 175
7. Gunpowder 200
8. Gunfounders 218
9. Modern Reconstructions 227
10. Conclusion 245

Appendix I: Annotated Drawings 247
Appendix II: List of Artillery Required to Support an English Army 249
Appendix III: Lists of Artillery Dated 1535–63 250
Appendix IV: European Standardisation of Ordnance 251
Appendix V: List of Ordnance from Bourne's *The Arte of Shooting* 252
Appendix VI: English Ordnance *c.* 1590: John Sheriffe SP 12/242 253
Appendix VII: A Table of Proportion for the Weight and Shooting of Great Ordnance 254
Appendix VIII: List of Ordnance from *The Gunners Dialogue* by Robert Norton (1628) 256
Appendix IX: French Ordnance from The Art of Warre 257
Appendix X: Details of the Ordnance Supplies for the Invasion of France 1544 259
Appendix XI: The Artillerie more or lesse, that is used to be conducted with an Army 261
Appendix XII: The Properties, Office and Duetie of A Gunner 262
Appendix XIII: Useful Data and Formulae 265
Appendix XIV: Engravings on the Art of Casting Cannon, from Diderot's *Encyclopédie* 267
Appendix XV: The Field Guns Carriages of the Musée d'Art et d'Histoire, La Neuveville 274

Colour Plate Commentaries 279
Bibliography 287

Preface

Five years ago, I began a very foolish project to build a reconstruction of a small(ish) early sixteenth century cast bronze gun, together with a suitable field carriage, loading tools, gunsights and all the lesser paraphernalia associated with it. There has been a lot to learn, a process that began when my son and I took detailed measurements of two octagonal bronze falcons belonging to the collection held by the Royal Armouries. What soon became apparent was that these guns were designed upon a similar system of proportion, relating to the diameter of the bore. Thus began a very steep learning curve and the acquisition of much arcane knowledge.

Making the hardware was one aspect of the project; discovering how it was managed and served was quite another. It was important that the procedures that we used for serving this piece were as accurate as the tools that were so carefully reconstructed. If we were to present the gun and her crew to the public, we should understand how the gun was employed and what its capabilities were. We should be also able to answer the myriad technical and human questions that it would raise. This book is in part the result of that research.

I have not attempted to describe the contribution of artillery in the many battles on land and sea that took place in this long period; that deserves a separate and much longer study. The book does address itself, at least in part, to the concept of an 'artillery revolution' which has been dated quite precisely to the invasion of Italy by Charles VIII of France in 1494. I would like to propose instead the idea of an evolution in the design and development of 'new' artillery, from the last quarter of the fifteenth century to the first quarter of the sixteenth, after which any improvements were incremental until the major technological transformations of the mid nineteenth century. I have also not attempted when considering artillery at sea to deal with galley-mounted pieces but limited myself to more northern waters.

Much of the book is devoted to the 'knowledge' that a gunner must have of his piece. I have drawn heavily on a number of contemporary guides for the 'modern' artillerist which I fear are often repetitive with little original to add other than personal preferences. They all show a very natural concern for safety and reliability. In these manuals the examination of the individual piece is explained and the correct loading, in terms of both techniques and weight of powder and shot, are stressed. When it comes to aiming point-blank, they concur on practical methods of assessing this but then falter when presented with the difficult task of establishing the *randon* or maximum range of a piece. One of the fundamental problems they faced was that they shared an Aristotelian understanding of motion rather than a Galilean, as his experiments and discoveries in this area had little impact on contemporary ideas.

I hope to explain how artillery was used on land, in the field and at sieges, as well as at sea. Using the extensive trials of reproduction weapons as well as contemporary accounts and evidence, I have tried to establish what the effects of artillery were against different targets, a topic rarely discussed. I will not dwell on the horrible damage to human targets, an aspect of the subject which would only be of ghoulish interest. The issue of the rate of fire and the effective range that could be achieved on land and sea has led me to draw some conclusions that will perhaps surprise the reader.

One of the most difficult issues to address were establishing the principles underlying the design of artillery and whether a typology could be defined in a period of rapid technological progress. I found that a typology defined by purpose and proportion, rather than simply scale, was feasible. As with so much in design in this period relatively simple rules of proportion predominated and certainly by the end of this period, they seem to have gained general acceptance with some variations. I have described the manufacture of guns in forge and foundry although focusing rather more on the latter, as it was the cast gun that would come to predominate in this period. Having experienced the joys, perils and frustrations of casting my own cannon, I appreciate how great were the achievements of the gunfounders, who did not benefit from our technology and scientific understanding. I have included a short chapter on the history of some of the individual as well as 'dynasties' of gunfounders in England and on the Continent.

The design and development of the gun carriage on both land and sea is an important topic that seems rarely to be considered. I have tried to show, drawing upon a large number of contemporary illustrations, how and why gun carriages developed through this period, responding as they were to the changes in the design of the gun barrel itself. I have also included a short section on the limber which with the trunnion was one of the most important and least considered aspects of the 'artillery revolution'.

The manufacture of saltpetre, gunpowder and pyrotechnics were issues of great importance to the gunner, and I have given them suitable prominence. The humble gunner may not have been expected to make his own gunpowder, but he clearly had to understand how it was made and how to assess its qualities, for the sake of his own safety and the accuracy of his gun. I have added an important chapter on modern reproductions of artillery pieces and their trial results. These have been of enormous value in establishing the ballistic performance of representative pieces from this period. The trials have enabled an accurate and realistic assessment of the actual 'impact' of such weapons to be made for the first time.

I hope that the resulting book will be of interest to those intrigued by the history of ordnance and of use to those who still bravely serve black powder cannons.

Acknowledgements

Many people have helped me directly and indirectly in the preparation of this book. I would like to thank Dr Fabrizio Ansani for allowing me to share his research on the development of artillery in Italy. Dr Bastian Asmus has been very helpful in explaining to me the science and technology involved in the casting process. Peter Hansen has proved invaluable in providing images of the ground-breaking work carried out by the HO Group. Dr Fred Hocker of the *Vasa* Museum has been very generous in sharing his many insights in the practical use of cannons, drawing on the many lessons from the trials of the *Vasa* reproduction gun. Dr Axel Muller has as always been of great assistance in helping me understand the manufacture of gunpowder.

I would like to thank and commend the staff at the Royal Armouries who have provided me with access to their collection and the librarians of the Shakespeare Studies Centre in Stratford the Royal Artillery Museum as well as the archivists at the Musée d'Art et d'Histoire Neuveville, for their unfailing politeness and helpfulness. The Master of the Great Hospital in Norwich and especially Niki Tansley, were very generous in allowing me access to their splendid falconet for which I am most grateful. Pauline Asher has yet again painstakingly proofread the drafts, and any wayward or absent punctuation is my own responsibility. Charles Singleton of Helion has proved to be an editor with a light but deft touch, who has again guided me through the labyrinthine processes of publishing.

I would like particularly to thank the many members of the Ordnance Society who have generously made their research available through the Society's publications. I have drawn heavily and gratefully on their scholarship when preparing this book and I hope the result will not be a disappointment to them.

I fear that as always, this book is both too long and too short, and as always, I have relied heavily on the work of others to guide me. I have attempted in this book to summarise current knowledge and understanding. If I have done any a disservice or misrepresented the ideas and opinions of others, I most humbly apologise. My intention is to bring to a wider audience the excellent work that has been carried out by many experts, whose depth of knowledge and breadth of experience far exceeds my own.

Introduction

> The most experienced gunner might hesitate to predict whether when he next fired it his gun would send its shot directly to the target, drop it with a sort of discouraged burp a few hundred feet ahead, or blow up at the breech, probably killing him and his crew.[1]

Robert Smith,[2] one of the leading authorities on medieval and early modern artillery, has produced a withering condemnation of those who shared the perception of artillery expressed by Mattingly in the quotation above.

> Without any misgivings it is true to say that, certainly for the period before the seventeenth century, the early history and developments of artillery has been extremely badly served by historians. The resultant corpus of published work is, in the main, of inferior quality, poorly researched, often containing many inaccuracies, mistakes, assumptions and omission, the whole dosed liberally with guesswork and hindsight, and all too frequently perpetuating stereotypes and myths.[3]

He continues:

> Among the common modern theories of early artillery there is a basic and pernicious underlying assumption that it was basically inaccurate, largely ineffective, often unsafe, always difficult to transport, and suffered from deficiencies in standardisation.[4]

I hope that my analysis will not outrage Robert Smith and if it helps, I must say at this point that I do not share many of the pernicious assumptions he outlines above, and which were so clearly expressed by Garret Mattingly. It is certainly true that artillery in this period could not match the dreadful efficiency of the massed batteries of Napoleon or the broadsides of Nelson's

1 Garret Mattingly, *The defeat of the Spanish Armada* (London, Jonathan Cape, 1960), p.105. He was the winner of the Pulitzer Prize for this book, which was considered 'A first class history and a literary work of high order'.

2 Robert Smith, now Kay Smith.

3 Robert D. Smith, 'All Manner of Peeces: Artillery in the Late Medieval Period', *Royal Armouries Yearbook*, Vol. 5 (2002), p.130.

4 Smith, 'All Manner of Peeces', p.130.

navy. On the other hand, there was very little to distinguish the guns at Waterloo from those fielded by the English army at Pinkie Cleugh three and a half centuries before.

Robert Smith, in the same article where he castigates us sloppy historians, also draws attention to the dangers of adopting modern preconceptions relating to development and progress. As he points out: 'there was seldom, if ever, a direct and upward progress of changes nor a direct and easily definable series of improvements, each superseding the last, in a clear and rational striving towards an ideal'.[5] The history of the development of artillery is a confusing one made more so by the variations in terminology employed by individuals and states over time. There are general trends that can be identified, areas of undoubted progress and also occasions when apparently obsolescent technology was still being used and even manufactured. It is important to appreciate that those responsible for the making and use of armaments in this period were as acute, competent and creative as we are or are not. We should attempt to understand the decisions that they made, rather than deride them for their foolishness. The historian has to demonstrate humility when dealing with those he is studying, not the arrogance that benefits from hindsight and an awareness of 'superior' technology.

One issue that the author would like to address is the assumption that is so readily made that early ordnance was 'unsafe', 'leaky' and 'lacking in power'.[6] Anyone who has spent any time reading contemporary gunnery manuals, such as those of Bourne and Norton, would appreciate that the first duty of the gunner was to ensure the safety of his 'peece' and to establish what a safe powder loading was. It was in his interest to ensure that his gun shot straight and would not blow him up! The casting and forging of artillery were inevitably prone to serious flaws. This is made apparent when there have been modern attempts at reproducing these guns, even with the benefit of the most modern technology. Anyone who has seen the extraordinary achievements of the founders and smiths of this period cannot but be impressed by their technological prowess, as well as appreciating the true magnificence of these weapons of war as works of art. Although sometimes imperfect early guns were certainly not 'lacking in power' as demonstrated by the recent trials of recreated guns.

Although the official start date for this study is 1494 there was an obvious continuity in both materials and technology in the preceding decades. I have chosen the date of Charles VIII's invasion of Italy, as it has been identified as the beginning of a new period in European warfare and especially in the history of artillery. It is a convenient starting point because it has aroused such strong opinions. The end date of 1628 was chosen for two reasons. Firstly, because it was the year of publication of two of Robert Norton's guides to contemporary gunnery practice. Secondly, it was also the date of the tragic but for us fortuitous sinking of the Swedish warship *Vasa* with its modern armament of magnificent 24-pdr bronze guns. One of these has

5 Smith, 'All Manner of Peeces', p.130.

6 Smith, 'All Manner of Peeces', p.130a.

been recreated and test fired by the *Vasa* Museum team and has produced definitive results for the impressive performance of what was 'cutting-edge' technology.

The period between those dates was one of transition, innovation and exploitation. It was also one of continuity, with wrought iron guns continuing to play an important part in the mixture of armaments. Older weapons continued to be used long after it was thought they might have been scrapped and replaced. This in previous years has been seen as a source of criticism by those who had witnessed the replacement of tanks, artillery and aircraft at a bewildering rate between 1930 and 1960, but we are witnessing weapons and weapon systems still in active service for 30, 40 or, as in the case of the venerable Boeing B52, more than 70 years. We should find it easier to understand why older weapons were still in use long after it might be thought that they would have become obsolete.

The 'Lazy Maid' is a remarkable survival of a wrought iron gun that was remounted to continue its role as a weapon as well as a talisman for its city. Medieval guns were sometimes discarded and abandoned, as in the case of the *Michelettes* at Mont St Michel. Fortunately, in their case after centuries of neglect, they were honoured and restored. This gun is unusual in being continually in favour and protected from the scrap heap of history. (Von Jan Rehschuh – Eigenes Werk, CC BY-SA 3.0 <https://commons. wikimedia. org/w/index. php?curid=261172>)

The Folly of the 'Lazy Maid' and the Tears of the Baba Merzoug

One of the most unusual surviving pieces of artillery to more than span the period covered by this book is *Faule Magd* or 'Lazy Maid'. It is a large wrought iron bombard made around 1420 probably for the city of Leipzig. It was of the usual hoop and stave construction with a sub-calibre powder chamber. It has a barrel 2.205 m long and shot a stone projectile of 345 mm diameter weighing about 48 kg. It was, as customary, mounted on a wooden platform or bed. In 1511 it was thought worthwhile to take this obsolete weapon and fit it to a 'modern' field carriage. To enable the gun to be elevated a pivot was made to the front of the wooden bed. Why was the expense thought worthwhile? Was it because of its formidable stone hurling capacity or had it become a talisman defending the city? It is noticeable that as much care was taken in the decoration of the carriage as would normally be taken with the decoration of the barrel of a bronze gun and it was painted in its city's colours. It was clearly a weapon which was held in high regard. It was also not merely a silent trophy, as it was reported that *Faule Magd* was used by Leipzig in its defence against the Swedes in 1637 during the Thirty Years' War. It has been preserved for successive centuries. Individual guns could undoubtedly develop a 'personality' and status of their own, as did *Mons Meg* in Scotland, virtually from its arrival.

In Algeria they are seeking the return of *Baba Merzoug* or 'Blessed Father', a 12 tonne and seven-metre-long barrel cast in 1541 with the aid of a renegade Venetian gunfounder. For three centuries she protected the city of Algiers, in 1683 being employed to fire the slim-hipped French consul, one Jean le Vacher, towards the French fleet. She was seized as a trophy by the French in 1830, who with black humour christened her *Le Consulaire* and mounted the piece vertically, with a crowing gallic cockerel now placed atop. The gun was placed in the naval port of Brest where she remains in exile, despite calls for her return from Algeria. The French have refused, with Michèle Alliot-Marie the Defence-Minister declaring in 2006 that it was 'part of the French military's heritage' to which 'it', presumably France, was 'very attached'.

Belkacem Babaci, a historian who has long called for the return of this iconic gun, recorded his own thoughts and emotions on the 'Blessed Father':

> Since I was 12 my grandfather told me stories of a legendary cannon. These stories inspired me to conduct research and study history ... When I touched the cannon, I felt moisture on my hands. In my imagination they were tears ... Baba Merzoug takes us back to the age when Algeria was dominating the Mediterranean. This cannon tells about the real history of Algeria.[7]

A cannon can represent many different things over its long life and arouse many strong emotions.

7 <https://www. bbc. co. uk/news/world-africa-54432571>, accessed 22 September 2021.

1

Medieval Firepower

The Early Years

The first definitive depiction of a cannon is to be found in the Holkham and Milemete manuscripts dated to 1326–1327. The earliest cannon that exists is the Loshult cannon, found in Loshult Skåna in Sweden in 1861 and dated tentatively to around 1340. It is a very small but very loud gun of cast bronze and originally like the Milemete guns shot a garros or gun arrow. Early guns seem to have been relatively small for several reasons, including the cost of saltpetre, the principal constituent of gunpowder, and the cost of materials, especially bronze.

The determining factor in the use of firearms of all sorts was the availability and price of saltpetre sometimes referred to as the 'master'. The two other constituents, charcoal that provided the fuel and sulphur that moderated the process, were already available, although sulphur was obtainable from only a few European sources. Saltpetre was on the other hand an expensive import. In 1347, for example saltpetre was purchased at 18 pence a pound from a spice merchant called William Staines on behalf of the English King's Privy Wardrobe.

Saltpetre could be collected and refined from naturally occurring efflorescences to be found in stables and privies, in fact wherever waste matter was found. What would make gunpowder much cheaper was the domestic manufacture of saltpetre in saltpetre *plantations*. The consequence was a massive reduction in the cost of gunpowder from 41 florins per hundredweight to 16 on the market in Frankfurt between 1381 and 1410. In England the price fell by more than three quarters between 1385 and 1417. In addition, the use of liquid, usually water in the combination of the three ingredients, produced a 'dumpling' of powder that was far less prone to damp and when crushed produced a more powerful powder than the dry mixture that that had been used before.

The availability of cheaper gunpowder meant that larger quantities could be used in larger guns and wrought iron guns were manufactured in greater number and increasing size. The development of the Catalonian furnace meant that blooms of 300 kg could now be produced rather than the 15 kg previously. Wrought iron was more plentiful and much cheaper than

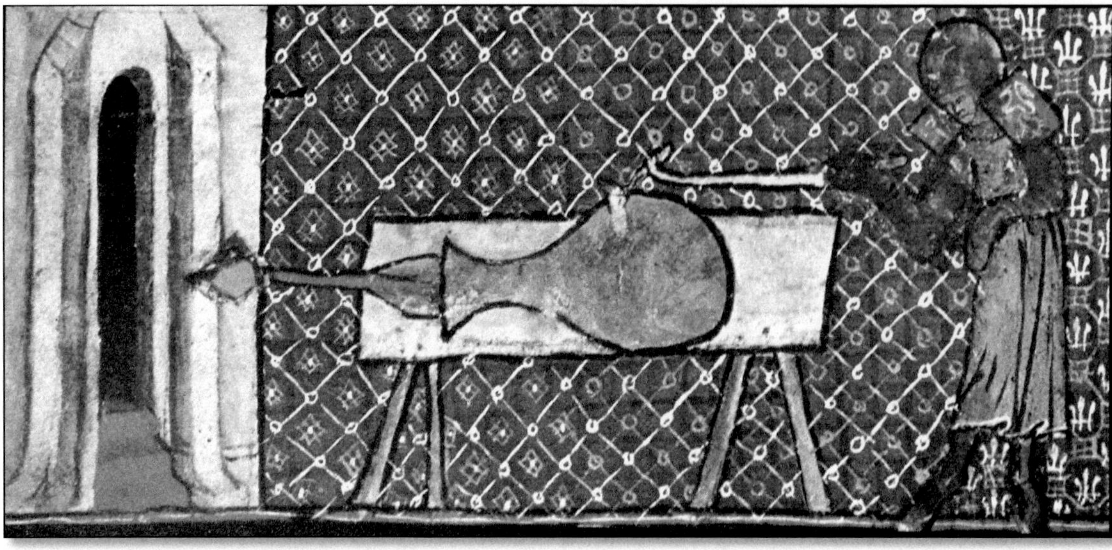

The illustrations from the Milemete Manuscript have been subject to various interpretations. It seems likely that they were in reality smaller than those images. A 'full-scale' reproduction weighed in at 800 kg and even then, only had a bore a few millimetres larger than the Loshult gun weighing 10 kg. Although the Milemete guns were clearly of cast bronze, the use of 'pots de fer' to describe other pieces would suggest that bronze and wrought iron pieces served side by side. Manuscript illustrations show both materials being used throughout the Middle Ages, the lack of surviving cast bronze guns from the early period need not necessarily suggest that they were not there: more likely they were recast either as swords or ploughshares! (De Nobilitatibus Sapientii Et Prudentiis Regum Manuscript, Christ Church MS 92 fol. 70v. Wikimedia Creative Commons World Imaging)

expensive bronze. The attractions of wrought iron as a material for the hoop and stave-built guns are clear:

> Wrought iron has a very low carbon content of less than 0.1 percent, fibrous slag inclusions which make up about two percent of the mass give it a 'grain' and assist the process of heat welding and resists corrosion. Wrought iron can be heated again and again and reworked. The material is ductile with a strong tensile strength making it an ideal material for the manufacture of artillery as it is easily worked and strong. If the metal is brought into the range 1,150–1,350 degrees centigrade then by hammering it can produce almost perfect welds as the slag inclusion melts and covers the join, which stops the oxidation process which would make the weld fail.[1]

Hoop and stave manufacture was suitable for a large barrel, but the breech, which was subject to higher pressures, had to be reinforced or made of a single piece of welded wrought iron. The weakest point was the joint between the barrel and the breech. In large guns where the breech was integral it was joined to the barrel by a tongue and groove arrangement. This was the point most likely to fail, as it did with *Mons Meg*, which is why the

1 Jonathan Davies, *The Medieval Cannon 1326–1494* (Oxford: Osprey, 2019)

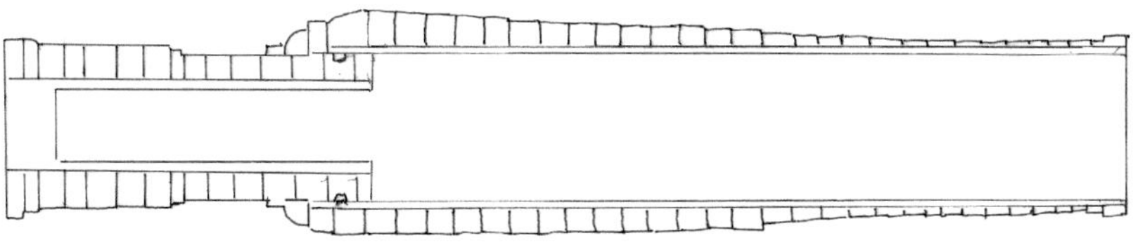

This cross-section of *Mons Meg* demonstrates the key features of the medieval wrought iron gun, if on a massive scale. The sub-calibre powder chamber was heavily reinforced, while the barrel is much thinner as the pressure there will be much lower. In effect the barrel only really acts to direct the projectile. The junction of the powder chamber with the barrel is heavily reinforced. This is where high gas pressures emerge from the gun to be 'impeded' by the projectile, before pushing it into the much larger space of the barrel where the pressure will be much reduced. (Author's drawing).

gun was reinforced with extra thick rings or hoops at this point. This also produced the characteristic shape of the large bombard.

The key to the success of the wrought iron gun was in its sophisticated construction and powerful propellant but also ironically in its principal ammunition. Stone was much lighter than the equivalent weight of iron and required the development of less pressure to force it down the barrel. The high-pressure breech was strongly constructed, the barrel was the low-pressure section of the weapon. In the case of the Middelaldercentret Cannon trials the barrel pressures were characteristically less than a third of the breech pressures. The result of this combination of materials and propellant was a weapon that could be constructed on the same principle to almost any dimension and which was safe and would shoot a stone projectile at a much higher velocity than any traction, torsion or counter-weight machines. Smaller wrought iron pieces were strong enough to shoot projectiles of lead and in the sixteenth century iron, but the larger guns almost exclusively shot stone balls.

Wrought iron was a wonderful material but even better but more technically challenging was bronze. There is an assumption that bronze was used for the earliest vase-shaped cannons but was rapidly superseded by cheaper wrought iron. Iron was certainly used from the earliest period; Edward III employed five iron as well as five copper cannons, in the defence of Cambrai in 1339. Bronze was far more expensive than iron; in London in 1375 iron was sold at 2½ d per pound and bronze at 8 d. Even so, the casting of bronze guns of increasing size continued throughout the Middle Ages. So strategically valuable was copper that in 1354 King John of France banned its export. One of the attractions of bronze was that ploughshares could be turned into swords, as when in 1414 William Woodward and Gerard Sprong cast guns from the pots and pans in the King's kitchens. No doubt these guns were quite small, as was certainly the case of the 49 guns that were melted down to cast a single gun of 3,904 lb given to the Count of Saint-Pol in 1433. This was by no means the largest bronze gun of the period.

Although bronze and wrought iron were very different materials, the gun made with them used the same ammunition and the same design, both employing a sub-calibre powder chamber. Bronze guns could have reinforcing bands on the barrel, although unlike the wrought iron guns where they were an essential part of the structure on a bronze gun, they were decorative rather than functional. The overall design, with a thickening between the powder chamber and the barrel was also mimicked, again to little purpose in bronze guns. By the 1480s the bronze foundry could cast weapons of considerable size with great skill, what was needed was an innovation in design to take advantage of the material and the skill of the founders.

The major innovation in artillery took place in the last quarter of the fifteenth century. This was not a change in the bronze but a change to the internal architecture of the gun and its use of new cast iron ammunition. The large calibre barrel with a sub-calibre chamber would now be replaced by a constant diameter tube, (with some variations in the end of the breech shape). The stone ball would be replaced by an iron one and the much higher pressures induced could be managed by the use of cast bronze, exploiting the full potential of this wonderful material. This new weapon did not supersede all the others, in practice wrought iron guns continued in use on a very large scale. In addition, some bronze guns continued to be cast with sub-calibre chambers and thin-walled barrels, for shooting stone balls. Even the new bronze guns could be divided into two complementary types, a long-barrelled narrow bore gun for longer-range and field work and a shorter-barrelled larger bore gun for close and siege work.

1494 – A Turning Point?

Geoffrey Parker has identified the moment when the military world was turned upside down with the new artillery of Charles VIII:

> The catalyst of major change was the French invasion of the [Italian] peninsula in 1494–1495. Charles VIII brought to Italy an army of 18,000 men and a horse-drawn siege-train of at least forty guns. Even contemporaries realised that this marked a new departure in warfare. In 1498 the Venetian Senate declared that 'the wars of the present time are more influenced by the force of Bombards and artillery than by men-at-arms' and frantically set about acquiring arms.[2]

This revolution is apparent in the illustration of the French army entering Naples in 1494. It shows horse-drawn field pieces followed by light wagons loaded with 'pale de fierro', and 'sache de farina', which were undoubtedly iron shot and ready-made cartridges. Following the guns and limbers is a pack train of animals, loaded with casks of what one can assume was powder and more shot. What we have here in all its essentials is a powerful train of mobile artillery, capable of rapid movement both to and on the battlefield and prepared with a large amount of ready-use ammunition for rapid, accurate and sustained fire. The artillery train that Charles had in Tuscany in 1494 included at least eight cannons, four large culverins and 30 demi-culverins, all cast in bronze.

Francesco Guicciardin in his oft-repeated description of the new artillery, provided a brilliant summary of the state of artillery before 1494 and the very tangible impact that such weapons had on Italy, during the French campaign:

2 Geoffrey Parker, *The Military Revolution: Military Innovation & the Rise of the West, 1500–1800* (Cambridge: Cambridge University Press, 1996), pp.9–10.

Two of the French guns that accompanied Charles VIII to Italy. The gun on the left is clearly a wrought iron piece, while its companion is a fine bronze gun. The bronze piece has a complex faceted barrel, while the cascable is in appearance very similar to the guns still to be found at Neuchatel. Wrought iron guns could have trunnions fitted to them on an additional collar, enabling them to accompany and complement bronze guns. Both are mounted on up-to-date carriages, the one on the left having folded trails to which a horse could be put. There was much debate as to the disadvantages or benefits of a long or short 'tail'. (Fasciculus temporum Italy, Naples, *c.* 1498 MS M. 801 fol. 122r, The Morgan Library. (Author's drawing)

The largest pieces of the artillery we are speaking of, were called bombards, and from that time were spread through Italy, and made use of in sieges. Some of them were made of iron, and some of bronze, but they were of so large a size, that, on account of the little experience of the gunners, and clumsiness of their carriages, they were moved from place to place very slowly, and with great difficulty, and for the same reason were very unhandy when placed against the walls of a town. The intervals between the firings were so long, that a great deal of time was lost, and little progress was made in comparison to what we see these days. This gave time to the besieged to cast up ramparts and fortifications behind the breaches at their leisure. But, notwithstanding all these impediments, the violence of the saltpetre, of which gunpowder is made, was such that, when these weapons were fired, the balls flew with so horrible a noise, and stupendous force, even before they were brought to their present perfection, that they rendered ridiculous all the instruments, so much renowned, invented by Archimedes and others, and used by the Ancients in sieges of towns.

But now the French brought a much handier engine, made of bronze, called a cannon, which they charged with heavy iron balls, smaller without comparison than those of stone made use of heretofore, and drove them on carriages with horses, not with oxen, as was the custom in Italy. And they were attended with such clever men, and on such instruments appointed for that purpose, that they almost always kept pace with the army. They were planted against the walls of a town with such speed, the space between the shots was so little, and the balls flew so quick, and were impelled with such force, that as much execution was done in a few hours, as formerly, in Italy, in the like number of days. These, rather more diabolical than human instruments, were used not only in sieges, but also in the field, and were mixed with others of smaller size. Such artillery rendered Charles's army very formidable to all Italy.[3]

3 Francesco Guicciardini, *The History of Italy* (Princeton: Princeton University Press, 1984), pp.147–149.

Paolo Giovio described the movement of the artillery train through Rome. Here it was deployed successfully to intimidate the Pope, who had immured himself in the Castel St Angelo:

> More than thirty-six guns, mounted on carts [carriages], induced astonishment and fright among everyone. They were drawn by horses, at incredible velocity, over flat ground and on slopes. The biggest of these, two and a half metres in length and two thousand kilograms of bronze in weight, were named cannons, and fired an iron ball as large as a man's head. After them followed the culverins, half as long again, but with a narrower barrel and a smaller shot. Falcons of different dimensions ended the parade. The smallest of them threw a ball similar in size to an orange ... The pace of cavalry could even be equalled by artillery carthorses, across plains, due to the tremendous skill of carters and masters.[4]

The powerful impression given by the artillery was more than enough to intimidate the Pope and the population without firing a shot. As Marin Sanudo explained, the 'French soldiers do not draw their swords for fighting, but only for scaring'.[5]

The key features of this new artillery could be summarised as follows:

- That they used lighter, more manoeuvrable bronze guns (rather than built up wrought iron pieces or cast bronze bombards)
- Iron balls were substituted for stone (iron has three times the density of stone) and shot at higher velocities
- They were hauled and fired from their own gun carriages (rather than carried, dismounted and placed on wooden beds)
- The use of trunnions had become common: these simple devices allowed a gun to be laid quickly and accurately using a quoin or wedge between the barrel and a transom in the trail of the gun
- Horses were used as draught animals in contrast to the oxen preferred in southern Europe
- The consequences were that such weapons gave artillery much greater firepower and increased mobility

Christopher Duffy in his study of siege warfare strikes a note of caution when considering these events:

> Machiavelli, Guiccardini and almost all people who have written since about Renaissance warfare are surely right to stress the revolutionary impact of the French and their new artillery. What the authorities are talking about was

4 Giovio, 'Le historie del suo tempo' (Venice: Appresso Domenico de' Farri, 1555), 59rv. In Fabrizio Ansani, '"This French artillery is very good and very effective". Hypotheses on the Diffusion of a New Military Technology in Renaissance Italy', *The Journal of Military History* 83 (April 2019), p. 361.

5 Sanudo, *La spedizione di Carlo VIII in Italia*, in Ansani, 'French Artillery', p.362.

> essentially a Blitzkrieg which depended as much for its effect upon speed, energy and the potential for destruction, as the actual scale of physical damage.[6]

Robert Smith and Kelly DeVries in their magisterial study of Burgundian artillery share Duffy's scepticism when they write that:

> The mistake that Guiccardini made, which has been followed by all those who have relied on his statement … was to attribute French success to revolutionary types of gunpowder weapons, iron cannonballs, and light, moveable carriages drawn by horse. In fact, none of these were new or revolutionary in 1494. The French gunpowder weapons were the same as those which had been in existence for half a century'.[7]

Their knowledge and expertise are unmatched, and they are undoubtedly correct in their chronological analysis of the technology. From the third quarter of the fifteenth century the use of trunnions had become common. Cast iron projectiles were readily available from the 1460s and bronze guns were certainly part of the equipment of Charles the Bold. There is also much to support the argument that French success in sieges can be attributed to a number of factors other than the availability of this 'revolutionary' artillery.

There is a strong argument made by Simon Pepper that the early French successes had little to do with their use of artillery.[8] The strategically important and powerful fortresses of Sarzana and Sarzanello were bypassed, as Florence abandoned any thought of resistance. The massacres of garrisons and civilians that took place at the capture of Favizano, Pontremoli, Mordano and Montefortino did more to ensure the collapse of resistance elsewhere than the use of artillery. When the French faced the partially modernised Neapolitan defences of Castelnuovo, they expended much of their stock of ammunition without achieving any significant damage, despite employing a battery of up to 70 guns. The castle surrendered only after its own gunpowder magazine exploded, probably accidentally. When the French held the same castle and were besieged by a substantial Aragonese siege train they held out for five months. It was only after resupply by sea had become impossible and the barbican outwork had been devastated by a gunpowder mine that they surrendered. On both occasions, artillery seem to have proven indecisive in determining the outcome.

6 Christopher, Duffy, *Siege Warfare* (London: Routledge Kegan Paul, 1979), p.11.

7 Robert Smith, Kelly DeVries, *The Artillery of the Dukes of Burgundy* (Woodbridge: The Boydell Press, 2005), pp.43–44.

8 Simon Pepper, 'The Face of the Siege: Fortification, Tactics and Strategy in the Early Italian Wars', in Christine Shaw (ed.), *Italy and the European Powers: The Impact of War, 1500–1530* (Leiden: Brill, 2006).

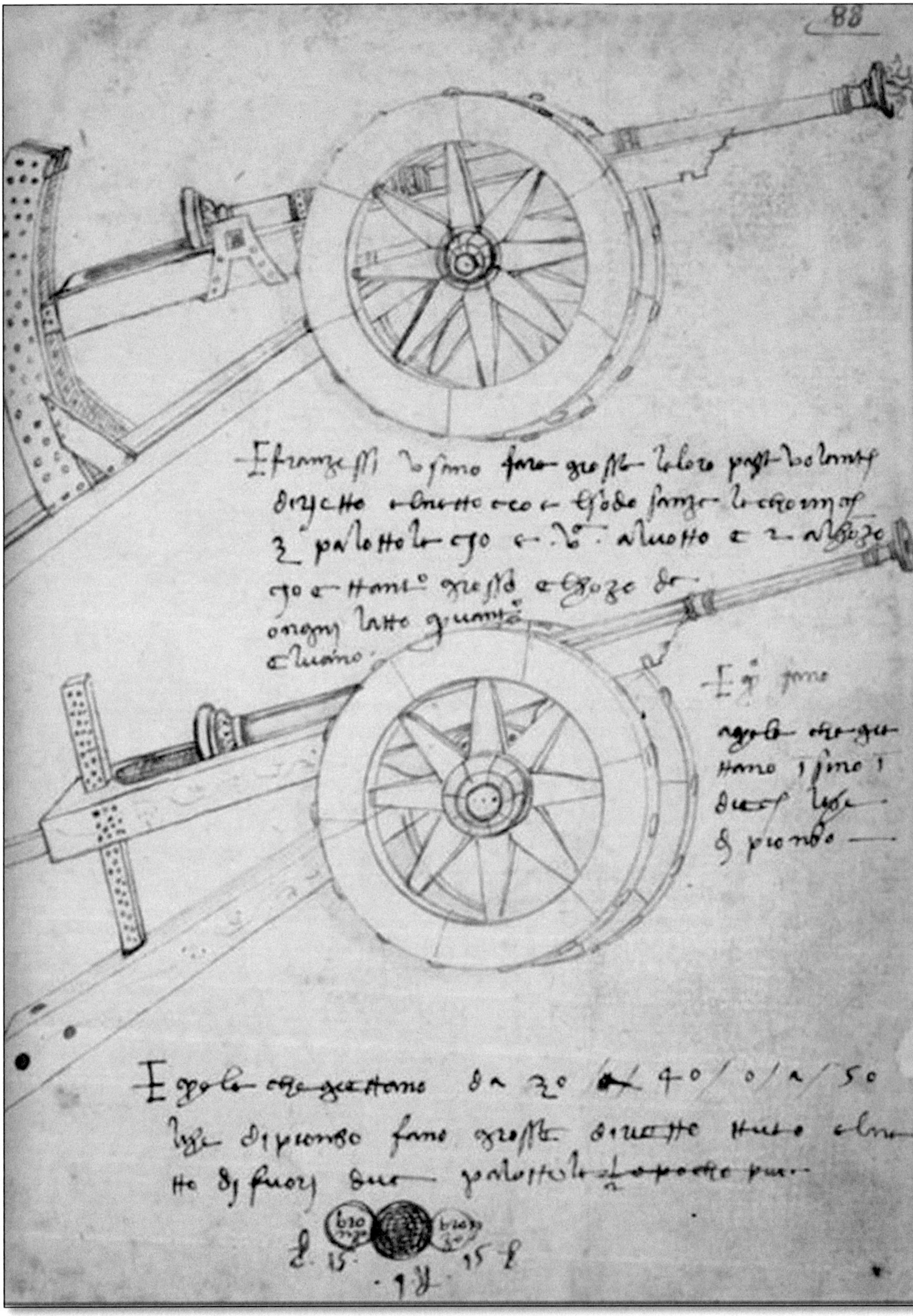
88

Facing page: The *Zibaldone* was a collection of drawings made by Bonaccorso Ghiberti (1451–1516) and is generally dated to the 1480–1490. The notebook contains drawings of bell and cannon casting, as well as technical drawings of systems of levers and screws. The lever, wedge, pulley, screw, and windlass were the five simple machines described by Heron in Roman Alexandria and were the playthings of Renaissance mechanicians. The guns in the illustration on the facing page are transitional types. Clearly of cast bronze but without trunnions, which required them to be mounted in wooden beds in the same manner as wrought iron pieces. The elevating system with a curved or straight arc was clearly the detail Ghiberti was interested in. The large gap after the flat breech in the mounting would suggest that they were breech-loading guns, again a system popular with wrought iron pieces. Of course, they need not be illustrations of real guns, merely a designer's doodle! (Banco Rari 228, in BNF, 88r. With the permission of Ministero dei Beni e delle Attività Culturali e del Turismo, Biblioteca Nazionale Centrale di Firenze)

Mallett and Shaw in their study of the Italian wars have pointed to the division between and within Italian states to explain the failure to resist the advancing French/Swiss force. Florence, Milan, Sienna and the Papacy were either collaborators, neutral or the victims of what seemed an unstoppable army. Alfonso of Naples and his successor possessed neither the loyalty of their people nor their nobility and although the defences of Naples held out for a short period their cause was lost before the French arrived. Isabelle d'Este wrote to her husband that the collapse of the Aragonese monarchy in Naples 'should be an admonition to all rulers to esteem the hearts of their subjects more than fortresses, treasure and men-at-arms, because the discontent of subjects wages worse war than the enemy in the field'.[9]

The Italians experienced a strategy of 'shock and awe' at the hands of the French. At Monte San Giovanni, one of the few fortresses loyal to Alfonso rejected the demand for surrender and mutilated the French envoy. In a few hours the fortress was stormed and in the following hours a massacre ensued:

> The garrison of Monte San Giovanni numbered three hundred men. Having confidence in the thickness of walls, and disbelieving the force of French artillery, they dared to harm the king's emissary … This act inflamed Gallic pride. Without any delay, soldiers assaulted the castle, driven by rage. In short order, they charged in. Defenders were cut to pieces … The castle was sacked and burned. This bitterness terrified the surrounding villages. Considering the cruel fate of Mordano and Montefortino, everybody believed that neither barrier nor warriors could stand up to the French army and the fury of its artillery without severe damage.[10]

Guicciardini reported that this use of terror was a strategic ploy to discourage further opposition, in which it succeeded. It was barbarity rather than bombardment that struck him as being a new obscenity in the history of Italy:

> This method of making war, not used in Italy for many centuries, filled all the kingdom with the greatest terror, because in victory, in whatever way it had been acquired, the farthest point of the cruelty of the victors had ever gone was to plunder and then free the captured soldiers, to capture towns captured by assault and take

9 Michael Edward Mallett and Christine Shaw, *The Italian Wars 1494–1559: War, State and Society in Early Modern Europe* (London: Routledge, 2012), p.26.

10 Giovio, 'Le historie del suo tempo', 70rv. In Ansani, 'French Artillery', p.363.

> the inhabitants prisoner in order that they would have to pay ransom, but always sparing the lives of men who had not been killed in the heat of the combat.[11]

The implications for Italy of this artillery Blitz were obvious to all those who had witnessed it from near or far. The historian Fabrizio Ansani summarised the lessons that were learnt:

> The rapidity and the efficiency with which French gunners could move, aim, and fire their guns represented an absolute, frightening novelty. The practice of positioning numerous guns near the walls, the simultaneity of the explosions, the saturation bombardments, the use of iron cannonballs, and the immediate infantry assaults were unprecedented in Italian Renaissance warfare. According to de Commines, Venetian senators were incredulous about the French way of seizing towns, which was simply 'new, in Italy'.[12]

The response was immediate with both the allies and enemies of France seeking to update their ordnance through urgent programmes of re-equipment.

In March 1495 the Florentine government recorded that: 'this French artillery is very good and very effective'; clearly something needed to be done.[13] The alliance between Florence and France in 1495 led to close inspection of French artillery by Francesco di Bartolomeo Telli and Lorenzo di Giovanni on the orders of the Florentine state. They produced drawings and measurements which would be used to design their new artillery. In Florence an entirely new foundry was ordered to be built in January 1495. The first cannon, *ala franzese* (in the French style), was cast and was sent directly to the Pisan front line in March of the same year, a commendable effort. In February of 1495 another potential technological transfer took place when a Pistoiese merchant presented to the Florentine state several patterns for 'French' guns:

> Here is a man who has beautiful models for making various kind of guns. He has acquired them from some Frenchmen. These models are perfect for crafting bombards, passavolanti, mortars and other pieces. They are fourteen or fifteen pieces. We could buy them for a reasonable price. I exhort your lordships to search for this man, and your lordships could see these models and grasp if they are just what you need. I think so.[14]

---

11 Guicciardini, *The History of Italy*, p.73.

12 Ansani, 'French Artillery', p.364.

13 Fabrizio Ansani, 'The Life of a Renaissance Gunmaker: Bonaccorso Ghiberti and the Development of Florentine Artillery in the Late Fifteenth Century', *Technology and Culture*, July 2017, Vol. 58, p.760. Dieci di Balìa, Missive, in Archivio di Stato di Firenze (henceforth ASF), 32, 79rv: 'Queste artiglierie franzesi sono molte buone et fanno grandi effecti'.

14 Ansani, 'The Life of a Renaissance Gunmaker', p.761. Dieci di Balìa, *Responsive*, in ASF, 38, 244r

The Venetians were well aware of the capabilities of the French artillery as their ambassador Zaccaria Contarini had already provided a detailed report before the invasion.

> The [French] king's bombards fire about thirty kilograms of iron shot. They are mounted on small carts, and, thanks to an admirable system, they can shoot easily, without any other support. It is said that Charles has countless spingards mounted on carts. The French use these guns in two ways. When the army encamps, the soldiers group these carts into an impenetrable fortification. When the army besieges a city, or a castle, these pieces demolish walls much more smoothly and, in less time, than our heavy bombards. People say that more than thirty thousand horses were necessary for the transportation of the ordnance of Louis XI. Now, the monarch [Charles VIII] takes twelve thousand horses with him.[15]

The Dukes of Ferrara who always had a personal interest in the development of artillery were able to take advantage of their alliance with the French for the transfer of expertise. Giovanni Alberghetti together with other Ferrarese gunfounders cast dozens of long-barrelled culverins between January and November 1495 at the Castello Estense foundry. Similar exertions in this field took place in Florence, Milan, Venice, and the Papacy as the peninsula re-armed itself. The rapid spread of the new technology gave no individual state a significant advantage. In 1499 a Florentine army besieged Pisa and demonstrated how much they had been taught by the events of 1494.

The Florentine army numbered about 15,000 infantry and cavalry. They also carried 80 guns. Culverins and cannons, mounted on carts, accounted for 50 pieces and there were seven heavy bombards. Florentine gunners fired about 200 shot per hour against the city walls and in the city. The entire world seemed to be getting destroyed.[16]

The lessons that the French had taught had been learnt well. As Fabrizio Ansani records, the technology could be applied so quickly because of the advanced state of Italian industry and commerce.

The technical transformation was possible thanks to the steady improvements matured in the broader, receptive milieu of the Italian Renaissance. And the technological 'bursts', above all, would have been only a part of the changes in the political landscape in Italy.[17]

Excluding the guns held in fortifications the French king could dispose of 150 artillery pieces but the number of 'barrels' was not the whole story. As impressive as the quantity and quality of the materiel, was the organisation that underpinned it. There were two Grand Masters of Ordnance, Jean de La Grange and Guyot de Louviers. They were ably supported by a lieutenant, and a captain of supply together with a Treasurer and commissioners responsible for the purchase, payment and storage of ordnance. The Ordnance was divided into five regional *bandes* in Lower and Eastern Brittany, Brabant, Western Flanders, and Normandy. Attached to these were the carters and

15 Ansani, 'French Artillery', p.359.

16 Portoveneri, 'Memoriale', pp.341–342, in Ansani, 'French Artillery', p.374.

17 Ansani, 'French Artillery', p.378.

carpenters, blacksmiths and foundrymen who could be supplemented in wartime but in peace provided a cadre of experience and expertise. This infrastructure was essential to sustain what was a highly expensive and specialist organisation. The problems that Italy faced were considerable. The infrastructure and expertise that made the French assault possible was initially lacking. Even the new strengthened gun carriages struggled on her mountainous roads, a situation exacerbated by a shortage of draught horses and a continued reliance on oxen. Casting cannons was one thing, sustaining such an elaborate martial construct as a siege train quite another.

In the light of the apparent and rapid developments in artillery there was surprisingly little innovation in the area of fortification. There had only been five sieges in total during the invasion and none had been especially arduous. Incessant warfare meant that Italian cities, towns and villages were almost universally in a permanent state of defence. Yet as Dr Ansani records:

> The French artillery, in fact, did not have an immediate impact on the construction of fortifications. Military engineers resorted to the traditional 'anti-gunpowder defences' experimented with throughout the course of the fifteenth century, that is, the reduction of the height and the thickening of walls, the increase in the number of gunports, and the building of ravelins. In the Tuscan documents of the time, the term 'bastion' still meant a level space fortified with wooden structures and numerous guns, located on the outside of the town perimeter. The customary round towers continued to proliferate until the third decade of the sixteenth century, while the bastioned front was still evolving from traditional forms.[18]

The artillery pieces that Charles VIII took to Italy in 1494 were the product of a culmination of French technological and administrative progress over the previous decades. They did have a profound impact, but this was on Italian morale rather than on their fortresses and city walls. They were 'terror' weapons. What was little understood but soon came to be obvious, was that the hardware alone was not enough to grant victory, especially as the technology spread rapidly. Guns without competent gunners, carpenters, carters, horses and pioneers were little more than decorative items suited only for a ducal courtyard. Effective artillery was the manifestation of a powerful state that had to commit many men and much coin to its development and maintenance, long before it could be brought into action.

18 Ansani, 'French Artillery', p.377.

2

The Gunner's Art

Zurich was in need of a specialist to manage their ordnance. Master Gunner Spross laid out his competencies in his letter of application to the Zurich council. He declared that he was more than able to conduct, for example:

> Reconnaissance of the target and preparation of a dummy battery of his siege pieces to mislead the enemy; positioning his real siege guns in a suitable position; development of his gun positions to enable him to cause optimal damage to the enemy and protect his own guns; the building of gun beds, gun shields, the building of hedgehogs, and the carriage building of his remaining guns. He will maintain order and obedience among his subordinates. The breaches which his guns opened during the day he would keep open during the night with the fire of his guns to prevent repairs by the defenders. The preparing of red-hot shot which he will be firing with slings[1] at the gun shields of the defender to set them alight.[2]

He went on to explain his mastery of all types of fuses and shot, the construction of field gun carriages, the building of roads and bridges as well as the manufacture of gunpowder and pyrotechnics of all sorts. Master Spross got the job but soon after was killed by enemy artillery fire at the failed siege of Stockach in 1499. No doubt this was to the great relief of the enemy and even greater relief of his colleagues. Spross was clearly a paragon and his letter of application points to the breadth of competence that was expected from a master gunner.

If Spross was a master of his craft, the simple and humble gunner responsible for his piece of artillery still had to display a wide range of competencies and was a well-paid 'professional'. Although guns may have been of similar length, material and diameter of bore they all had to be treated as individual items. In practice no two guns, even if cast as part of a series, were ever truly identical. It was the responsibility of the gunner to first establish the safety of his gun, its trueness (whether the axis of the bore matched the axis of the metal) and its safe powder loading. He was then responsible for hitting his allocated target.

1 I assume this is a reinforced wrought iron gun of the *sling* type, used to fire iron shot, and not a catapult or similar pre-gunpowder artillery.

2 Rudi Roth, 'The Renaissance Man', *Journal of the Ordnance Society*, Vol. 6 (1994), p.6.

Inspecting the Piece

An exterior inspection would identify blemishes, pinholes, variations in colour and superficial cracks but what was required was an investigation of the integrity and trueness of the bore. The dangers were principally honeycombing and a bore that was not true, resulting in there being too little sound metal around the breech for safety. Honeycombing was caused by bubbles of air and gas being frozen into the metal during the process of solidification, producing a Crunchie Bar effect! In the absence of a sophisticated X-ray system, which the author used to check the interior of his falconet, the alternative was simple and cheap. Robert Norton provided a detailed description of several practical methods that could be used by a careful gunner:

> But to know whether any Crackes, Flawes, or Honycombes be within her Chase or Chamber, put in an usuall Searcher, made with two or three springs with points like great pinnes heads, bending outwards, which being bended together with your hand, untill it will enter into the mouth of the Peece, and that put up to the bottome of the Bore, and turned round in the motion of pulling it in and out of the Chase; so if any Honycombes be therein, it will sticke with the points in the said Flaw, Crack or Honycombe. Also in a sun-shine day, with a looking-glasse, the Sunne beams may so be reflected into the Chase, as that all flawes, crackes and spungie Honycombes may be discerned. Or if the Sun shine not, a Wax candle lighted and put upon the end of a halfe Pike, or a bright Sword, will also reflect the Sunnebeames, as afore-said. Now if no flaw be found, you may proceed to her proofe, but if any be therein, that Peece is dangerous for breaking, and second lading, and had need to be carefully proved, and respectively handled, lest in Lading the second time, any fire remaine therein to enflame the second Charge. Therefore, if any such Peece must needs be used, you must not neglect to spunge her well first, with a wet spunge. And also after spunging, advisement taken, not to allow such Peeces the ordinary lading of powder, but with discretion an abatement must be made, more or lesse, according as the fault or defect is found to be more or lesse.[3]

He clearly assumed that the presence of honeycombing did not necessarily make the piece unusable but that the powder loading should be moderated. Great care should also be taken to ensure that the wet mop was used to extinguish any embers that might be caught up in the uneven surface of the bore.

This was highlighted in the Diary of Thomas Dallam in 1599 when aboard the ship *Heckter*. An impressive salute, conducted at Constantinople before the Sultan or 'Great Turke' in his kiosk, began and ended with tragedy:

3 Robert Norton, *The Gunners Dialogue. With The Art of Great Artillery, Robert Norton, Enginier and Gunner* (London: John Tap, 1628), pp.9–10.

Anker was wayed, the Daye verrie calme and fayere. Althinges beinge reddie, our gonores gave fiere, and discharged eighte score great shotte, and betwyxte everie greate shott a vallie of smale shott; it was done with verrie good decorume and true time, and it myghte well desarve commendations. When all was done the Grande Sinyore sent tow men abord our shippe to se how many greate peecis thare was, for he thoughte there hade bene four score, and there was but 27.

But one thinge I noted, which perswaded my simple consaite that this great triumpte and charge was verrie evile bestowed, beinge done unto an infidell. Thare was one man sicke in the ship, who was the ship carpinder, and wyth the reporte of the firste greate peece that was discharged he died.

Lyke wyse at the verrie end of this sarvis an other man, who was one of the stouteste saileres in the shipp, and all this whyle had plyed a great peece in the beake heade of the shipe, as he was raminge in his cartridge of pouder, som fier being lefte in the bretche of the peece, the pouder touke fire and blew that man quite awaye in the smoke; aboute 3 dayes after all his lower parte, from his waste downward, was founde tow myle from that place, and his heade in an other place.[4]

Nye suggested two simple methods for testing the integrity of the gun:

So soon as you have discharged a Piece of Ordnance, cover the mouth of the Piece close, and stop the touch-hole at that instant; by this you may know if any unknown cracks or flaws do go through the mettal, for if they do, a visible smoke will come through those cracks and flaws.[5]

Also, if (in striking a Piece of Artillery upon the outside of the mettal with an iron hammer) you shall at any stroke hear a hoarse sound, then without doubt there are honycombs, &c. but if you so striking the Piece shall at every stroke hear a clear sound, assuredly the Piece is clear of any dangerous flaws cracks or honycombs.[6]

The potentially porous nature of the bronze casting is made clear here by Nye.

Checking that the bore was central to the barrel was vital not just from a point of accuracy but also safety. In the first casting of the author's falconet the core had slipped and although the bore at the muzzle was central, the bore at the breech had only 10 mm of metal above it at the touch hole, not the 40 mm it should have. If the core slipped due to 'negligence or default' the bore would not be true it was therefore essential to discover the 'true' bore and the thickness of metal around it. For us it meant the melting down of the gun and a later but more successful recasting.

To check the bore the simplest tool was the double staff (illustrated overleaf). It would be placed in the bore and the gap between the outside staff and the metal from breech to muzzle measured at several points. The staff could then be rotated in the barrel and further measurements taken at the same points. If the measurements were the same, appearing as a series of concentric circles,

4 J. Theodore Bent (ed.), *Early Voyages and Travels in the Levant* (New York: Burt Franklin, 1893), p.59.

5 Nathaniel Nye, *The Art of Gunnery* (London, 1624), pp.47–48.

6 Nye, *Gunnery*, pp.48–49.

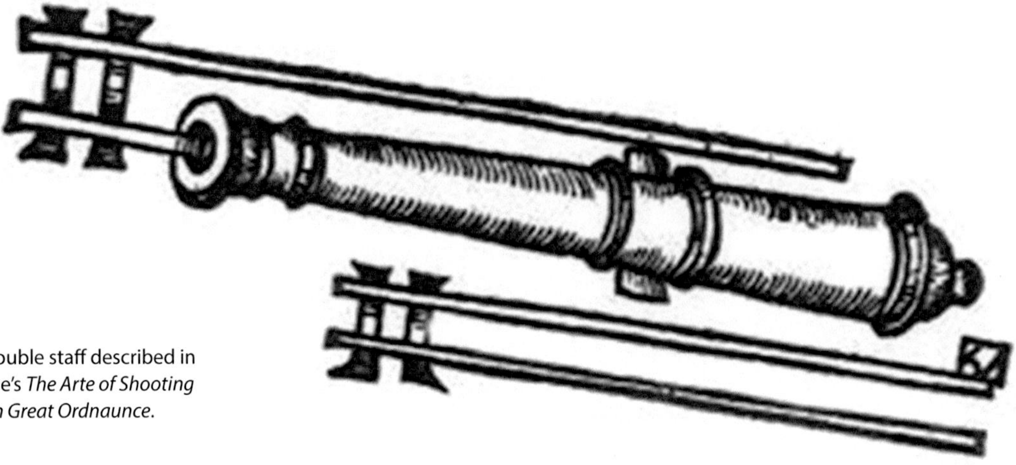

The double staff described in Bourne's *The Arte of Shooting in Great Ordnaunce.*

The 'double calibres' described in Bourne's *The Arte of Shooting in Great Ordnaunce.*

Below: The demi-culverin found in 1970 from the Armada ship *El Gran Grifón* was a fine example of poor workmanship. The broken muzzle end showed that the metal was porous with numerous voids visible in the bore. These would have made the gun dangerous to shoot and burning debris could have lodged in the pores. Worse still was the bore itself, which was badly off centre. At the breech the thickness of metal was not a uniform 10.35 mm but as much as 14.7cm on one side and as little as 6 cm on the other. The design, as well as its prosecution, was deeply flawed, being a very lightly metalled piece for a culverin, especially in the breech. This is one of the few occasions when one can legitimately say of a piece that it was probably as dangerous to its crew as to the enemy, even though it probably passed proof! (Author's drawing)

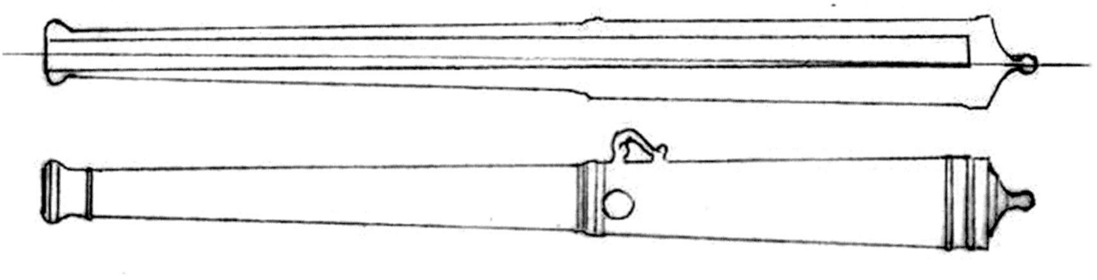

then the bore was true. Norton suggested putting pins into the staff which would just touch the metal, which would show a difference in the thickness of the metal by touching the metal or not. A variation of this system was to fit a rammer head to the staff placed in the bore, with the portion of the staff at the muzzle fitted with a mop collar that would enable the contraption to rotate.

A final method of establishing the trueness of the bore was to use the 'double calibres' (see illustration on facing page):

> Lastly, there may also for that purpose be an Iron or Brasse Instrument with a joynt in the midst called Double Calibres with foure legges like the figure here described, which by putting them into the concave of the Peece, and turning it round within the bore, clasping the other legge without to the mettall of the Peece, the distance betweene the other two legges without will shew the thicknesse of the Peece in each part, and shew, with applying all the openings vnto an inch Rule or scale of equall parts, whether the Peece bee thicker upon one side then on another, being alike distant from the mouth of the Peece, and how much and where.[7]

You can make of the above what you will!

Gunnery Instruments

> Question. What Instruments are most fit for a Gunners use?
>
> Answer. The Callibre Compasses, Height-boord, Inch sight-rule, Gunners Scale, Gunners Quadrant diuided into ninetie degrees, and into twelue points and their minutes, with a Geometricall square to take mountures, leuels, heights, breadths, and distances, and a paire of straight pointed Compasses.[8]

The very model of the modern gunner was no rude mechanical but a technician who had at his command the latest instruments, as well as the books and list of tables that made his craft a modern mystery. A reasonable but not high standard of numeracy might be expected from the putative gunner. It is significant that Smith and Norton spend much time explaining simple arithmetic, including addition, subtraction, multiplication and most challenging of all division. Numeracy was clearly expected and literacy implied by their manuals. How far the workaday modern gunner was the paragon that Bourne and Norton would have wished him to be, is another question.

Some of the instruments were simple and cheap. Bourne uses a ruler and a plumb line for making measurements and ensuring alignments. The instrument that was most often shown in illustrations and referred to in books of instruction was the quadrant. This was an instrument known to astronomers, astrologers and navigators in the form of the astrolabe. In this form it could be an object of some mystery and beauty and potentially great

7 Robert Norton, *The Gunner Shewing the Whole Practise of Artillerie* (London, 1628), p.82.
8 Norton, *The Gunners Dialogue*, p.11.

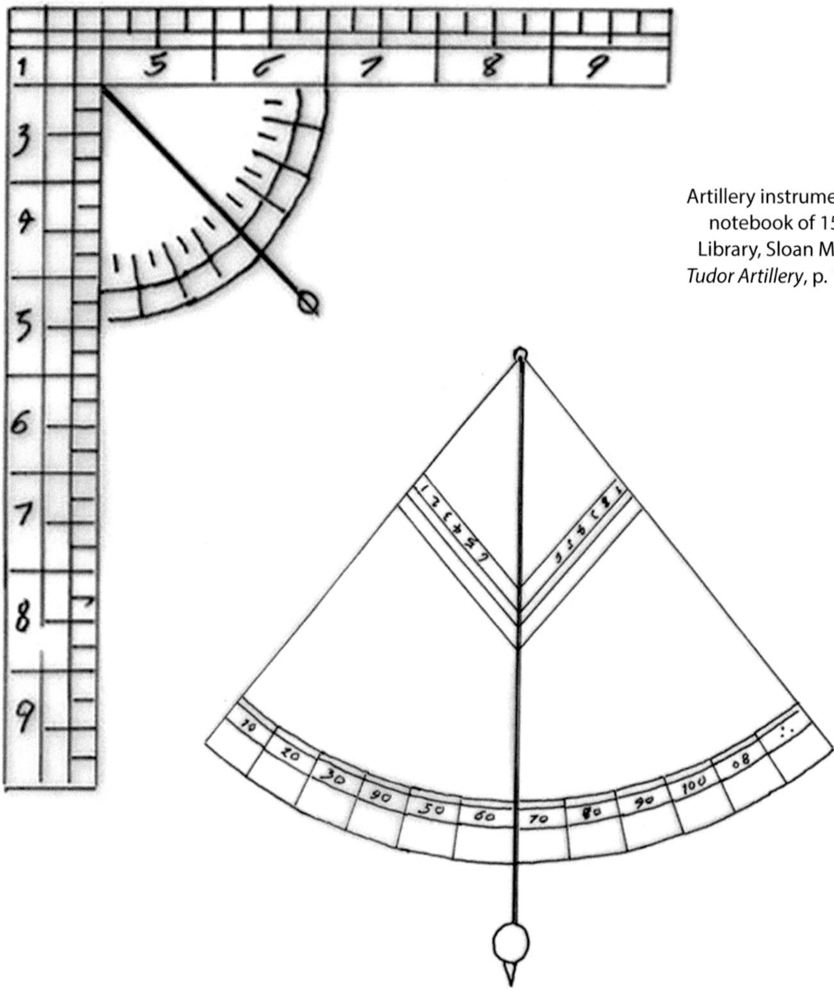

Artillery instruments from an unsigned notebook of 1540. (Source: British Library, Sloan MS No. 2497; Caruana, *Tudor Artillery*, p. 16. (Author's drawing)

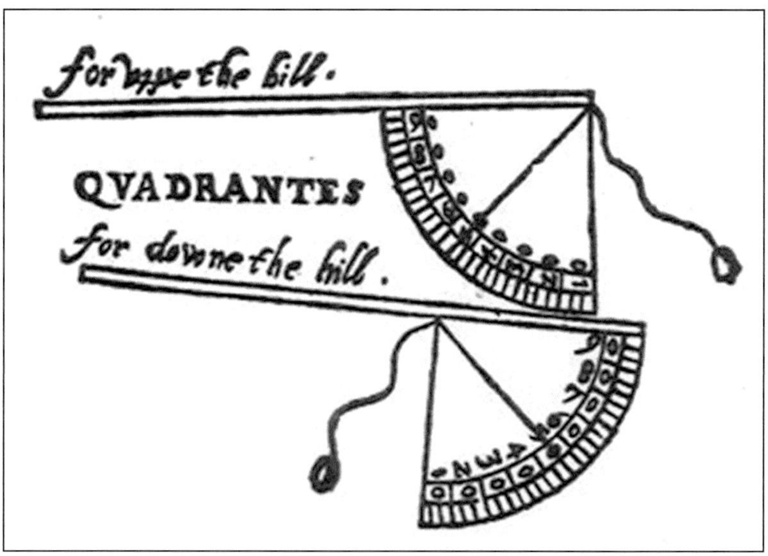

A gunner's quadrant from Bourne's *Arte*. The long arm could be inserted into the barrel. Shooting downhill could be a problem when using these examples but was resolved in some examples by extending the arm of the quadrant.

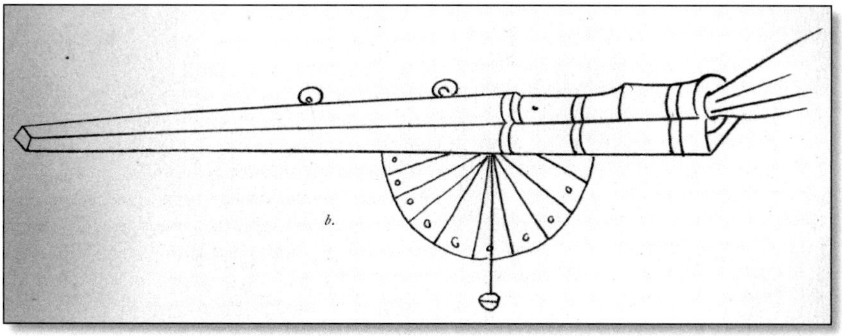

An illustration of a long-stocked handgonne provided with a quadrant, to be found in Essenwein and von 1873, table A. XIX. Dated to 1410.

expense. It was for the gunners' use a much reduced and simple tool, used to measure angles of elevation, not to navigate oceans or investigate the universe. Bourne considered that worse perhaps than the failure to possess this vital piece of equipment were those that had instruments, and 'yet be utterlie voide of the uses of them'.[9]

The gunner's quadrant is a work of genius, and its use is simplicity itself. Its invention was sometimes attributed to Tartaglia, but its origins were much earlier. It is probable that the quadrant was used during the Hussite wars 1419–1434. An example of a quadrant is found in an illustration in the *Ex ludis rerum mathematicarum* by Leon Battista Alberti (*c.* 1452), showing how to 'measure of the sight of a bombard with the equilibra'.

Bourne instructed the gunner: 'take your Quadrant, and looke through the two sight[s] of the Quadrant, the plummet hanging at libertie, till you may see the mark justly upon the hill, winking with one of your eyes, then looke upon what degree and place the plummet line doth hang along'.[10] Norton, as always, describes the instrument in greater detail:

> The Gunners Quadrant is a Geometricall Instrument, made of brasse or fine grained wood, containing in the circumference one quarter of a circle, divided into 90 equall parts or degrees in the outmost limbe, and in the second limbe within into the 12 poynts of the Gunners quadrant, having within the same also a Geometricall Quadrate, with each side divided into 12 equall parts, and those each of them subdivided by meanes of paralels and Dyagenalls into 10 equall parts, so that each side will be thereby found distinctly devided into 120 equall parts, fitly serving to take all Geometricall Mensurations, of distances, heights, breadthes, and depthes, accessible and inaccessible, by the directions hereafter mentioned … each Quadrant … hath a strait Ruler about 2 foot long, ioyned either firmely, or by a Grove [groove] dove-tayled vnto one of the sides of Semidiametres thereof.[11]

The quadrant could be inserted into the bore to level it or to ascertain to what degree it was elevated. It was also of use when establishing the angle between the observer's eye and his mark, by which the gun could be adjusted and with the use of some trigonometry heights could also be calculated. If

9 William Bourne, *The Arte of Shooting in Great Ordnaunce* (London, 1587), page unnumbered.
10 Bourne, *The Arte of Shooting*, p.50.
11 Norton, *The Gunner*, pp.95–96.

Left: How not to use a quadrant! What may be obvious to a modern observer was not necessarily so to those ignorant of its use. (Unknown author, public domain)

Below: An illustration from Tartaglia's *Nova Scientia* showing the correct use of a surveyor's set square, a form of quadrant. Knowing the angle to the castle and the distance to the base of the cliff, the height of the castle could be calculated. Adding the height of the observer was a factor that should not be overlooked. Tartaglia, *Nova Scientia*, 1558, third book, 26r.

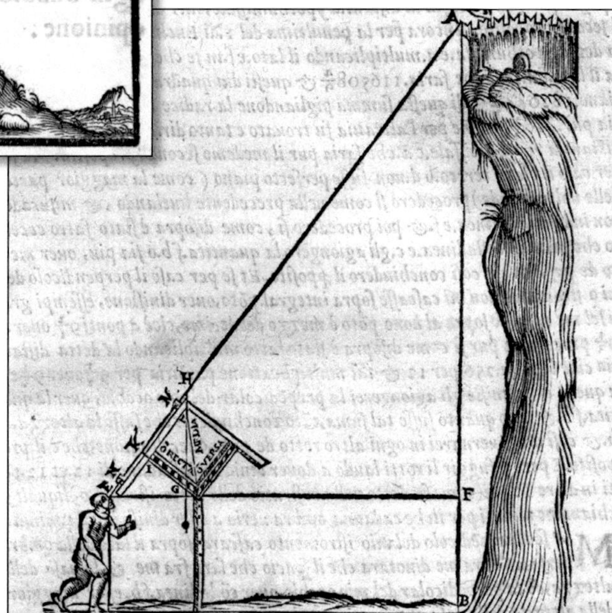

Below: Another valuable yet simple instrument was the square or Egyptian level. This was an A-shaped frame from the apex of which hung a plumb bob. With legs of the same length, when the plumb bob line was in the middle of the crosspiece, the frame was level. It could also be used as a form of quadrant if the crosspiece was calibrated. It could be a humble instrument made of wood used by carpenters and masons; it could also be found in fine brass and gilt versions, folding and calibrated, suitable for the discerning gunner of means.

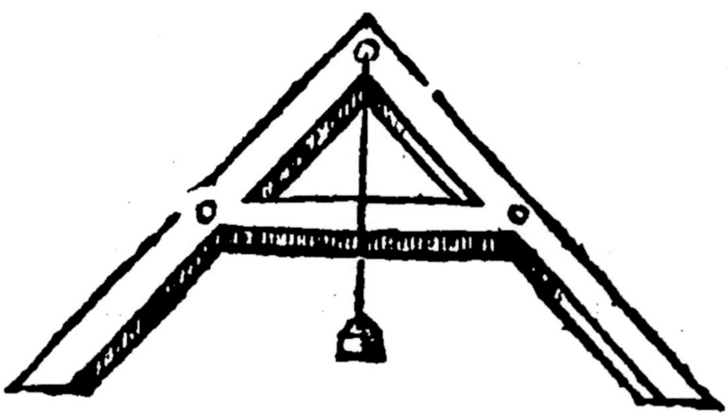

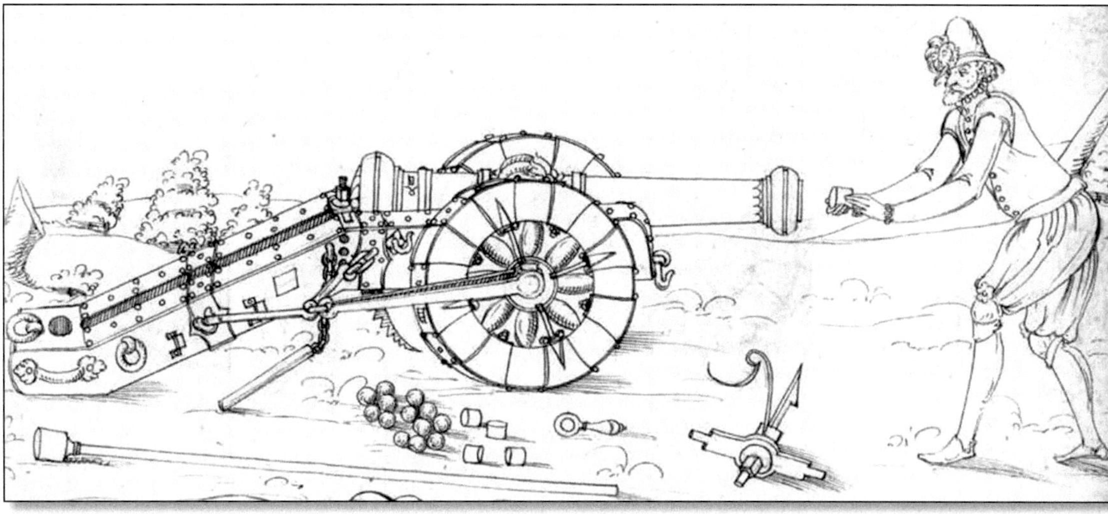

This foolish gunner is approaching the muzzle of the gun in such a position as to ensure his bloody extinction if anything went wrong! Gunners are enjoined never, ever, to place themselves in front of the muzzle when loading. Loading tools have to be handled in such a way as to avoid a premature explosion causing their hands, as well as the loading tool, to fly off. As in so many illustrations the loading tools are left on the ground, a practice avoided by modern gunners in order to avoid picking up grit and other detritus. He appears to be loading 'clod' or cylindrical shot, a type that does not seem to have been used in England but is often illustrated by Litzelmann. (Walther Litzelmann: *Vortrab zu der Arckalay – Büchsenmeisterbuch* – Shelfmark, Fol. 909 p. 5. Courtesy of the Bavarian State Library)

the target was downhill then a variation of the quadrant where the scale was extended to register minus figures could be used.

The gunner's rule was a very useful little tool to enable the gunner to establish the height and weight of shot for his piece. In later years it became a decorative status object sometimes in the form of an Italian stiletto dagger. It was a ruler of 'Brasse, Boxe, or other fine grained Wodde', that could be used to measure the height of a ball or the diameter of the bore of the cannon. On the rule were engraved the weights of balls in iron, lead and stone that corresponded to that height. Lead seems only to have been used for small-bore pieces and even then, the projectile often had an iron or stone core. The scale was sometimes graduated for heavy lead shot. In some cases, the rule seems to have been used with a pair of calipers, which were used to measure the diameter of the shot. Norton considers that the rule was a satisfactory alternative to an accurate balance. Much effort has been taken to decode surviving rules, as there were so many variations in units of length and weight.[12]

12 For further information on this subject please see R. R. Brown, 'Comment on the Tal-y-bont gunner's rule', *Journal of the Ordnance Society*, Vol. 2 (1990), p.71; J. Green, 'Gunner's rule, Further information on gunner's rules or tally sticks', *Journal of the Ordnance Society*, Vol. 2 (1990), pp.25–32; Alex Hildred, 'A gunner's rule from the *Mary Rose*', *Journal of the Ordnance Society*, Vol. 15 (2003), pp.29–40; A. Konstam, 'A Gunner's Rule from the "Bronze Bell" Wreck', *Journal of the Ordnance Society*, Vol. 1 (1989), pp.23–26.

Loading Powder and Shot

Loading any artillery piece both then and now is not a task to be taken lightly. Hands, arms and even lives have been lost when carrying out this duty; safety was and remains the priority. The procedure for artillery pieces currently follows this pattern with some variations:

1. The barrel is searched using the worm, to ensure that the piece is unloaded and that there is nothing untoward in it
2. The vent should be cleared, for the same reasons as above, using a priming iron
3. A cartridge or loose powder should be loaded and rammed
4. A wad should be inserted and rammed
5. The projectile should be loaded and rammed
6. A further wad should be loaded and rammed
7. The gun should then be primed with powder in the vent
8. The gun should be fired using a linstock
9. The vent should be cleared and the vent stopped to induce a vacuum in the bore during stages 11 and 12
10. The gun should be searched using the worm, to remove any remnants of the cartridge
11. The wet mop should be used to clean and cool the barrel
12. The dry mop should be used to clean and dry the barrel
13. Reloading and firing procedure continues
14. On ceasing fire barrel should be searched and mopped, vent cleared and tools cleansed. Barrel may be cleaned out thoroughly with water and dried if no more firing is expected.

The loading tools were tied to the barrel when the gun was being moved, a highly practical solution to the problem of how to manage them. They have to be well-constructed and of a good length to be practical and reliable when in constant use. Norton provides one of the few detailed descriptions of these necessary accoutrements which are so often found in illustrations:

> The Button or head of the Ladle-staffe must be one dyametre, and of such height or thicknesse, that it together with Brasse plate may bee equall to the height of the Shott (due vent being abated) for Spunges, their buttons or heads are to be made of soft fast wood, as Aspe, Birch, Willow, or such like, and to be one dyametre, and ⅔ in length, & not aboue ⅔ of a dyametre of the shots height: The rest being covered with rough Sheepes skinne wooll, and all, be nayled thereon with Copper nayles, so that together they may fill the Soule or Cavity [bore] of the Peece.[13]

13 Norton, *The Gunner*, p.111.

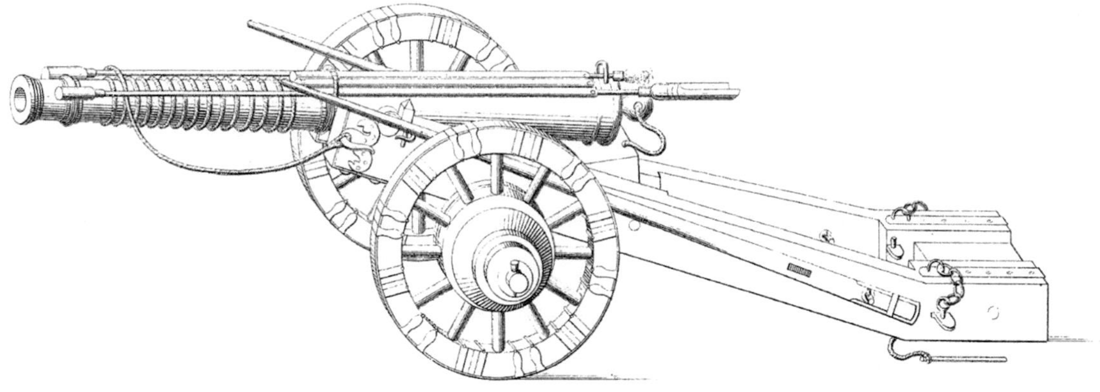

This light field piece has its loading tools tied to the barrel using the towing ropes fitted to the cheeks of the carriage, an admirable solution. Also well illustrated is the pair of trails used to draw the gun which have been turned forward so as not to impede the crew when in action. It is drawn from extant examples of guns and carriages captured by the Swiss from the Burgundians in 1476, at present in the Musee de Neuveville. Source: Napoleon III, *Études sur le passé et l'avenir de l'artillerie*, Volume 1 (Paris: J Dumaine, 1846)

It seems that Norton like Bourne assumed that the rammer will be on the reverse of the ladle or the sponge staff and not a separate item. The sponge heads recovered from Armada wrecks were tapered oak. Martin explains their use:

> Close to the face ends, where the heads are widest, the wool would have compressed tightly into the barrel sides and so have served to dislodge the firing residues, while the looser wool surrounding the taper behind it (which was normally soaked in urine) would have gathered and held the loosened dirt.[14]

The author would like to see if that worked especially well, as the parallel-sided lambswool ones his gun crew uses work perfectly well even under the direst of conditions.

The use of copper nails as a safety feature is noticeable but even more apparent is the absence of any reference to the 'worm' with which the bore could be searched and any detritus from a cartridge removed safely. Smith suggested the following combination of tools 'every rammer staffe ought to have a sponge at the one end, to cleanse the peece with, and a tampion of wood at the other end, to put home the bullet and wad with, in the centre of which ought to be a hollow screw wherein the Gunner may screw in a wad hooke to unloade any peece at his pleasure'.[15] There were certainly 'worms', fitted with a corkscrew-style end, found in the *Vasa* wreck of 1628. The purpose of the wad-hook was as its name suggested to remove the wad before unloading the gun and not for inspecting the bore for the remains of cartridges and removing them.

14 Colin John Mackenzie Martin, *The Equipment and Fighting Potential of the Spanish Armada*, A Thesis Submitted for the Degree of PhD at the University of St Andrew (1984), p.341.

15 Smith, 'All Manner of Peeces: Artillery in the Late Medieval Period', *Royal Armouries Yearbook*, Vol. 7 (2002), p.87.

The positioning of the powder barrel was not an afterthought, as more than once a gun battery was reduced to smoking ruins by sheer carelessness. Norton recommended that the careful gunner should: 'place the Boudge or powder barrell to wind-wards, and cover it safe with some Hide, garment, or cloath: alwayes avoyding to stand before the mouth of the Peece, but on the right side thereof in loading her for feare of further danger', excellent advice (Norton, The Gunner, p.38). This rare example found in the Tøjhusmuseet Copenhagen has a leather draw string cover. The keg is bound with hazel withies rather than iron hoops which meant a less secure fixing. (Author's photograph)

Of first importance when considering the safety of the piece was the quantity of powder to be loaded. This could be calculated in terms of volume or weight and related to the weight of the projectile and the strength of the gun. The gunner was as conservative as possible in determining the powder load, as his life depended upon not over-straining the gun. Breakages were likely to occur at the breech, especially the point where powder and shot met, also near the trunnions and at the muzzle.

Bourne assumed that an equal ratio of powder to shot would be the standard with the ratio of shot to the weight of metal determining higher or lower loads.

- 1 to 1½ cwt of metal to every 1 lb of shot requires the same weight of powder as shot minus 3/9
- 1½ cwt of metal to every 1 lb of shot requires the same weight of powder as shot minus 2/9
- Less than 2 cwt of metal to every 1 lb of shot requires the same weight of powder as shot minus 1/9
- 2 cwt of metal to every 1 lb of shot requires the same weight of powder as shot
- 3 cwt of metal to every 1 lb of shot requires the same weight of powder as shot plus 1/9

These measurements took no account of the length and profile of the piece, especially the crucial thickness of metal at the breech. The profiles of early guns were very different from later and could be lightly built and without reinforces. The quantities of powder were laid down in numerous published and private documents, but it was for the gunner to decide what the individual gun for which he was responsible could take safely. Bourne's rough guide was that the volume of powder would equal nine diameters of the ball. One noticeable difference between modern and contemporary powder appears to be that modern powder takes up significantly less volume for the same weight.

Loading could be done with powder taken loose with a ladle directly from a 'budge' barrel, which had a leather cover fitted that could be tied shut. There were innumerable detailed and very dull instructions on the dimensions of the ladles for the great variety of guns in use. Bourne's account will suffice as an example:

> Question: How you may Arithmetically know the true breadth of the plate of the ladle that is due for any peece of Ordinance by knowing the height of the bullet fit for the said peece.
>
> Answer: Take a line and compasse the bullet in the mids, laying the same measure to an inch rule, devide the same measure into 5 equall parts, 3 of those parts is the iust bredth you ought to make your plate of, which being orderly placed on the staffe, and bent circularly, serves to hold the powder in: the other ⅖ partes being cut and taken away, and so left open, serves to turne the powder into the peece.[16]

As it usually contained half of the powder needed the ladle would be loaded at least twice. Norton was meticulous in describing how to use the ladle in a safe and soldierly manner:

> then his assistant declining the powder or boudge barrell aside, he shal thrust in his Ladle to fill it, striking off the heaped powder, giving a shogge to shake downe the surplusse: and it being so filled and striked, put in the Ladle downe to the bottome of her Concave Cillinder vnto the Touch-hole, but at the first putting of the Ladle (so filled) into the Mouth of the peece, slide in the Ladle staffe, so that the upper side may keepe uppermost all the way, and when it is arived to the bottome of the Bore, then hee laying his right thumb upon the said upper side of Ladle staffe, neere the Mouth of the peece, and turning the said staffe so much, untill the said thumb unremoved upon the staffe be found directly under the same; then giving two or three shakes, and bearing up the Ladle, that the powder may bee turned out, or goe out cleane, and that the Ladle bring no powder back therewith, for that were a foule fault for a professed Gunner to commit.[17]

16 Bourne, *The Arte of Shooting*, p.27.
17 Norton, *The Gunner*, p.101.

Not only a 'foule fault' but a potentially lethal one. The loose powder now had to be rammed and wadded.

Norton very helpfully suggests marking the ramrod to indicate the full depth of the bore and also the depth of powder:

> First, put the rammer into the Peece up to the Touch hole, and marke, the staffe, even with the mettle at the mouth of the Peece, and then pull it out 3 dyametres, for the Cannon, and 3½ for the Culuering, and 4 for the Saker Falcon, &c. and marke there another marke, which is the place that the Powder must supply in the Chamber.[18]

How hard the powder should be rammed was an important question that would greatly affect the force of the explosion. Bourne thought that ramming home the wadding too hard would slow ignition of the main charge and result in much of the powder venting through the touchhole. In the case of the small Loshult cannon when too small a charge was used and a gun arrow rammed hard into the muzzle all the force of the powder was vented through the touch hole, leaving the projectile in situ! If the powder was wadded too loosely that would also result in a loss of pressure and range: 'The powder rammed too hard and the wad also, it will be long before the piece goes off … The powder too loose … will make the shotte to come off the mark'.[19] The consequences of wadding too hard was to provide a compressed propellant suitable for rockets. Norton on the other hand was not one for messing about:

> Then shall he put the powder home softly, with the Rammer that is at the end of the Ladle staffe; putting in a good Wadd, and thrust it home to the powder, giving three or foure hard strokes, which wil gather the scattered pouder together, and drive close the same, and the rest to the bottome of the Chamber.[20]

Biringuccio recommends hay as wadding and that the wad should be pressed in firmly with the rammer, the ball pushed into the wad and then: 'strike it with the ramrod with all your force and strength'.[21] In the *Vasa* trials old lessons were relearned:

> how much dead space was occupied by the wadding … has a surprisingly large influence on muzzle velocity) [as did] how much the charge was compressed in the ramming sequence. The Swedish naval gun drill of the seventeenth century specifically stated that the charge should be rammed home, and then given two good thumps. We discovered that the first thump has a mushy feel, as the air in the charge is driven out and the grains of powder are jammed into each other, while the second thump should feel hard and make the gun ring. Firing the charge

18 Norton, *The Gunner*, p.112.
19 Bourne, *The Arte of Shooting*, p.31.
20 Norton, *The Gunner*, p.110.
21 Vannoccio Biringuccio, *The Pirotechnia* (New York: Dover Publications inc., 1990), p.418.

Loading the *Vasa* cannon was made rather more difficult by the insistence of Bofors that no one could be in front of the barrel after a cartridge had been placed in it. The rather unwieldy rammer slowed the process but even so, as they demonstrated, a well-trained crew could work efficiently and safely in a relatively small space. (Photo by Richard Belles,Courtesy of the *Vasa* Museum)

> without these thumps creates very high variation in muzzle velocity, but once we had the knack of the thumps, our muzzle velocity settled into a narrower range.[22]

It was very important to make sure that a gap did not develop between the powder and the shot as this could produce disastrously high pressures at this point in the barrel. William Bourne warned 'if the shotte does rest any thing short, it will breake the peace (or else it is a chaunce) in the vacant place betweene the powder and the shotte'.[23] Norton was also clear about how the shot ought to be safely placed: 'And then put in the Shot which with a Rammer he must put softly home, and afterwards another Wadd of Hay, Grasse, Weedes, Okham, or such like: And againe, give two or three good strokes with the Spunges Rammer head'.[24]

The alternative to loading with loose powder was to use a cartridge. Cartridges were preferred to loose powder as the: 'peece is more sooner and easilyer laden and or charged, and hee shall keepe the length of the marke the better'.[25] Loose powder could also be easily spilled and the quantities of powder could be uncertain, resulting in the problems he mentioned. Norton has provided a detailed description of the cartridge and how to make it:

> Cartredges are either to bee made with Canvas Fustian, or other linnen cloath, or with thicke strong Paper, especially of Paper Royall: which prepared, take the height of the bore of the Peece, without the vent of the Shott [windage i. e. the full diameter of the bore], and cut the cloath or paper of the breadth of three such heights … and leaving in the midst at the top and bottome one other such height, at each place to make a cover and bottome for the Cartredge, cutting each side and end, somewhat larger, then the strict measures appointed for the sowing or glewing of the seames thereof, so much as will countervaile the same, having also a respect for augmenting and deminishing those measures, as the powder

22 Correspondence with Dr Hocker, 26 July 2021.
23 Bourne, *The Arte of Shooting*, p.1.
24 Norton, *The Gunner*, p.110.
25 Bourne, *The Arte of Shooting*, p.31.

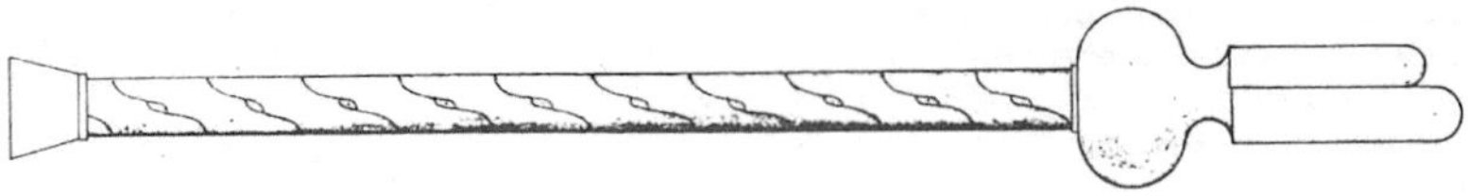

A powder ladle for a perrier. The bulb or grommet would enable the ladle to be centred for the sub-calibre powder chamber. (John Lad's Notes (1587) in Caruana, *Tudor Artillery*, p. 15. Author's drawing)

> shall bee better or worse then ordinary, and also abating with discretion, when as your Peeces shall be already heated in fight, least else you endanger the breaking or splitting of your Peece. Having resolved then for what sort of Ordnance your Cartredges are to serve, you are accordingly to have a Modell or Former of wood turned of the height of the Shot, and of a convenient length, longer then the Cartredge is to be. Then if you make them of Canvas, halfe a dyametre is to be allowed more in breadth for the seames: but if they be made of Royall paper, then having lapped it once about the Former, leave about ½ inch surplussage more then will compasse it, which with Starch, Paste, or mouth Glew, close about the said Former, having some part of the same substance, fitted upon the end of the Former; first for a bottome, which must also be pasted or glewed close, and fast to the side of the Cartredge, so that being dry, it may hold the Powder fast, and sure from spilling. And you must remember first to tallow the said Former, so that the Cartredge being so moulded thereon, it may be easily and without tearing, slipped off againe.[26]

If the gun was chambered, that is having a sub-calibre chamber smaller than the diameter of the bore of the gun, a special ladle was required which had a wooden or leather grommet, which would centre the ladle in the muzzle.

The next duty of the gunner would be to prick the cartridge and prime the vent with loose powder and then 'give fire' with a linstock. This procedure is well described by Norton:

> with the Rammer-head, to put it home, to the bottome of the bore of the Peece, with two or three easie stroakes: and then with a sharpe three squared [triangular?] Pryming Iron, to cut and pryme the Cartredge, that the Powder prymed at the touch-hole, may give fire to the quick powder thereby. In all other things for wadding before and after the Shot, and ramming home the Shot, you are to performe the vsuall manner taught in his proper place.[27]

When loading with a paper cartridge Biringuccio recommend ramming it so hard as to break the paper and spill the powder, this procedure could create problems. Some of the cartridge would remain in the bore, sometimes still smouldering. It was therefore vital for this to be removed and extinguished. Paper wadding could also be compressed in the rear of the breech, which in the past has caused serious accidents. It is surprising how often parts of paper cartridges remain in the barre smouldering or unburnt. The author's gun crew have used silk cartridges in their guns as this is completely burnt leaving no detritus; wool is thought to have similar properties, but the author

26 Norton, *The Gunner*, pp.127–128.

27 Norton, *The Gunner*, p.128.

A gunner is seen using his calipers to assess the roundness of shot and the diameter of the 'clod' shot. Clod or cast cylindrical shot must have been easier to produce to a precise diameter; round shot seems to have been something of a mixed bag. Whether the aerodynamic performance of the tubular shot was markedly inferior to round shot at point-blank range is worthy of further research. (Walther Litzelmann: *Vortrab zu der Arckalay – Büchsenmeisterbuch* – Shelfmark, Fol. Cgm 909 p. 298. Courtesy of the Bavarian State Library)

has no experience of using it. The idea of cartridge and shot as one round of ammunition is usually dated to its use by the Swedes in their leather guns in the 1620s. There is however a tantalising much earlier reference from Italy in the 1480s referring to small pieces known as *spingarde* and *passavolanti a cartoccio*. These were loaded with a cartridge made from a sheet of paper, a *foglio reale* which contained both the powder and the shot.[28]

Round shot, the principal form of ammunition in this period, was not necessarily round or of a precise diameter. Measuring the diameter and shape of shot was something of an obsession for gunners, if the numerous calipers and shot gauges both illustrated and still extant are a good indication. Wooden shot gauges are in practice easier to use than calipers as calipers require recalibration all the time while wooden gauges can be readily and quickly fitted around the projectile.

Stone shot was always problematic as these had to be hand carved and were more likely to be rough and uneven. Iron shot and lead shot would have been cast but there was again no certainty that it would be consistent in shape or diameter. The evidence from the *Vasa* is that although the cannons were cast with a remarkable degree of precision the shot was not. The bore of 12 *Vasa* guns was found to be 146 mm plus or minus 0.5 mm, while the balls varied from 135–143 mm in diameter, leaving a windage on either side of the ball of between 1.5 and 5. 5 mm. In the case of iron shot recovered in situ from pieces from the *Mary Rose* the windage varied from ¼ in to 1 in (6–25 mm). Shot that fitted too well or too loosely were not conducive to precise shooting. A shot that was too large was obviously to be avoided but a ball that fitted at the muzzle need not fit the bore if it was also irregular. It was again the duty of the gunner to know whether the inside of the bore was true.

The windage, that is the gap between the ball and the bore was very important when considering both velocity and accuracy and of course

28 Ansani, 'The Life of a Renaissance Gunmaker', p.757.

safety. If the gap was wide then much of the force of the explosion would be ejected around the ball before it left the barrel, as is apparent in slow motion photographs of guns firing. If the ball ricocheted down the barrel (a progress known technically as balloting) it would have a spin imparted to it. The result is called the *Magnus* effect and as the spin will impart different air pressure on one side of the ball from the other as it passes through the air, the ball will not fly true. This is something well known and oft experienced by golfers when they slice a ball! Bourne warned that a loose ball would fall short or *swerve*; the smaller the windage the less would be the problem. In England a standard windage of a quarter of an inch was supposed to be applied, in Italy the windage of 1⁄20 of the diameter of the ball was considered ideal. The English rule made little sense, for as Norton complained:

> the best experienced Gunners of late yeares disliking, have for a generall proportion chosen rather one twentieth part of the height of the Bore to be abated for a due height of the Shot vent, affirming that in the Faulcon and small peeces the abatement of ¼ of an inch is too much, and for the Canon ¼ of an inch is too little, and that 1⁄20 of the Diametre of the Bore will be a reasonable abatement for all Peeces, be they higher, or lower, which being by Arithmatick so easily found, will not require any example.[29]

Loading a mortar was a very different task, more akin to shooting a medieval chambered piece. The chamber should be according to Norton half the diameter of the bore and in length half that of the barrel. He stressed the importance of making sure that the charge was well wadded, and a wooden tampion fitted and to ensure that no air gap developed between shot and charge:

> First then the Chamber is to bee well sponged and cleansed before the putting in of the powder, whether you loade it with loose powder or Cartouch turning the Mouth neere upright; the powder being so put into the Chamber, ther must be a wad put in either of hay or Okam, & after a Tampkin [tampion] of Willow or other soft Wood; such as may, together with the powder that was first put in, fully fill up the whole Chamber thereof, that there may bee no vacuity betweene the powder and wadd, or wadd and shot; after which the shot shall be also put in at the Mouth with a wadd after it; especially if the Peece be not much mounted, least the shot goe out too soone, and the wadd between the Tampkin and the Shot, is not onely to save the shot from the Tampkins breaking of it, but also to avoide vacuities, which are very dangerous for the Peece by second expansions.[30]

Whatever the type of ordnance, once the powder, wadding and shot had been carefully loaded the charge had to be pricked, primed and ignited. The vent was sometimes provided with a hinged vent cover or a small lead apron pressed to conform to the shape of the breech; the weight of the lead held it

29 Norton, *The Gunner*, p.95.
30 Norton, *The Gunner*, p.60.

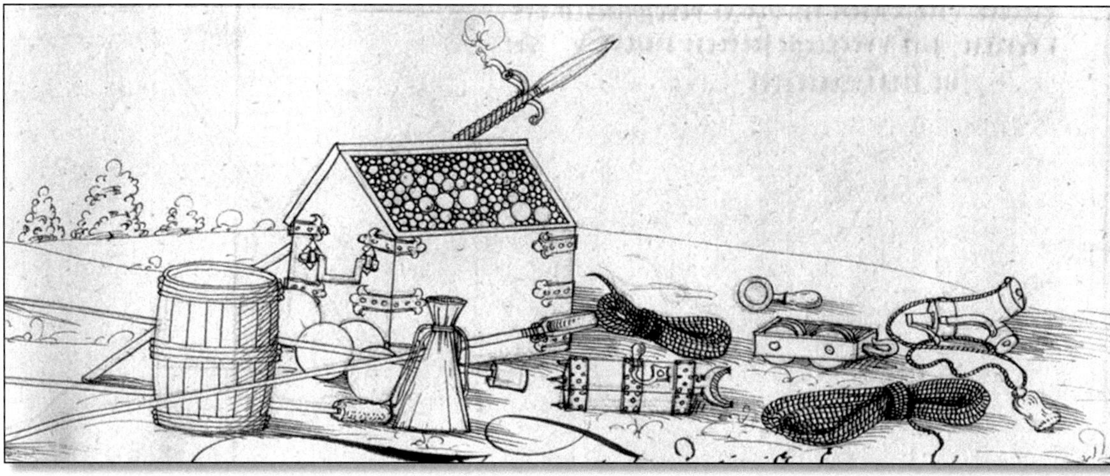

A delightful illustration of some of the accoutrements associated with gunnery. The small powder barrel is next to a shot-box, somewhat overfilled, in front of which is a ratchet jack and pulley block. There are some hanks of neatly coiled match cord and at the far right a nicely made horn priming flask. Loading tools lie on the ground and a partisan style pole arm has been modified to mount match cord, some of which can be seen happily fizzing away, which it can do at an alarming rate. (Walther Litzelmann: *Vortrab zu der Arckalay – Büchsenmeisterbuch* – Shelfmark, Fol. Cgm 909 p. 319. Courtesy of the Bavarian State Library)

in place in high winds and works a treat. Once the vent cover was opened a copper or brass pricker was used both to clear it, as it may become clogged, as well as to pierce the cartridge if one was being used. The prickers tended to be long and usually had a simple figure of eight twist on the top, although some others had finely decorated finials. Neither were designed to be thrust down with the flat of the hand, which if the charge went of prematurely would drive the pricker through it!

The vent should now be primed with fine corned powder using a powder horn. In the case of a powder horn found in the *Vasa* the wide end was fitted with a removable cap for ease of filling and if the powder in the horn was to explode then the explosion would blow backward out of the horn and not produce an extemporary grenade. In Spanish practice, at the time of the Armada, a small brass or tin funnel was used for the purpose, presumably useful in high winds and to avoid spillage. Ignition was achieved with cord soaked in saltpetre held in the jaws of a linstock.

At its inception in Europe, gunpowder was probably ignited using a red-hot iron wire. This was replaced by the wooden linstock holding match cord. The wooden linstocks found in the *Mary Rose* seem to have been the possessions of individual gunner who have lavished much effort on carving them with intricate patterns and zoomorphic features. They were on average 65–80 cm in total length, with a diameter of 20–25 mm. They would add a good arm's length to the gunner giving him plenty of clearance from the hot gases venting from the touch hole and safe distance from the recoiling gun carriage. Much longer linstocks were needed for field carriages which eventually led to the modification of pole arms such as the partisan to hold match cord in a pair of jaws. Linstocks could be fitted with an iron ferrule to allow them to be planted in the ground, which is handy and practical. the author's linstock also has a long leather thong attached to the end which enables me to twirl it at speed to make sure that the cord is burning nicely and avoids having to blow on it constantly. Whether this was a contemporary historical technique, the author does not know.

Current practice would be to worm the piece to dislodge any linen or paper cartridge, to clean the bore with the wet mop and then with a dry mop to ensure that no water remained in the bore. The cartridge and wadding would then be loaded. Dutch practice in the early eighteenth century seems to have been to mop the barrel out after every seventh or eighth shot in order primarily to cool the barrel. It was felt that the greater danger was that powder residue when wetted, produced a lumpy surface, that would tear cartridges when loaded, causing potentially premature explosions where 'frequently, a man may lose his hands, indeed his life, by it'.[31]

After continuous firing the gun would become too warm and in the case of long bronze barrels they may even visibly have drooped. 'A brasse peece made hote with often shooting, is more apt to breake then when it is cold';[32] To maintain the integrity of the piece it had to be cooled. Norton is again free with useful advice:

> If the service be such, as she may not have time to rest, in that case, she must be cooled with Spunges wet in Lee [wine dregs], which is best, or in urin and water, or water and vineger, or with the coolest fresh or salt water; bathing and washing her within and without, untill she be come to her due temper againe; but if I may, I will rest her one or two houres in twelve, and between whiles coole her every tenth or twelfth shot.[33]

Gunners do not seem to have always demonstrated the care that Bourne recommends in the handling of their pieces. The fascinating analysis of the surviving Tudor bronze guns of the Channel Islands,[34] reveals a sad tale of clumsy and careless gunners using shot too large and powder loads too great. The Sark falconet shows evidence of scoring in the bore of the barrel while the bore of the Castle Cornet falcon shows 'severe internal damage from tightly-fitting shot or possible extended period of firing, causing heating and softening of the metal'.[35] The bastard falconet has:

> severe scoring of the bore and a vertical crack in the right-hand side of the barrel about 50 cm back from the muzzle, suggest that oversized or irregular shot have been used in the past, to the gun's detriment. Indeed, viewed from above, the gun barrel appears to bend imperceptibly to the left.[36]

31 V. de Tollen, *Het nieuwe licht bosschieterye,* Amsterdam (1734), p.302 ff in Jan Piet Puype, 'Guns and their Handling at Sea in the 17th Century: A Dutch Point of View', *Journal of the Ordnance Society,* Vol. 2 (1990), p.22.

32 Smith, 'All Manner of Peeces', p.100.

33 Norton, *The Gunners Dialogue*, p.23.

34 Robert Waterhouse, Tudor Bronze Guns of the Channel Islands (2015), pp. 82–109, <http://www.christopherlong.co.uk/aanna/papers/WaterhouseR/TudorBronzeGunsOfTheChannelIslands.pdf>

35 Waterhouse, *Tudor Bronze Guns,* p.103.

36 Waterhouse, *Tudor Bronze Guns,* p.97.

At Brading on the Isle of Wight, an Owen cast gun has a bulged barrel which was probably caused by using a large turf wad when firing a 'blank' round. Turf wadding would expand sideways when a charge was fired, a danger pointed out to modern black powder gunners by Adrian Caruana. The muzzle of that gun also showed evidence of cracking and distortion from a possible overcharge. To conclude the survey, there are fragments from St Saviour Parish of a falconet which had suffered a catastrophic failure. The author concludes that:

> The bulged and cracked barrels, split and distorted muzzles and scored bores, all show a disregard for safety attested by the shattered state of the St Saviour fragment from Jersey, whose gunner, if not other members of the St Saviour militia, was probably severely maimed or more probably killed by this unrecorded explosion.[37]

Loading Guns at Sea

It is worth pointing to what seems to have been a universal naval practice which was to keep guns permanently loaded. The large numbers of loaded guns found during archaeological excavations of shipwrecks would suggest that this was common practice for merchant vessels as well as warships. The muzzle could be plugged with a tampion and the vent with a 'fid' which was: 'a little Okum made like a naile put in at the toutch hole'.[38] This was potentially a very dangerous practice, but it did enable guns to be brought into action quickly and in the light of the loading procedure used at sea may well have been the only shot fired by that gun.

Loading naval ordnance was very different from manning a gun in the field or a fortification. The limited space on the gundeck was one problem, the other was the movement of the ship and of course the gun, which with its carriage could weigh several tons. The truck carriage and the trail carriage will be discussed in the chapter on carriages and the issues surrounding them considered there.

Heavy medieval wrought iron guns were usually breech-loaded and mounted on a large and heavy oak bed, which would have limited their recoil. As breech-loaders they also did not have to be withdrawn to be reloaded. The cast bronze guns on wheeled carriages were sometimes it seems fixed to the hull sides and after firing withdrawn and reloaded. What would become common practice after 1625, using the recoil of the gun inboard to enable reloading, was part of a tactical transformation designed to maximise the firepower of ships. These would now engage in an artillery duel until a ship either blew up or was boarded and captured.

Guns could be drawn in and reloaded or reloaded in situ by the crew outside the hull, an activity which might seem at best foolhardy and at worst

37 Waterhouse, *Tudor Bronze Guns*, p.107.

38 John Smith, *A Sea Grammar* (London: Printed by John Haviland, 1627), p.68.

suicidal. There is a first-hand account of this practice from Jon Olafsson an Icelandic gunner serving in a Danish warship in 1622 when attacked by a corsair off Gibraltar:

> When I had loaded the first gun and was lying outside the gunwale on the next, a heavy wave struck the port side three times, whereupon the ship rolled all the starboard guns over, and me on my gun with them. I swallowed much water, and was nearly carried away, but the Grace of God enabled me to keep hold of the gun, so that I suffered no more than a wetting.[39]

Another account from 1686 also highlights the danger posed by such a practice and comes from an English merchantman assailed by five pirate vessels:

> their men [the pirates] loading their great guns without board (as is the custom of these West India gunner pirates) were cut off as fast as they appeared to do their duty, and this was the reason they fired but few great guns when they bore down upon us, for which we are beholden unto our small arms'.[40]

This account would also suggest that the engagement took place at a very short range. The Dutch artist Willem van de Velde (1633–1707) also made sketches of sailors sitting outboard, astride gun barrels, with rammer and sponge.

It has been argued that such a procedure was only carried out by merchant vessels, lacking the space or the crews to bring guns inboard for loading. Although dangerous it was certainly possible, especially if a ship was out of the action. It is worth remembering that loading tools are considerably longer than the bore of the piece making the procedure potentially very clumsy and slow when space is limited. John Smith refers to loading tools with a rope rather than wooden handle: 'Spunges and Rammers they vse now a stiffe Rope a little more than the length of the Peece, which you may turne and wind within boord as you will, with much more ease and safety than the other'.[41] Where, when and how extensively they were used is unclear.

One of the most imponderable questions is what would be the numbers and level of training to be found in the gun's crew? In the case of the Spanish Armada a reasonable assumption would be that a large piece would be manned by a gunner and up to six soldiers, those who were not already engaged in the defence of the ship. In English ships of the same period and before, the trained gunner commanded a single piece and sailors and supernumerary soldiers may have assisted. A gun crew requires considerable training if it is to perform safely and with expedition, a factor rarely considered. Explaining the principles of loading and the key safety aspects to a new gun crew takes little time but to conduct reloading with safety and under considerable pressure requires regular practice.

39 Martin, *Spanish Armada*, p.370.

40 Martin, *Spanish Armada,* p.370.

41 Smith, *A Sea Grammar*, p.66.

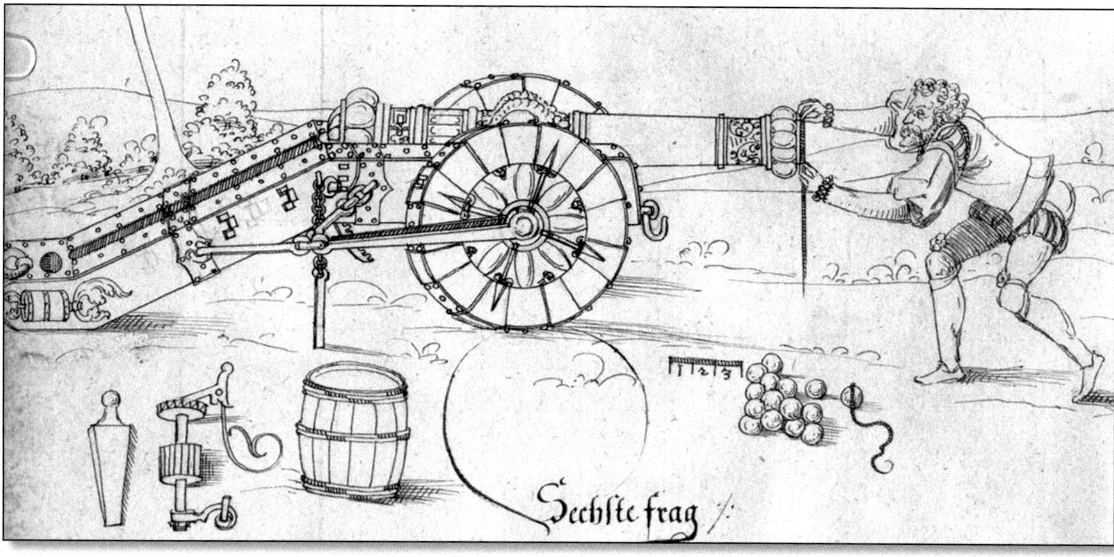

A gunner caught in the act of disparting his piece. He is measuring the height of the bore or 'concavitie'. There is a small powder barrel, a quoin and what appears part of an elevating system on the ground. One of the round shot appears to be attached with string, perhaps to test the bore for precision, as if it became jammed it could be pulled free? (Walther Litzelmann: *Vortrab zu der Arckalay – Büchsenmeisterbuch –* Shelfmark, Fol. Cgm 909 p. 296. Courtesy of the Bavarian State Library)

One factor which would have determined the rate of fire would have been the available stocks of ammunition, which were never extensive. Tudor warships seem to have had 20 to 25 rounds per gun and later up to 40, while the Spanish Armada ships had approximately 50 per gun. The author cannot imagine that any commander could have dedicated even a small fraction of those to actual training. How much useful training could have taken place without live firing is questionable. The gun crew could have been trained with 'false fire' or blank fire, a practice used with Tudor infantry whereby the only powder used was for priming with no main charge being loaded. This practice may have been used but again there is no evidence for this supposition. It would certainly not have prepared them for the blast, noise and recoil that they would experience with a full military load of powder and shot.

Disparting and Aiming

F. L. Taylor, whose book on Italian Warfare concentrates on Ravenna, the Renaissance artillery battle par excellence, asserted that: 'with the rough methods of sighting then in vogue it was very difficult for a big gun to hit a battalion of infantry on the march, while no gun was able with accuracy to hit a target on a different level from itself'.[42] The time and effort devoted in the various artillery manuals to aiming, would suggest that this was not considered to be the case.

Aiming is a technical craft. It requires the individual to understand the weapon that he is using,[43] what the correct sight picture is, what is his point of aim and how it may well differ from the point of impact. The gunner must appreciate the effects of wind and rain as well as the position and relative

42 F. L. Taylor, *The Art of War in Italy* 1494–1529 (London: Greenhill books, 1993), p.88.

43 It is fairly safe to assume that the gunner would always be a man.

movement of the target in relation to him. He might also be expected to understand the contemporary theories related to the trajectory of projectiles.

The first requirement of the gunner was of course to establish whether the axis of the bore matched the line of metal. This was of course essential for safety as well as for accuracy. Once it had been established that the axis of the bore was correct, the gunner should calculate the *dispart* of his piece; that is the difference between the external radius of the gun at the base ring and the muzzle. Because the barrel of a cannon is wider at the breech than at the muzzle, if you aim down the line of metal than you will not be looking along the true bore of the gun. This is a point that Robert Norton made when considering the business, for: 'In generall, you may observe, that in taking ayme by the highest of the mettall, without consideration of the disparture, if the visuall line thereof aspect the marke, the shott will bee alwayes too high within the right range, contrary to the Gunners diseigne'.[44] Simply put, if you aim the gun at a target using the line of metal, then in practice the bore will be pointing upwards. To counter this the gunner had to calculate the *dispart*, the difference between the true bore and the line of metal and fit what was in effect a foresight to the muzzle, so that the line of the gunner's eye to the foresight would be parallel with the bore of the gun.

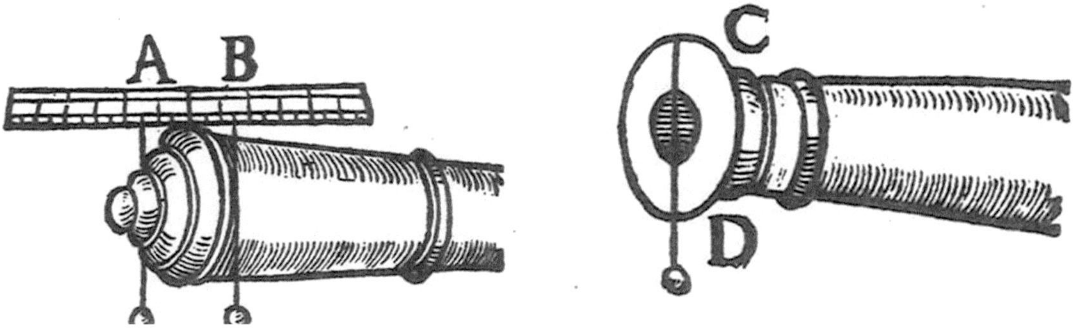

Two methods of disparting a piece. Illustrations from Bourne, *The Arte of Shooting in Great Ordnaunce*.

To achieve this, you can measure the diameter of the breech and muzzle, using a piece of string to measure the circumference, or the diameter by using a ruler and two plumb bobs. To calculate the diameter from the circumference, divide by 22/7 which is π, or if the gunner is mathematically challenged then simply by three. A better way, suggested by Bourne, was to take a ruler and placing two plumblines on either side of the metal at the breech A–B and then another plumbline at the muzzle C–D and measure the amount of metal. Taking the measurement of one from the other and dividing by two will give the dispart.

It was possible to use a vent pricker or *pryming yron* to measure the thickness of metal and the full depth of the bore at breech end. Then by placing the vent pricker at the muzzle again measuring the overall 'depth', the difference between would be the dispart. Once the dispart had been calculated Bourne's practice was simplicity itself: 'I take a straw, setting that perfectly upright, according to the dispart … and make it fast with a

44 Norton, *The Gunner*, p.108.

The gunner manning the rear piece is aiming along the line of the metal using his thumbs. Over the barrel can be seen an Egyptian level calibrated for use as a quadrant and a gunsight with several peepholes drilled for different elevations. The gun in the foreground is being loaded and barrels of powder and loose shot are visible. The containers have the appearance of the separate breeches of wrought iron guns but may be clumsy interpretations of cartridge cases. (*Pirotechnia* of Biringuccio, *c.* 1540)

little wax.'[45] Bourne despaired of gunners who did not calculate the dispart, especially sea gunners. 'What mad men be those gunners that will be of such an opinion, that they need not dispart their ordnance.'[46]

Having in effect created a foresight, what was now required was some form of rearsight. Aiming could be achieved by: 'sighting down the metal between upthrust thumbs',[47] a technique recommended by Vanocchio Biringuccio in his *Pirotechnia* of 1540, and still common practice in British naval gunnery in 1860. Biringuccio's explanation of shooting by rule of thumb follows:

> For ease, most people usually put a little wax in the middle of the rim of the mouth and they make a little hole like a sight with the ends of their thumbs brought together at the foot of the gun, and by looking through this they see the wax and, beyond the wax, the object. Thus by raising and lowering they adjust the breech in a straight line with the mouth and the object.[48]

Of course, the thumbs would be placed some distance over the metal and this would have to be taken into account. It would also only enable the gunner to aim with any degree of accuracy to the point-blank range, that is before there was any significant ballistic drop of the projectile.

A ruler marked in inches could be placed vertically at the breech end, the elevation required for the gun would be measured by using the rule as a rearsight and the muzzle as the foresight. The range achieved at different elevations could be marked on the rule. Commonly gunners would shoot

45 Bourne, *The Arte of Shooting*, p.37.

46 Bourne, *The Arte of Shooting*, p.45.

47 J. F. Guilmartin, *Gunpowder & Galleys: Changing Technology & Mediterranean Warfare at Sea in the 16th Century* (Cambridge: Cambridge University Press, 1974), p.165.

48 Biringuccio, *The Pirotechnia*, p.420.

using their judgement and then assessing the result using an inch rule. If the shot falls short at the eight inches mark, the gun would be elevated to the 10 inches mark and the impact of the shot recorded. If the shot went over the gunner should use the nine inches mark, with the shot (hopefully) hitting the mark. This to Bourne's mind was far too miss and occasionally, but not often enough, hit, for as he said: 'by this order of shooting, he shall never become cunning, although he shoot a thousand shots, for that there is no method or order in the doing thereof, but only with that piece at that mark'.[49] This approach betrays his early experience as a fortress gunner, where distances around the fort could be calculated and potentially ranging shots fired from the gun's actual position.

Bourne denounced the use of an inch ruler as a rearsight in the aid of gunnery. He gives the example of three culverins of 10, 12 and 14 feet. Using the inch rule, at 12 inches the 10 feet length barrel culverin will be elevated by five degrees, the 12-foot barrel less and the 14-foot barrel only four degrees. He does provide a useful conversion table of degrees to inches of elevation at the breech. Thus, the author's falconet with a barrel length of almost precisely five feet would be elevated by 1 1/22 inch (1.05 in) for an elevation of one degree and 5 5/22 inch for five degrees. Bourne was determined to establish the correct elevation in degrees and then to use a sight, graduated in degrees, to lay the gun at the correct elevation.

Thomas Smith explained a very simple way of establishing the height of a degree on any length of barrel. First the length of the bore had to be measured carefully. This would establish the radius of a circle formed by the bore. This number should be doubled to calculate the diameter and thereby the circumference of the circle. Circumference is calculated using the formula $C=\pi/d$ where π is calculated by multiplying by 22 and then dividing by seven (or 3.142). This number should be divided by 360 to give one degree. The author's falconet has a bore 61 inches long, the diameter of a circle with a radius of 61 would be 122, this figure is multiplied by 22 = 2,684 then divided by seven = 383, and then divided by 360 which is 1.06 inches or one inch near enough.

A quadrant placed in the mouth of the gun could establish the degree of elevation or a proper gunsight could be used. Biringuccio describes a simple but effective sight:

> In order to keep the shooting straight, a scale one palmo long is made of wood, sheet iron, brass or copper. This has many little holes at equal distances along the middle and also along the exact middle is a fine line along which a string with a lead in the form of a plummet hangs from the top when desired. To which is attached a stand so that it can be put to rest firmly over the middle of the surface of the breech of the gun. One looks through this for my fixed sight at the first, second, third, fourth, fifth, tenth twelfth hole, or any that is necessary to give its elevation.

49 Bourne, *The Arte of Shooting*, p.19.

An impressive multi-purpose sighting mechanism. There is a graduated gunsight, a pendulum that could be used to level the piece in both axes and what is probably a simple 'battle sight', set probably for point-blank fire. Note that the base is curved and designed to sit on the base ring of the piece. (Walther Litzelmann: *Vortrab zu der Arckalay – Büchsenmeisterbuch* – Shelfmark, Fol. Cgm 909 p. 47. Courtesy of the Bavarian State Library)

Bourne writing some decades later also describes a simple but perfectly effective rearsight:

> Prepare your rule, and let it be well divided into inches, half inches, and quarters of inches and half quarters of inches, and then let there be a slit in the middle of the rule, and in the slit, let there be made in brass or latten a sight, that it may be moved up and down at your pleasure, and then the rule is finished.[50]

The process of aiming could be somewhat time consuming if the rearsight and dispart had to be refitted on each occasion; the dispart was a rather flimsy item after all. He suggested that in a siege once the gun had been aimed satisfactorily the gunner could confirm the correct point of aim for that point of impact by using the old rule of thumb and use that method for successive shots until a new target was designated.

There were a number of factors that could seriously affect the accuracy of the piece, even after the business of taking aim with dispart and sight had been achieved:

> She will shoot wide also if the Trunions lye not directly the one against the other: as also if the platforme lye awry one side higher then the other: or if the Gunner take not his ayme, so that the visuall line that passeth from his eye to the marke: or if one wheele be higher then the other, or if one wheele be stiffe and the other jocond [moved freely], or be squatted with any rub in the reverse [recoil] of the Peece, or run upon soft, and the other upon hard ground; or if the Cariage-maker have made the Cariage leane, or hang more upon one, then the other side: If the Nave [gap between the cheeks of the carriage] be too wide, that the Peece shake therby in her discharge, and reverse: or if the tayle of the Cariage rest harder on the platforme upon one side more then on the other, when the shott is too lowe for the Peece, and at the delivery toucheth upon one side. And lastly the vehemency of the wind being sidewise, over-rules it to the contrary-side-ward, &c.[51]

To this impressive list he would add several more:

> There are also many other Accidentall alterations happening by reason of the winde, the thicknesse or thinnesse of the Ayre, the heating or cooling of the Peece, the different manner of charging by Ramming fast or loose the Powder, by close or loose rouling or lying of the Bullet … whereof no rules certaine can be prescribed, to reduce these uncertaine differences to any certaine proportions:

50 Bourne, *The Arte of Shooting*, p.35.

51 Norton, *The Gunner*, p.109.

An impressive pose from an equally impressively dressed gunner! He is sighting the piece through a rearsight and achieving a few degrees of elevation. There is a long but very narrow wedge or quoin on the ground near the gun. The breech and muzzle appear to have roughly the same diameter which would save the effort of having to calculate the dispart. Also well displayed is the heavy ironwork fitted to the substantial carriage, essential if it was to deal with the rigours of contemporary roads. (Walther Litzelmann: *Vortrab zu der Arckalay – Büchsenmeisterbuch* – Shelfmark, Fol. Cgm 909 p. 292. Courtesy of the Bavarian State Library)

> but all these are by Practise, Discretion, and Judgement to bee considered, and uniformely guided and performed in their best perfection.[52]

It is surprising in the light of these potential pitfalls that the poor gunner hit anything. There were certainly impressive examples of precision. At the siege of Thérouanne Henry VIII personally directed *St Bartholomew*, one of the famed Apostle guns, to shoot at the towers of the cathedral of Notre Dame, hitting the bell-tower with the first ball![53] Benvenuto Cellini the artist and artillerist snapped an Imperial officer in two with a shot from a falconet during the Sack of Rome in 1527. The range was so great that the achievement led to his being presented to Pope Clement himself to be rewarded for his large-bore sniping. At the siege of Montreuil in 1544, one round fired from the besieged town killed Sir Edmund Carew while he sat at table in the army council in Lord Herbert's own tent. This was a mile from the town, which meant that the unfortunate knight was the victim of a very lucky or rather unlucky shot. The most remarkable evidence of sharpshooting comes from Henry Hexham when recording counter-battery fire in the Netherlands in Elizabeth's reign, where after a cannon had been dismounted things became serious:

> one of your [Spanish] Canoniers made so good a shot, that he shot his bullet just into the very bore or month of our Canon, ours being charged, your bullet and ours striking together in our Canon, from that violent blow flew sparkes which gave fire to some loose cornes of powder which were not well driven home, and so our peece giving fire to it selfe, sent you backe your owne bullet, and ours to

52 Norton, *The Gunner*, p.15.

53 C. G. Cruickshank, *Army Royal: Henry VIII`s Invasion of France 1513* (Oxford: Clarendon Press, 1969), p.145.

Although Hexham and Collado may appear to have been exaggerating the precision of artillery, above are two photographs of guns which clearly have been struck by cannonballs. The left-hand gun is in the Heeresgeschichtliches Museum (Vienna) and the other a captured Russian gun from the Crimean war in the Royal Armouries Fort Nelson collection. The Russian gun was hit by at least four balls, one of which entered the barrel and partially tore out the bottom of the chase. (Author's photograph)

> accompany it, without breaking or doing any harme to our peece, and this is most true, for there are some Officers yet living that saw it.[54]

Hapless Henry proceeds to tell more tales of lucky escapes for some and less luck for others with arms and legs thrown into the air and bowels shot to 'pash'. He also describes the damage to the muzzle from a cannonball, an event which can be seen for real in numerous examples to be found in museums. Gunners clearly thought that remarkable precision should be acknowledged and rewarded:

> Louis Collado in his Manuall Practise of Artillery, writeth, that at the Siege of Siena there was a Peece lodged upon the great Church, from which the besieging Army received much domage: but in the end a Germaine Gunner made a shott thereat, who at the first did not onely dismount the same Peece, but also made the Gunner thereof, and those that were about him, flye together in the ayre to their destruction: the which Marquis de Martinian, the Generall of the Army seeing, it pleased him to take a Chayne of gold from his owne necke, and to give it unto the Gunner that made the shot, for his reward, which did not only rejoyce him, but encouraged all the rest of the Gunners to endevour diligently, so to purchase the like honour and reward, when like occasions were presented.[55]

54 Henry Hexham, *The Third Part of the Principles of the Art Military Practised in the Warres of the United Provinces* (Rotterdam, 1643), p.27.

55 Norton, *The Gunner*, p.110.

How to Shoot Straight

Bourne's instructions were quite straightforward. The gunner should correctly dispart the piece and fix his dispart to the mouth of the piece, level the carriage and ensure that the recoil would be even. 'Bring the myddle of the tayle of your piece to the top of the dispart upon the mouth of the piece, and that the mark that you shoot at all three upon one right line, by the light of one of your eyes'.[56]

Once having established how to aim, the gunner's task was to establish point-blank range. This is the distance achieved by the: 'direct fleeing of the shot, without any descending from the mouth of the piece unto the mark'. The gun should be mounted on perfectly flat ground, Bourne recommends a marsh where the water level would assist in ensuring correct measurement. The gun when placed on flat ground would then be levelled using a quadrant and fired at an earth butt. If the shot fell below or above the height of the barrel then the gun would be moved until the point of impact was at exactly the height of the barrel. This would be the point-blank range. In the absence of a convenient marsh an alternative system was suggested by Norton:

> If many Papers or Cloathes be set in the right line or way of the Shotte, betweene the Mouth of the Peece and the Marke, then making an orderly shott with a full Bullet directed to some Marke, setting a sticke in each place, with the top thereof right in the Centre of the hole which the shott went through at every paper: You shall thereby perceive where the shott did first begin to decline, and how much at each distance, which being measured, will very much informe the Gunner both for levell Batteries, and shooting at randon [maximum range].[57]

To establish the range at different elevations Bourne recommends again using marshland for practice. This time the gun should be sited in a (no doubt watery) pit, with its muzzle just above ground level. The quadrant would then be used to measure one degree, the gun fired and the point at which the ball first touched the ground would be measured. In 1813 a 9-pdr cannon achieved a point-blank range at 200 yards, the first graze was at 400 yards and when fired with an elevation of only one degree the range achieved was 700 yards. The closest equivalent Tudor piece of artillery would have been a demi-culverin. John Sheriffe in his list of ordnance of 1590 gives a point-blank range for a 9.5 lb demi-culverin of 400 yards. Bourne's calculation of the relationship between elevation and range was as follows. Using the *right line* as he refers to it as the base unit of measurement, he calculated the following.

+5° of elevation equals right line range +2/9 increase in range
5–10° of elevation equals right line range +1/9 increase in range
10–15° of elevation equals right line just
15–20° of elevation equals right line range +½ increase in range
20–42° of elevation equals right line range +⅔ increase in range[58]

56 Bourne, *The Arte of Shooting*, p.47.
57 Norton, *The Gunner*, pp.104–105.
58 Bourne, *The Arte of Shooting*, p.27.

He did calculate that the maximum range could be achieved with an angle of 42° but counselled that: 'neither is there any service above the mounting of any piece of ordnance above 20 degrees except it be a mortar piece'.[59] Later he asserts that: 'there is commonly no service to be done, above the mounting any piece of ordnance, more than ten degrees.'[60] His table converting inches into degrees of elevation was dependent upon barrel length but only applied to angles up to 10°. It is worth noting that even in 1914 the maximum elevation of the British 13-pdr and French Modele 97 75 mm field guns was 16° and 18° respectively. It was assumed that only direct fire would be used so that high angles of elevation would be unnecessary.

Targets not helpfully positioned on the same level but above the gunner produced a problem but one that it was suggested was readily solved. The gunner would first establish the range to the target and knowing his piece would elevate to the required degree. To reach the target he would then:

1. Raise their piece the number of degrees it would take to reach the target on level ground
2. Determine with their quadrant the number of degrees the target on the hill stood above the piece
3. Raise their piece this additional number of degrees
4. Align target, dispart, and rear sight[61]

Shooting downhill towards a target required a similar calculation but with the angle between the target and the muzzle being subtracted.

To establish the distance to a target Cyprian Lucar proposed the following relatively simple method in his Appendix to Tartaglia. It was a means:

> How you may certainely know by the Gunners Semicircle whether a ship uppon the Sea, or an Armie uppon the land, or any other thing seene a farre of, doth come towardes you, stande still, or goe from you: and how you ought to discharge your great ordinance of diuers sortes against a ship, or an Armie comming towardes you.[62]

One of the gunners first placed his quadrant on a level spot, #I (see figure overleaf) and sighted the target, #5.

Second, without moving his quadrant, he sighted back along it several hundred paces to another point, #3, which he marked.

Third, he turned his quadrant 900, and sighted a point, #2, several hundred paces from point #1, which he also marked.

59 Bourne, *The Arte of Shooting*, p.27.

60 Bourne, *The Arte of Shooting*, p.35.

61 Summary from Henry J. Webb, 'The Science of Gunnery in Elizabethan England', *Isis*, The History of Science Society, Vol. 45, No. 1 (May 1954), p.18.

62 Cyprian Lucar, *A Treatise Named Lucar Appendix, collected by Cyprian Lucar Gentleman, out of divers good authors in divers languages*. In Nicolò Tartglia, *Three Bookes of Colloquies Concerning the Arte of Shooting in Great and Small Pieces of Artillerie*, trans. Cyprian Lucar (London, 1588), Page unnumbered, chapter 67.

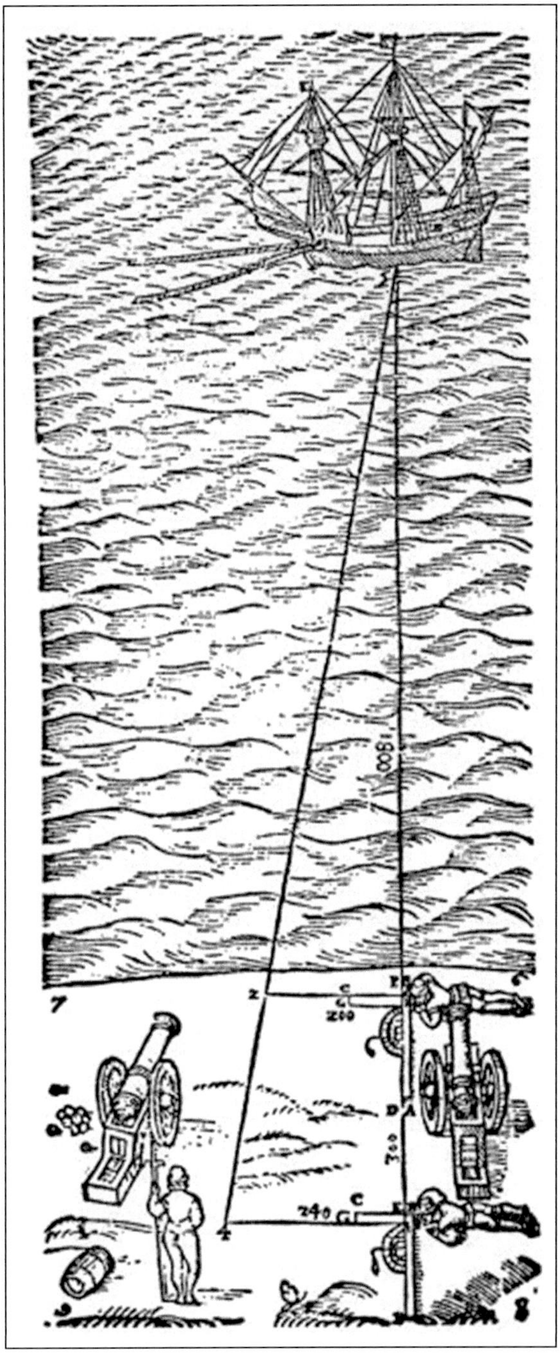

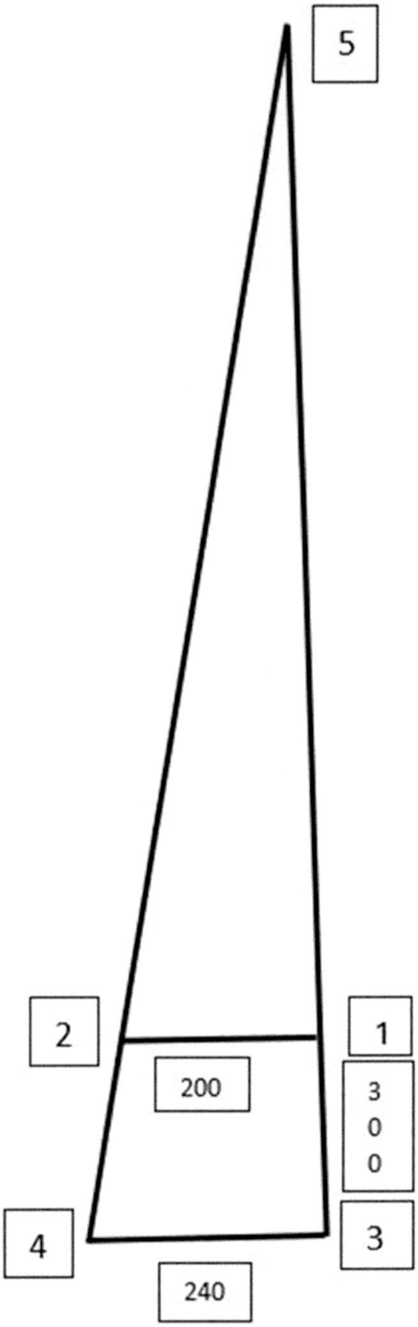
5
2
1
200
300
4
3
240

Fourth, he took his quadrant to point #3 and, after aligning it with points #I and #5, turned it 900 and marked a point, #4, which was in alignment with points #2 and #5.

By measuring distances between points #I and #2, and #I and #3, and #3 and #4, he had enough information to calculate the range to his target. The formula used was as follows:

$$\frac{\text{(distance from 1 to 3) X (distance from 3 to 4)}}{\text{(distance from 3 to 4)} - \text{(distance from 1 to 2)}}$$

For instance, suppose (as Cyprian Lucar does in the accompanying diagram) that the gunner found the distance from #I to #3 to be 300 paces, the distance from #3 to #4 to be 240 paces, and the distance from #I to #2 to be 200 paces. He would then find that the distance to his target was 1,800 paces, as follows:[63]

$$\frac{300 \text{ X } 240}{240 - 200} = \frac{72{,}000}{40} = 1{,}800 \text{ paces}$$

Although no doubt practical for establishing the range of fixed points it would prove fairly useless at sea or against a moving target. Judging distance by eye was and is a common accomplishment for a well-trained observer, while the standard measurement of that period would have been a 'bowshot' of 220 yards or an eighth of a mile.

Gunnery, an Art or a Science?

In *The Geometry of War*, the authors ask the questions: 'was this a practical science or a polite one? Did it belong at war or at court?'[64] They write of the instruments that: 'they were supposed to be employed in battle, but it is clear that their purported military value also had other functions – in justifying textbook geometrical problems,for example, and in attracting patronage'. The clear impression that they give is that it was an art and not a science,[65] dependent upon experience and luck rather than books of tables, beautifully crafted brass instruments and instructional manuals. Guilmartin states his case with authority and conviction when he writes that:

> Experience, basic intelligence, and a 'feel' for the business paid off more surely than precision in following a rigid set of procedures. Sixteenth Century gunnery was an art not a science ... Its finest practitioners were men of unusual ability and daring who were attracted to gunnery by the mystery of the esoteric knowledge involved, to the danger, and to the prestige.[66]

63 Webb, 'The Science of Gunnery', p.19.

64 Jim Bennett and Stephen Johnson, *The Geometry of War 1500–1750* (Oxford: Ashmolean Museum, 1995), <http://www.mhs.ox.ac.uk/geometry/content.htm>

65 Bennett and Johnson, *The Geometry of War*, p.9.

66 Guilmartin, *Gunpowder & Galleys*, p.174.

Colin Martin argues that in 1588 English gunners, compared to their Continental counterparts, actually benefited from a lack of printed gunnery manuals:

> Most of this theoretical knowledge, however, we now know to have been unnecessarily complicated, usually inaccurate, often spurious, and generally of little practical help. Not to have easy access to or interest in such information may have been one of the English gunners' greatest strengths.[67]

A. R. Hall, the principal authority on the history of ballistics in the early modern period, has presented the argument that:

> ballistic theory provided a scientific 'veneer' in support of attempts to gain patronage from rulers and military leaders who were anxious to gain an advantage in the new cannon warfare that played a crucial role in the development of the emerging European nation states from the end of the fifteenth century.[68]

'Science' in his opinion did little to further the 'art' of artillery. France argues that although science had little effect on the practice of artillery it was important in encouraging the practical and theoretical in scientific enquiry, the union of the 'mens' [mind] and the 'res' [thing]. Tartaglia, who was the most influential mathematician in this period, played an important role, for as France records:

> Tartaglia's more important contribution to a science of motion is his vision of what could be achieved and how it could be achieved. He was the first to attempt to build a science of motion on systematic, axiomatic, mathematical foundations as the only basis for precision in gunnery practice, and in his first definition, of an equally heavy body (corpo equalemente grave), he attempted to abstract air resistance from motion, thus breaking decisively from the Aristotelian milieu theory which asserts that a projectile is kept in motion by the air. The separation of motion from its impediments was the crucial prerequisite for its mathematical analysis.[69]

It was important in the history of science even though it contributed little to the accuracy of artillery. Gunnery was in practise neither an art or a science but a craft, a combination of innate skill, experience, common sense and cunning.

Tricks of the Trade

One problem facing an unlucky gunner was if his piece had been spiked, that is having a nail or several nails driven into the vent making it unusable.

67 Martin, *The Equipment and Fighting Potential of the Spanish Armada*, p.376.

68 Catherine Ann France, *Gunnery and the Struggle for the New Science (1537–1687)*, University of Leeds School of Philosophy, Religion and History of Science, September 2014, p.iv.

69 France, *Gunnery and the Struggle for the New Science*, p.24.

Norton as always knew exactly what to do. He disregarded the opinions of others who suggested putting oil in the vent and hitting the muzzle or dismounting the piece and heating the breech until red hot. He proposed drilling another touch hole or loading the piece with a full charge of powder and shot but no wadding and igniting it from the muzzle using a *fusee*: 'I will there make to fire the charge, to avoid danger as much as I can'.[70] This task would clearly require a willing or very gullible 'volunteer'!

To remove a 'rusty pellet', that is a shot rusted in position, Lucar recommended pouring vinegar down the barrel and shifting the ball with a few hefty wallops with a rammer. If still stubborn the main charge of powder could be flushed out through the touch hole and a small powder charge, loaded via the touch hole, used to drive it out. This process implies that guns could be left loaded for a prolonged period, most likely at sea or in a fortification. If this did not work, hot oil and more poking and prising accompanied by vibrating the whole barrel with some hefty whacks on the muzzle should do the trick.[71]

To shoot further with the same gun and ammunition Norton recommended the following procedure after the useful careful loading had been done. Firstly, the bore should be carefully oiled and the gun should be braced against a wall or similar to stop its recoil or make the 'Peece reverselesse'. If time permitted, he recommended drilling the touch hole further forward as this would: 'make the Powder all to fire, and sooner together, if it be rightly done'.[72] Drilling new touch holes was quite common practice when the original had become too wide and was plugged, the idea of initiating the charge from its middle rather than the breech end may have had some virtue but does not seem to have been practised.

The application of steam power to artillery was not something the author thought to discover in this period. The practice, if conducted at all, was a 'merry conceit' but deserves recording here.

> A merry conceit, how to charge a Peece of Ordnance without Gun-powder. This may be done with aire and water onely, having put cold water into the concavity of the peece, filling one quarter of the Cylinder, put in after it a Tampion of wood as long as broad, perfectly made to fit the peece, with an oyled cloath about it, doubled either more or lesse, that it might be two mens strength to put it home, this done, put in the shot, elevate the peece a little, and make a fire under the hinder part thereof, the touch-hole being very close stopped, then put the spunge of the peece in oyle, and wet all the Cylinder of the peece with oyle, that it may passe out the glibber: for when the fire hath rarified the water, it will burst out suddenly. The experiment which we have in long Trunks shooting out pellats with aire onely, proveth this also.[73]

70 Norton, *The Gunners Dialogue*, pp.23–24.

71 Lucar, *Appendix*, p.24.

72 Norton, *The Gunners Dialogue*, pp.17–18.

73 Nye, *Gunnery*, pp.30–31.

3

Artillery in Action

Deployment in the Field

By 1494 artillery had been a feature of warfare for over 160 years. For just about a century artillery had been the essential accompaniment to all sieges, supplementing and then replacing the trebuchet. Although its first appearance on the field of battle was probably at Crécy in 1346 it was still, even by the end of the fifteenth century, not appreciated as a potentially decisive force. This was partly because although the possession of a grand artillery park was a matter of both pride and envy, there was little glory to be found in manning the artillery piece, far more in riding to victory on the field or mounting the breech artillery had made in a siege. The guns were manned by craftsmen and labourers, the gentry and nobility found their place on the battlefield amongst the ranks of the cavalry and at the head of the pike block.

European monarchs developed an obsession with acquiring the largest and most modern artillery park. The Venetian merchant Bavarin was so impressed by the scale of Henry's artillery sent to besiege Therouanne that he wrote that: 'Hell itself could not resist so powerful a force'.[1] In 1518 when a small Venetian fleet arrived in England their ambassador recalled that: 'the king chose to have all the guns fired again and again, marking their range, as he is very curious about matters of this kind'.[2] The Dukes of Ferrara were noted for their devotion to this new aspect of warfare, even going so far as becoming directly involved in their casting. Despite the passion, almost an obsession with artillery in this age, it was still assumed that artillery had little part to play in 'modern' warfare. James Raymond in his study of the subject, asserts that: 'artillery salvos would continue to be valued as much for their psychological effect, as for their practical efficiency for much of the sixteenth

1 *Calendar of State Papers and Manuscripts Relating to English Affairs, Existing in the Archives and Collections of Venice and in Other Libraries in Northern Italy*, Vol. 2 1509–1519, ed. Rawdon Brown (London: Longmans, Green, Reader and Dyer, 1867). 116, September 1513. Dated 2nd and 12th August 1513.

2 James Raymond, *Henry VIII's Military Revolution: The Armies of Sixteenth-century Britain and Europe* (London: Tauris Academic Studies, 2007), p.29. Quoting H. A. Dillon, 'Arms and Armour at Westminster, the Tower and Greenwich, 1547', *Archaeologia or Miscellaneous Tracts relating to Antiquity* (London: 1888), vol. 51, p.228.

century',[3] and Hogg goes further when he writes that: 'noise smoke and all the features of a grand parade was their idea of firepower'. [4] Ian Heath in his study of artillery in the sixteenth century pointed to its weaknesses and concluded: 'though by the 16th Century artillery was present on the battlefield in most major engagements its contribution to the fighting remained minimal'.[5]

Contemporary opinion was also unimpressed by the performance of artillery on the battlefield. Machiavelli, whose expertise in military matters was admittedly questionable, gave little prominence to artillery in the field. He assumed that after a short bombardment, the artillery would either be overrun by the enemy or become irrelevant as friendly formations blocked its field of fire. His account of the very limited contribution of artillery was written in 1516:

> Doe you not heare the artillerie? Ours have alredie shotte, but little hurte the enemie: and thextraordinarie Veliti [infantry], issuying out of their places together with the light horsemen, moste speadely, and with moste merveilous furie, and greateste crie that maie be, thei assaulte the enemie whose artillerie hath discharges ones [once], and hath passed over the heddes of our footemen, without doing them any hurt, and because it cannot shoote the seconde tyme, the Veliti, and our horsemen, have nowe gotten it, and the enemies for to defende it are come forwarde, so that neither our ordinaunce, nor thenemies, can any more doe their office … the armie … with the men of armes on the sides … marcheth in good order, to give the charge on the adversarie. See our artillerie, whiche to give them place, and to leave them the space free, is retired by the same space, from whens the Veliti issued.[6]

Thomas Digges, at the end of the century, was equally dismissive of the role of artillery on the battlefield when he wrote that:

> Farther all men know (that are of skill or experience) that great Artillerie verie seldome or never can hurt any footeman, that vpon giving fire onely doo but abase themselves on their knee till the volew bee past, being much more terrible to vnskilfull and vnexpert new Besoignes [green recruits], than any way hurtfull to trayned Souldiers.[7]

Whitehorne made the same point but added the problem faced when the armies actually engaged:

> footemen easely avoide, either with gowing covered by the nature of the situacion, or with falling downe vpon the grounde, when they shoote: the which also by

---

3 Raymond, *Henry VIII's Military Revolution*, p.27.

4 O. F. G. Hogg, *English Artillery* (Woolwich: Royal Artillery Institution, 1963), p.23.

5 Ian Heath, *Armies of the Sixteenth Century* (St Peter Port: Foundry Books, 1997), p.151.

6 Nicolò Machiavelli, *The Art of War*, in *Machiavelli: The Chief Works and Others*, Vol. II, translated by Allan Gilbert (Durham, NC: Duke University Press, 1989), fol. xlvij. The Veliti were light infantrymen.

7 Thomas Digges, *Four Paradoxes, or Politique Discourses Concerning Military Discipline,* 1604, p.63. Digges died in 1595, and the work was published by his son.

> experience hath beens seene not to be needefull, in especially to be defended from greate ordinaunce, the which cannot in such wyse be levelled, because if they gowe highe, they touche thee not, & if they gow lowe, they will not come neere thee. Then the armies being come to hande strokes, this is more cleare, then the lighte, that neyther the greate, nor the littell can after hurt thee: for that if the same, which hath the artillery, be before, it becommeth thy prisoner, if it behind, it hurteth the friende before thee.[8]

Matthew Sutcliffe an ardent supporter of military archery, was less dismissive of the part that artillery could play but he did point to the circumstances which would negate its effect and he never considered it as playing a decisive role:

> Great artillery against troupes standing thicke and in even ground worketh great effectes. But in other places, and against men ranged otherwise, the sound is greater than the hurte. At Moncontour the Kings ordonance beating among the horsemen ranged hedge-wise, did not in twentie shot hit once. Neither did it anie hurte to the footemen by reason of the vnevenesse of the ground. For eyther it fell lowe, or high. If the same be placed in the front of our armie, yet can not the same be employed past one or two volies. For at the ioyning of the battell, it ceaseth. And if our men march forward it breaketh their arrayes. If the same be placed vpon some hill, yet lying out of levell and shooting downeward, it doth no great hurt. But let it doe the vtmost that it may, being employed by iudicious gunners: yet never was victorie obtained by great ordonance in open field, nor the force of the enemy comming resolutely to the charge thereby stopped.[9]

Artillery appeared to be a one-shot-wonder, an impediment to martial progress on the battlefield and as much if not more of psychological value than material. It is therefore surprising to see so much effort and gold spent on the acquisition of massive artillery parks. These were seen as statements of power across Europe and especially in the England of Henry VIII. In practice, how successful artillery would be on the battlefield depended upon the skill of the gunners, their commander and the context in which it was used.

Contemporary opinion was essentially very conservative about its use. In *The Arte of Warr*, Thomas Audley established a few fairly obvious pointers to the deployment of guns, placing them on the flanks so that they would not interfere with the deployment of the infantry. This should give them a clear field of fire until the infantry and cavalry engaged, hiding the enemy from further bombardment. Placing artillery on the flanks also placed them in danger from being overrun. William Garrard wrote:

> In the fielde, when soever any day of service is, it is the office of the great Maister of the Orednaunce, to select a convenient place to plant his Ordenaunce, as well

8 Peter Whitehorne, *Certaine Wayes for the Ordering of Souldiours in Battelray* (London, 1588), p.37.

9 Matthew Sutcliffe, *The Practice, Proceedings and Lawes of Armes* (London, 1593), pp.190–191.

> to annoy the enemie, as also to be in such sort garded and fortified, that it be not surprised of the enemie.[10]

Whitehorne also stressed the value of field fortifications of one sort or another when deploying artillery:

> artillerie mynding to have it worke, hath needed to be garded eyther of a wall, or of ditches, or of banckes. And when it lacketh one of these gardes, it is taken, or becommeth vnprofitable.[11]

Thomas Digges shared this concern but seems to give a greater value to artillery:

> Neyther is there any Martiall Commaunder of iudgment, that will obiect his Battailes or Battalions as a But [the butts, traditional archery target] for the great or small Artillerie to play vpon: But will alway advance forward their owne shot and lightest armed loose men to encounter their enemies shot, and surprize his Ordinance before the battailes or battalions come within daunger of the Artillerie great or small, to bee plaied on at Poinct Blanke. And thereby shall the enemie bee enforced eyther to put forward his Light-horse or [in]Fanterie before his great Ordinance to guard and defend them: Or else they shall bee possessed or cloyed.[12]

Audley's clearest statement on the matter is a little gnomic to say the least: 'And the Artillerie to be newly charged, and set in good order for doubt of any chance that might happen'.[13] Styward in his *Pathwaie to Martiall Discipline* rather unhelpfully suggests that: 'the great ordnance [should] … be placed in the fronts of the battailes, or in such convenient place as maie most terrifie the enimie'.[14] The objective of the commanders seem to have been a simple one, that of placing the artillery where it could engage the enemy most easily, so that it could be on emplaced either to the front or on the flanks, or from a position that commanded a strategic or tactical advantage.

Norton helpfully summarises the accepted wisdom of the previous century when he advised:

> There are divers opinions therein; some say it is best to plant them in the Front, to disorder the Enemie, and cause them to breake their Ranks at first; which others disliking, say, that will hinder their own party from many advantages, and difficult to retire them: some would plant them in the Flanckes: And some would have some of them in the Front, & some in the Flanke: And lastly, some would have them behinde the Armie, that vpon a signe given, their owne Squadrons may open a way

10 William Garrard, *The Arte of Warre Beeing the Onely Rare Booke of Myllitarie Profession: Corrected and Finished by Captaine Hichcock* (London, 1591) p.281

11 Whitehorne, *Certaine Wayes for the Ordering of Souldiours*, p.38.

12 Digges, *Four Paradoxes,* pp.63–64.

13 Thomas Audley, 'A treatise on the Art of War', *Journal of the Society for Army Historical Research*, Vol. 6, No. 24 (April–June,1927), p.70.

14 Thomas Styward, *The Pathwaie to Martiall Discipline* (London: 1581), p.136.

To hit a large formation was not easy and shooting over the target was a common mistake to make. Skipping shot short was a sensible choice, as long as the ground was reasonable dry as otherwise it might simply swallow them up. If the ground was stony then the impact of the shot would create 'shrapnel', which is why those filling gabions or constructing earth fortifications were enjoined to make sure that they did not include stones in the filling. (Walther Litzelmann: *Vortrab zu der Arckalay – Büchsenmeisterbuch* – Shelfmark, Fol. Cgm 909 p. 330. Courtesy of the Bavarian State Library)

> for the Gunner to play vpon the Enemies Troops or Squadrons. But wheresoever they are planted, care must be had, that the Musketiers, appointed for guards, nor any other fire armes, approach within 100 Paces of the Ordnance, lest untimely firing them worke mischiefe to the party. But having viewed before-hand the place appointed for the Battell, if any eminent rising-place be there neere, to lodge the Ordnance thereon, they may thence most safely play upon the Enemie in Front, Flanke and Travers, with wonderfull advantage and hope of victory.[15]

Henry Hexham, writing in his imaginary dialogue between a general of Ordnance and a new captain, published in a German edition in 1614 and later translated into English, provided a thorough discussion of the placement of artillery in the field. He presented the usual conundrum that the best position in front of the army would endanger them and also make it impossible to support the army once the infantry engaged. What is especially interesting is the stress on mobility and rapidity of fire on the battlefield:

> when they are fastened to them by drawing-Ropes and yron Rings, that upon an instant they [guns] may be remooved and turned for the advantage of our own Troops, and give fire as fast as they can charge and discharge among the Enemies Troops, which is a matter of great moment for the obtaining of a Victorie.[16]
>
> Wherefore I do advise every Captain of the Ordnance, and Master-Gunners, to use some light field peeces, and small Drakes, which upon every occasion may be remooved from one place to another, and couragiously advanced to the places of most advantage, which might offend the Enemie.[17]

Hexham provides a wonderfully realistic appraisal of the problems of siting artillery on a real battlefield, rather than the paper one promoted in printed publications:

> But it happens very seldome that such convenient places can be found for the planting of Ordnance in a Battell, as were to be wished, for oftentimes one shall be driven of necessity to make choice of such places as the ground will afford, by reason of woods, hills, marish grounds, and such like disadvantagious places as you may meet withall, and for which one can give no certain rule.[18]

Bourne provided sound tactical advice when he suggests that a good point of aim would be at a turn in the road, as this is where any advancing formation is most likely to be delayed. It would also be useful to range in some weapons on identifiable points before battle was joined. Bourne recommends using heavy iron shot at a distance and then use falcon, falconet and base shot as the enemy closes. He also refers to the use of chain and clive (cleve or split and articulated) shot. Norton also provided some practical

15 Norton, *The Gunners Dialogue*, pp.22–23.

16 Hexham, *The Third part of the Principles of the Art Military*, p.49.

17 Hexham, *The Third part of the Principles of the Art Military*, p.50.

18 Hexham, *The Third part of the Principles of the Art Military*, p.49.

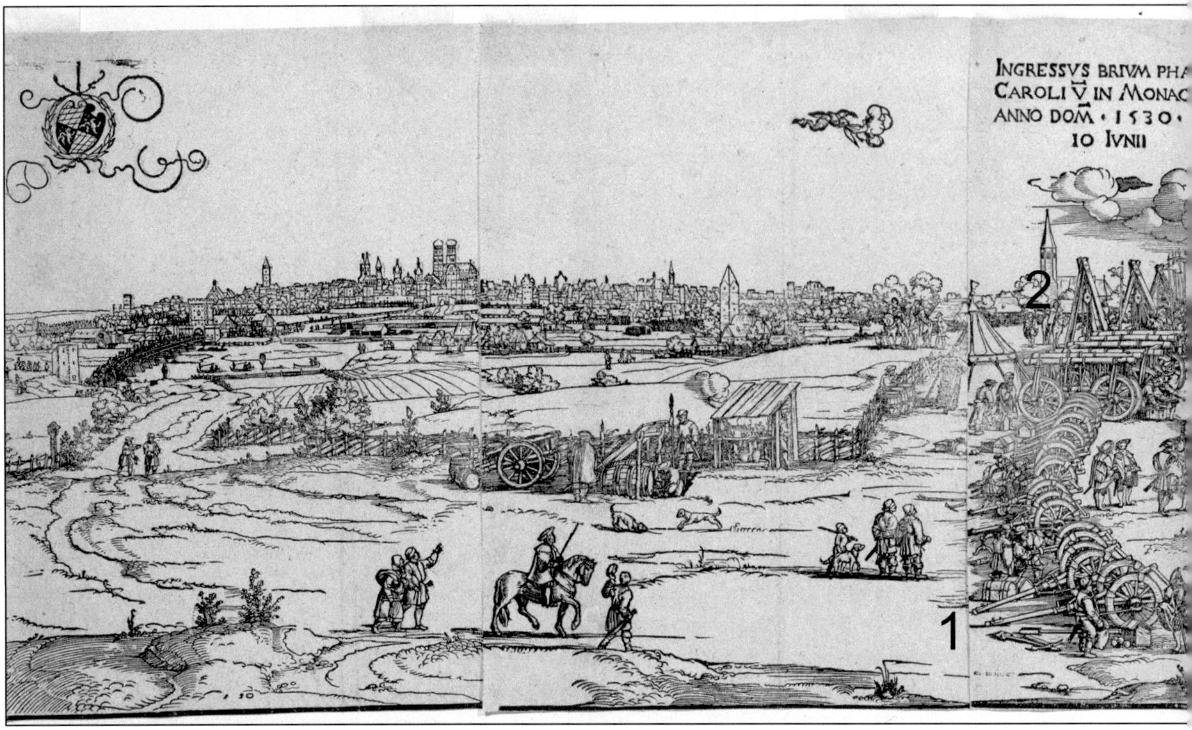

Entry of Charles V into Munich, 10 June 1530 by Sebald Beham. This splendid testimony to Imperial power and contemporary firepower shows almost the full range of artillery and accompanying equipment of the time. (Metropolitan Museum of Art, Open Access)

1. The limber first appeared at the end of the fifteenth century and was crucial in developing the mobility of artillery.

2. These large wagons were used for carrying ammunition and also the essential ancillary equipment, including the large three or four legged hoists which were indispensable if the gun barrel was to be dismounted. Carrying the barrel separately in a wagon or on a block trail would significantly reduce the wear and tear on the carriage.

3. These guns are firing a ceremonial salute from left to right. They are large siege pieces mounted on very similar carriages. The closest gun appears to be unique in this battery as it is clearly of wrought iron rather than cast bronze construction. Note its similarity to one of the French guns taking part in the French invasion of Italy in 1494.

4. Mortars are providing supporting fire for an assault on a fortress as part of a military demonstration. Mortars were able to provide the high angle fire that the flat-shooting field pieces could not. They were therefore essential accompaniments to a siege as was the case at Boulogne in 1544.

5. Two small multi-barrel guns are not taking part in the performance. Quite large in calibre, their individual barrels could be fired, or a train of powder could fire them all simultaneously. The trail is designed to allow one or two horses to draw them swiftly about the battlefield. It is surprising that more use was not made of these weapons. Loading a number of closely placed barrels today is a health and safety nightmare, for obvious reasons.

(Courtesy of the Metropolitan Museum of Art, Open Access)

advice and stressed the importance of enfilading fire in order to maximise casualties, something that is well illustrated in contemporary texts:

> but singly to shoote at a Troope of Horse or Squadron of Foote, and especially in gravelly or stony places it were not amisse purposely to shoote short to light upon the stones, to beate them so up, as that the raysing of the grauel and stones may do the more execution, entering traverse-wise amongst the Enemie. But in a plaine or levell Champion, and amongst Battalions of men the Peece should be so bended and directed, as that her shot may passe by the midst or thickest of them, guirdle [waist] height: and so it may cut off a hundred, or whole Ranke or File of Men at once, and breake their Orders and Rankes, whereby those that fall on their Faces will hardly escape, but shooting over, profiteth nothing at all, but is meerely lost.[19]

Although smaller pieces were clearly more likely to be used in battle this did not preclude guns that would otherwise be considered for use in the siege. Henry VIII, one of the most enthusiastic proponents and purchasers of artillery, employed his famous *Apostles*: 'for closing of the field'.[20] During the engagement of the English forces by the French at Tournehem, the Chronicle of Calais records that: 'the kynge with his [h]ost kept theyr ground,

19 Norton, *The Gunner*, pp.108–109.
20 Cruickshank, *Army Royal*, p.75.

The Papal fortress demonstrates the steps taken towards the full *trace Italienne*. Round towers were added to Hadrian's circular mausoleum in 1447. These were encased by octagonal gun towers in 1492–5 and all were enclosed by Francesco Laparelli's angled bastions between 1561–5. Note how the guns emplaced on the bastion provide flanking fire, the key to the new form of defence. (*Speculum Romanae Magnificentiae* (*The Mirror of Roman Magnificence*): Castello Sant' Angelo 1557 by Giulio de Musi. Courtesy of the Metropolitan Museum of Art, Open Access)

and shot thyr grate ordnaunce, and broke the array of the Frenche men'.[21] As late as 1588, when the distinction between field and siege artillery was well understood, the army 'to incounter the enemy' included in its train four demi-canons, four culverins, three demi-culverins and only seven smaller pieces (sakers, falcons and minions). It seems from this evidence that large pieces could be and sometimes were used in the field.

Technically artillery was potentially very effective. Its limitations were those that it would face for generations to come. It was predominantly a direct-fire weapon and topography and friendly troop deployments could shield an enemy force. It was also an expensive and relatively cumbersome component for any army whose commander had to appreciate its possibilities and limitations if it was to be properly deployed.

21 J. G. Nichols (ed.), *The Chronicle of Calais: In the Reigns of Henry VII and Henry VIII to the Year 1540*, Vol. 35 (London: Camden Society 1846), p.13.

The English battery, referred to as the *Monte*, at the siege of Boulogne in 1544. At Boulogne the full force of England's artillery was to be directed at what was a rather miserable and mostly medieval town. The bombardment was certainly impressive and the preparations were monumental, as suggested by this detail. Vast quantities of powder and shot had to be transported, magazines for the storing of powder built and even the earth for the *Monte* had to excavated and moved from elsewhere. Masons can be seen finishing stone balls for the cannon perriers and bombards. Unlike iron projectiles these could not be recovered but were inevitably smashed to pieces on impact. Detail from the Cowdray engraving. (Dr Dominic Fontana, with kind permission)

Artillery at the Siege

Sieges were the 'bread and butter' of Renaissance warfare, whereas battles were rare birds indeed. Sieges were usually relatively low intensity and low risk for the besieger and the besieged. Battles were risky ventures for both participants and only to be ventured if the point was forced or if one side was certain of victory. In this period there was arguably a transformation in the technology of fortification which, it has been argued, had a profound effect on not only warfare but the future of the state and the transformation of Europe into a global force. This is a debate which goes well beyond the remit of this book and the patience of its author.

What is unarguable is that the medieval castle and city walls were giving way to the *trace Italienne*, a system of fortifications that relied upon thick low walls, angled bastions and artillery to produce a system of defence that would dominate war for centuries. Medieval fortifications could be modernised according to these new principles and still could present a difficult target even for the most modern artillery. As its name suggests the new system emerged in Italy over a long period of development but by the 1530s had become a pan-European phenomenon. It was the star-shaped forts of the Dutch rebels that were to provide the backbone of resistance to Spain in the Eighty Years' War. It was to achieve its ultimate form in the fortifications of Vauban whose methods were to be copied wholesale by competing military engineers for centuries.

The perspective of the humble gunner was a simple one. He had to batter through the walls of stone and earth in the quickest and most efficient way possible. The positioning of the guns was a crucial part of the role of the

Master Gunner. In 1542 Christopher Morres the Master Gunner of the Tower was criticised for attempting to bombard Edinburgh castle from the town. On this occasion after a two-hour long bombardment of the Scottish guns 'nothing impaired them' and the English gunners suffered heavy losses. It was advisable to fire from a position of advantage not simply to open fire at the earliest opportunity on arrival.

At Boulogne in 1544 another strategy was used by the Duke of Suffolk to gain a position of advantage, which was to build an artificial mound: 'The Monte' upon which to place his guns. The collection of enough earth was made very difficult by the hard and thin soil. This was a tactic employed by the Ottoman forces at the siege of Rhodes in 1522, where two such mounds were built to enable their artillery to shoot over the Hospitaller defences. The Scottish guns at the siege of Haddington in 1548–1549 were first mounted on the 'cragg towards Aberlady', probably on the lower slopes of the Garleton Hills, and were capable of firing into the town. They were soon forced to retire, under counter battery fire from the English guns mounted on Bowes bulwark. In the siege of Steenwijk in 1592 a similar mount was constructed to enable the artillery of the besiegers to shoot into the town. The use of large mortars as at Boulogne made this tactic less necessary. Mortars firing shell as well as shot assisted the besiegers in making the lives of the garrison and population miserable. At both Haddington and Boulogne, the fire was so intense and so prolonged that almost all the houses suffered damage.

Bourne is the earliest contemporary author to suggest how artillery should be used in the context of a siege. He recommends that crossfire be directed at the base of the wall: 'as the bottom being beaten away, the top will fall away of itself'.[22] He would allocate guns to bombard at a specific height a foot above or below another siege piece, which seems wholly impractical unless the distance was extremely short. He also encourages the use of volley fire to increase the impact so that the guns might: 'shake the wall at one time together',[23] increasing the likelihood of the wall collapsing. Thomas Smith also believed in the value of concentrated firepower:

> To batter the broade side or curtaine of any wall, Castell or Fort, the middle Ordinance placed on the middle mount or platforme, directlie against the object to be beaten downe, are called the peircers, and are onely to shake and beate the wall, and the Ordinance on the two other side mounts, or platformes shooting something slanting, are to coyne or cut out that which the Ordinance from the middle platforme doth shake or loose.[24]

Continuing the bombardment through the night was necessary if the besiegers were not to allow the enemy to make good at least some of the damage inflicted during the day. Before the placing of the guns a daytime reconnaissance was required. With a sledgehammer, two iron stakes one foot in length and a large astrolabe, the intrepid gun positioning officer would first

22 Bourne, *The Arte of Shooting*, p.65.
23 Bourne, *The Arte of Shooting*, p.65.
24 Smith, 'All Manner of Peeces', p.69.

The gunners are concentrating the fire of their pieces on the flanks of the tower. Flanking fire was at the heart of the defensive system known as the *trace Italienne*. They are very carefully laying their pieces using a sophisticated combination of gun sight and level. The graduated crosspieces on the two trestles would enable the guns to be aligned at night, although this was a rather clumsy method. (German edition of Ramelli's *Le diverse et artificiose machine* of 1588 published in 1620)

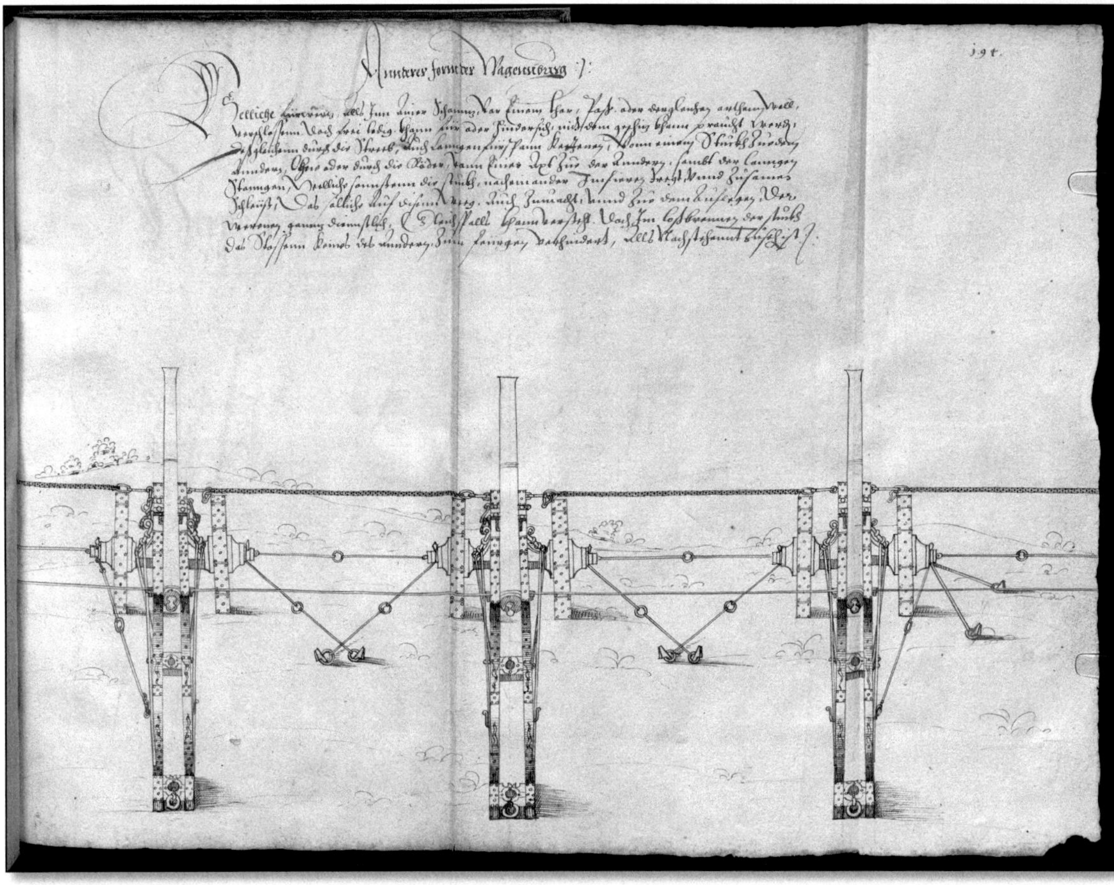

Litzelmann illustrates in great detail a battery of guns joined with chains and rods, pegged accurately to ensure precise alignment. The destruction of fortifications required not a shot to be wasted. A serious siege was as much, if not more, a matter of logistical competence as martial bravery. (Walther Litzelmann: *Vortrab zu der Arckalay – Büchsenmeisterbuch –* Shelfmark, Fol. Cgm 909 p. 411. Courtesy of the Bavarian State Library)

drive the stakes into the ground aligned with the target. He would then use the quadrant to measure the angle to it: 'then remember to write that degree, and place it in some book or paper'.[25] Bourne recommended for this task the use of a very large quadrant with a half diameter of two feet (radius of 1 foot) with the degrees being divided into fourths in order presumably to ensure precision.

At night three guns could be brought forward, the base line would be used to set two parallel lines on either side, 10 feet apart. These three lines would then be used to align the guns, elevation would be set and accuracy was apparently guaranteed! If greater precision was needed to concentrate fire on a specific point, for example an enemy cannon mounted to provide flanking fire, the same procedure would be carried out as above, except that the two lines flanking the centre base line would need to be angled inwards slightly.

Bourne's example of concentrating battery fire worked as follows:

- Distance to target would be 600 feet or 10 score yards
- Base line under the gun 20 feet 1/30 distance to the target.
- The two other guns in the battery would be set 10 feet from the base line

25 Bourne, *The Arte of Shooting*, p.83.

The gabion was a strong and flexible structure which any countryman would be able to make with the coppiced material that would have been readily available. They could be constructed away from the fighting and brought forward to be filled with earth, preferably at night. The jointed wooden structure between the two gabions would be far too complicated and expensive for general use and would also produce dangerous jagged splinters when hit. (Walther Litzelmann: *Vortrab zu der Arckalay – Büchsenmeisterbuch* – Shelfmark, Fol. Cgm 909 p.42 (Courtesy of the Bavarian State Library)

- One inch of lateral movement of the gun would be magnified by 30 times at the target. The gun's trail if set four inches of centre line would then aim 120 inches or 10 feet (4 times 30 equals 120) off, to point at the mark.

What was sauce for the goose was sauce for the gander and it was essential that the besieger should protect himself from the weapons of the besieged. Smith showed his concern for the poor gunners and pioneers who faced the inevitable counter-battery fire with: 'Baskets ramd full of earth being placed betweene each peece of Ordinance [which] are to defend the Gunners and Laborers from hurt of them that are besieged',[26] The gabions, according to Lucar would either be four feet in diameter or eight feet if 'double':

> Also in the very middest of the Gabbion drive a strong peece of oke into the ground, & lay other peeces of tymber joyned fast togeather overthwarte the gabbion from one of his sides to the other. After all this, fill vp the said gabbion with good clay, or with blacke, small, and wet earth well rammed downe, or with greene turfes digged out of meddowes, or with bagges full of earth or fine sande, or with bagges of wooll which is the best thing of al for this purpose and take heed that you do put no stone into any gabbion among those things which do fill it vp.[27]

The author has never felt satisfied that the illustrations of gun positions protected by a single 'rank' of earth-filled gabions would have been adequate, especially as there was also often nothing to protect the gun itself. Henry Hexham provides an excellent and convincing if convoluted description of what was actually required:

> If the Battery be to be made with Gabbions, they being filled with earth without stones,[to avoid stone shrapnel] moistned, and rammed 7 foot a peece in dyametre,

26 Smith, 'All Manner of Peeces', p.69.

27 Lucar, *Appendix*, p.41.

> 3 rankes betwen 2 peeces, if the place will permit or 2 at the least, and 3 rowes also one before another, setting one betwen two, so if the 1 ranke have 3 the second will have two, and the third one, but it will be hard to make a safe Battery with Gabbions, Canon, or Culvering proofe: And each platforme is to have 30 foot for the reverse [recoil]the Canon, and 27 foot for the Demy Canon, he ought to see that it be levelled, or rising 1 foot for 20 backwards the better to stay the reverse and facility, the bringing the peece being loaded to the Loop.[28]

Smith considers that even three gabions constituting over 20 feet of rammed earth would hardly be enough to protect the guns and the gunners. What is also apparent is the space needed to allow the piece to recoil, preferably on a wooden platform that was sloping downwards to assist bringing the gun back into its firing position.

If gabions were not available Lucar provides an engaging and imaginative set of alternatives to protect and hide the gunners, which may or may not have proved up to the task:

> where no baskets of earth are set to defende them, they may shadow themselveswith cables, or bigge ropes, with wet strawe or wet hay, with dead bodies of enemies, with mattresses, shippe sailes or canvas hanged stiffe vp so as the enemie shall not see any marke or man vppon the platfourme to shoote at, and with a line drawe vp the nearer or loost end of the same mattresses, ship sailes or canuas, so often as they shall have cause to shoote at their enemies.[29]

Garrard suggests the use of 'wool packs' or wool sacks. He is probably referring here to a sarpler, the standard English woolsack which was a very large canvas sack weighing approximately 364 pounds, which was equivalent to 240 wool fells. The wool had been compressed and would have absorbed the impact of projectiles very well:

The Impact of Artillery on Fortifications

> 'The violence of the artillerie is suche, that there is no wall, how greate so ever it be, which in few dayes it battereth not doun'.[30]

There were numerous detailed descriptions of ideal modern earthen fortifications, explaining their design, method of construction and dimensions. The following is reasonably representative description of the massive scale of the works if they were to be effective.

> That the rampart of the Bullworks be 15 foot high on the plain of the citie, and that of the curtain twenty five, without reckoning the parapet: briefly that it be so

28 Hexham, *The Third Part of the Principles of the Art Military*, p.14.

29 Lucar, *Appendix*, p.41.

30 Whitehorne, *Certaine Wayes for the Ordering of Souldiours*, p.35.

> high, as it may keep the houses from being battered and ruined. Let their parapets be from 15 to 20 foot thick, and 8 foot high.[31]

David Papillon considered that to be cannon proof ramparts should be 40 feet broad at the base and 20 feet wide at the top. This thickness of well rammed earth could be faced with brick or stone. The earthen structures constructed by military engineers were not simply piles of loose soil. Giovanni Battista Beluzzi described in some detail the materials and methods of making a strong earth structure. Firstly, it required substantial foundations formed by wooden piles driven deep into the ground and the gaps between them filled with rubble. Heavy timber uprights would then be positioned on a five-foot grid. A chain of lateral timber reinforcements would then form a carpet every three or six feet. This would be filled with two to three inch layers of earth and twigs well-rammed. All stones or pebbles had to be removed as they might create shrapnel if hit by a cannonball. The exterior of the wall would be contained by fascines (bundles of twigs), which would be tied to the internal timber structure. The outside could then be provided with a turf wall laid horizontally, secured with vertical pegs. The deck could be made from rammed clay and if necessary additional layers of clay could be used to act as damp-proof courses. The parapet and gun emplacement would then be provided for, using gabions (large wicker baskets), which would be filled with well-rammed earth.

What breadth of earth could contemporary artillery penetrate? The figures for the artillery train of Charles VIII given to Ferdinand by a spy seem extremely unlikely, especially as they are for penetration of a solid wall (see Table 5, p.110). The most precise and near contemporary figures for penetration tests are from 1651. These were conducted by the Professor of Geometry at Gresham College, one John Greaves. The target was made up of three composite oak and elm butts each 19 inches thick. The gap between the first and second butt was 42 feet and the distance to the last was a further 24 feet. What is not known is the distance from the guns to the first butt. The guns fired projectiles weighing from nine to 32 pounds. All guns penetrated the first two butts. Only the balls fired by cannons and surprisingly not culverins struck the third butt. This would support the conclusion of the *Vasa* trials that contemporary warships were easy meat for the cast ordnance of the time. Unfortunately, it is difficult to equate these results to impact on earth defences.

It has been customary to use much later evidence collected systematically by military institutions. European and American figures from 1836 are fairly representative of such exercises. They record penetration of almost seven feet for a 6-pdr shot, 15 feet for 12-pdr and almost 20 feet for a 24-pdr. Prussian artillery experiments carried out at different ranges against freshly made earthworks in 1802 produced some intriguing results:

31 Sieur Du Praissac, *The Art of Warre, or Militarie Discourses of Leavying, Marching, Encamping* (London, 1639), p.36.

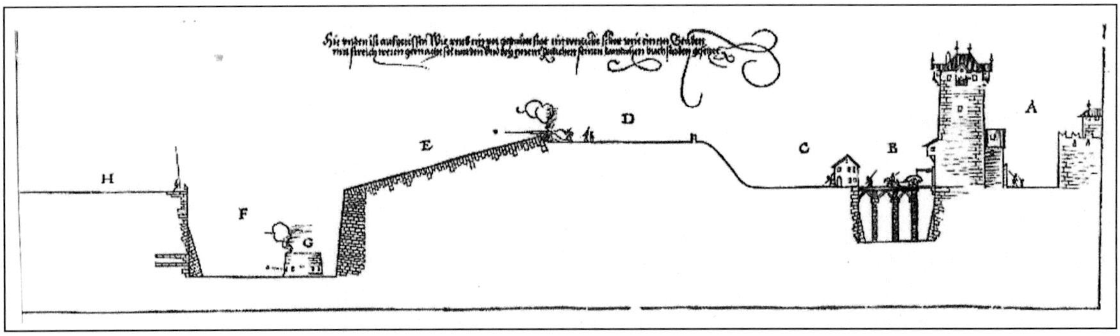

This illustration in Dürer's *Treatise* demonstrated how a feeble medieval town wall might be fronted with a massive earthwork fortification. The earth is protected by a stone carapace to secure it in place. The glacis is well sloped and would provide a glancing surface against incoming fire. The wide rampart would permit guns to be withdrawn into dead ground to avoid unnecessary exposure, or moved easily to a better position. The moat has a small circular pillbox which would develop into the caponier at a later date. What is impressive and daunting is the sheer volume of material that would need to be excavated. Such a fortification was neither cheap nor quick to construct, as Siena discovered to its cost. Even so the difficulty of displacing such a volume of compacted earth with artillery would present immense problems, especially the provision of powder and shot. (Courtesy of the Library of Congress)

Table 1. Prussian Artillery Experiments Against Earthworks

Gun	400 paces	600 paces	800 paces
24-pdr	7–12 ft	3–7 ft	2.5–6 ft
12-pdr	4–7 ft	3–6 ft	2–5 ft
6-pdr	2.5–4.5 ft	2.5–3.5 ft	1–5 ft
3-pdr	3–6 ft	2.5–8 ft	1–2.5 ft

Accepting the Prussian precision of these figures and the very wide range to be found within individual results, the conclusion is that what really mattered to the gunner was his distance from the target and the closer the better. It is also worth stressing that these were tests against loose earthworks erected as field fortifications, not the carefully made and rammed earth fortifications that would be met in a siege.

The most reliable figures that the author has found for the effect of artillery on 'proper' fortifications are those of Sébastien Le Prestre de Vauban, the grand master of military engineering. They are admittedly from the end of the seventeenth century but they have the weight of his authority behind them, as he was after all designing fortifications to absorb the impact of such weapons. He distinguishes between the penetration achieved against different qualities of earth, so that a cannonball: 'pierces two Fathoms [12 feet] of clos'd Earth. If it be not well setled, it pierces two fathoms and a half; and if the Earth be poor, and hungry, it pierces four fathoms'.[32]

32 Du Praissac, *The Art of Warre*, p.36.

Aston Hall in Birmingham was besieged by Parliamentarian forces in December 1643. One round shot passed into the house, shattering the staircase, causing large fragments to fly off. The damage was never repaired, kept as a testimony to the loyalty of the family to the royalist cause. As can be seen the neat entry hole disguises the massive damage to the newel post. The diameter of the hole would suggest that a demi-culverin was the culprit. (Author's photographs)

During the English Civil War St Bartholomew's Church in Tong was struck by musket and cannonballs. What is surprising is the limit of damage to a relatively small portion of the stonework around the impact point, with the substantial masonry wall resisting what was probably the standard field piece of the day with apparent ease. (Photo by 'The joy of all things', shared under Creative Commons license CC BY-SA 4.0, via Wikimedia Commons)

Table 2[33]

Type	Ball weight	Penetration of 'clos'd' earth
Cannon	33 lb	12 ft at 200 yards
Dutch cannon	48 lb	20 ft at 400 ft
Culverin	24 lb	12 ft at 100 yards
Falcon	12 lb	7 ft at 200 ft

What is clear is that heavy artillery could flatten whatever fortification it met, as long as the supply of ammunition was adequate and that the gunner knew his business, although at a range of no more than a few hundred yards at most.

The French siege of the English fortification at Haddington was very successful in reducing some of the earth bastions so that some were: 'beaten down to flatte so that a man might ryde in and out at the breach'.[34] Haddington did not fall, it was abandoned later but was not captured. It is important to stress that the outcome of a siege was dependent upon many factors of which the competence of the artillery was but one.

33 Sébastien Le Prestre de Vauban, 1633–1707. *The New Method of Fortification, as practised by Monsieur de Vauban, Engineer General of France with an explication of all terms appertaining to that art made English* (London: printed for Abel Swall, 1691), p.104.

34 Fulwell Ulpian, *The Flower of Fame: Containing the bright renowne, & moste fortunate raigne of King Henry the viii* (London, 1575), p.51.

Artillery at Sea

Bourne begins his disquisition with a blanket criticism of naval gunners who did not know how to dispart their pieces. This was especially relevant for naval guns as they were 'Great Ordnaunce' with a lot of metal in the breech, which would make them difficult to aim accurately by looking down the line of metal. When shooting at a ship Bourne considered that the gunner should take into account the heel of the ship and the rise and fall of the enemy vessel, shooting as it rises rather than descending. What he considered essential for accurate gunnery was a good helmsman, someone who was nimble and steady who could steer a good course, keeping the enemy vessel at the same range and with the benefit of the weather gauge. It was principally the ship that aimed the guns, the gunner would fire when his gun came to bear. This was especially true as the guns were not arranged in straight lines but mounted along the curved shape of the hull.

Bourne was especially aware of the problem of having a point of aim well over the target, caused by the 'loopy' trajectory of artillery. He suggested that using a cloud as the point of aim might suffice, which points to the problem of accuracy at long range. The *Vasa* trials produced some fascinating results. At only 20 metres the gun achieved a group of approximately 50 cm. At 1,000 metres if these results were extrapolated it would have achieved 11 hits on the hull and eight more in the rigging, all out of 22 rounds. This was impressive but with both gun and target stationary. The *Vasa* team calculated that the angle subtended by the ship's hull at this distance would only be 0.2 of a degree, in addition the ship may be rolling some six degrees, with the point of aim at that distance having to be well over the topmast. These results would suggest that at sea, under most circumstances, reasonable accuracy could only be achieved at comparatively short ranges, far less than the range at which they could still do serious damage.

The alternative to striking the ship directly, was to skip the shot across the water to the target (see illustrations overleaf). This was achieved by shooting low on the target, for: 'if that [shot] cometh short of the ship, it will graze in the water and rise again, and speed well enough, so that it be not too short of the ship, for too much too short both kill the shot in the sea and especially if that the distance be any thing far off'.[35] The practice he described is well understood by anyone who has wasted hours skipping stones. The height of the gun above the water was crucial. When: 'the piece shall be little higher than the water, then shall the shot run grazing in this form to rise again by that proportion that it doth hit the water, and to run til that the great force be decayed'.[36]

If the angle was too great the shot would simply plunge into the water or might skip over the ship. The Dictionary of Marine of 1780 considered that ricochet firing was practical with elevations from 3° to 6°. In Bourne's experience the same practice of 'skipping' shot was true for shooting on land, 'but the water is more certainer and truer'. He considered that with a

35 Bourne, *The Arte of Shooting*, p.55.
36 Bourne, *The Arte of Shooting*, p.81.

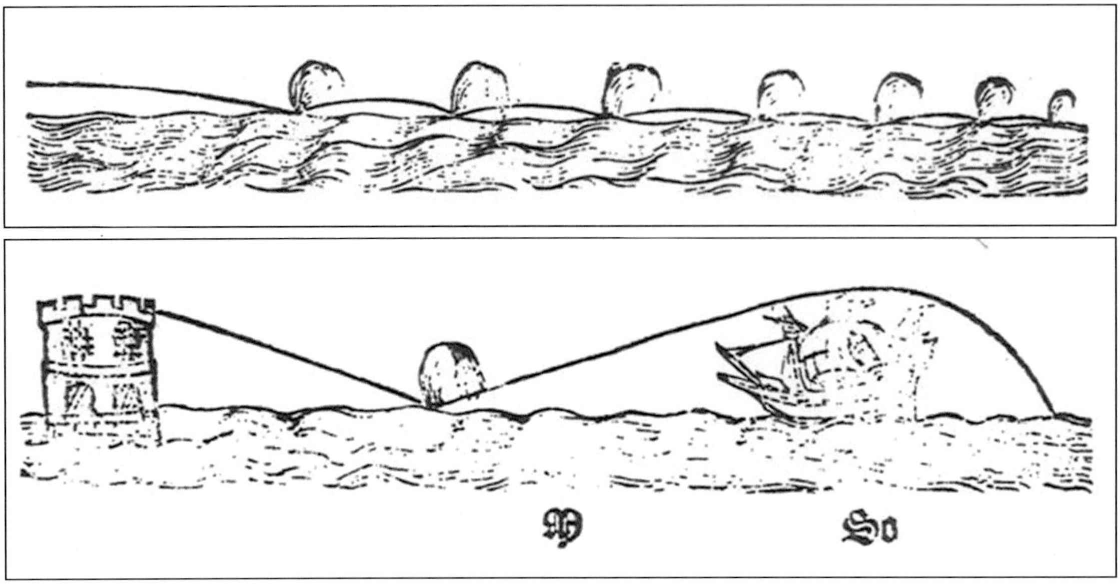

Skipping a cannonball across the to a target. From William Bourne, *The Arte of Shooting in Great Ordnaunce*

low angle of strike the ball could travel up to three quarters of its possible maximum range, whether on land or water.

William Johnson when considering the historical accounts of the ricochet, pointed to the problems exploiting the phenomenon at sea, which was:

> limited because it could have to take account of the steep sides of waves which at sea could be a few feet high; velocity was very rapidly lost if a cannonball had to penetrate water to some extent. Ricochet firing at sea was clearly only effective in circumstances where it was relatively calm and this was frequently only so close to a shoreline.[37]

Experiments aboard HMS *Excellent* in 1838 demonstrated what could be achieved under ideal conditions. An 18-pdr. with a barrel some 5' 4" above the plane achieved a first graze at 250 yards when firing horizontally, with the last of 29 grazes at 2,660 yards from the ship. At two degrees the first graze was at 926 yards and the ninth at 2,293.[38]

In *The Arte of Shooting* Bourne considered that the 'great' guns aboard ship should be used to dismast an opponent or damage the rudder but not, it seems, for sinking. For repelling boarders or for preceding the boarding of an enemy vessel, fowlers, bases and slings, short range anti-personnel weapons, would prove of use. Bourne did not consider that these smaller weapons required much in the way of training or even the measurement of their dispart, as their effective range would be measured in tens of yards. He thought that the gun mountings or 'truckes', of the heavy pieces, should be as low as possible to place the guns as close to the ship's hull and reduce recoil.

37 William Johnson, 'Ricochet of Non-Spinning Projectiles, Mainly from Water. Part I: Some Historical Contributions', *International Journal of Impact Engineering*, 9, Vol. 21, Nos. 1, 2 (1998), p. 19.

38 Johnson, 'Ricochet of non-spinning projectiles', pp.19–20.

The barrel when placed in the gunport should be dead centre, permitting equal angles of elevation or depression.

To shoot at a ship at sea from a fort or bulwark at night was also dealt with by Bourne. In this case the guns could be aligned with lights set on the other side of an estuary or fortification, or if needs be a small boat provided with a lantern. Any ship passing such a light would obscure it and at that moment it would be fired upon by the ever-vigilant gunner. The shot would be fired to graze the water rather than striking the ship directly and a depression of the barrel between one and three degrees was recommended; much of course was dependent upon the state of the tide and sea.

Impact of Artillery on Ships

How effective were cannons against wooden walls? The Royal Armouries constructed a *Mary Rose* port-piece and a demi-culverin.[39] These were tested against a wooden target corresponding to the hull of the *Mary Rose* at deck level. More recently the *Vasa* Museum has built a reconstruction of one of the ships 24-pdr cast bronze gun, which was then shot against a reconstruction of a portion of the warship.

The impact of the *Mary Rose* iron port-piece firing 12 lb stone shot on the 'ship' target was impressive. One shot for example substantially damaged: 'all three layers of the structure, the outer planking, frames and inner stringer'. Undoubtedly, at short range with stone shot the hull of a major warship would have been very vulnerable. In another test shot an original stone ball from the *Mary Rose* was used. The projectile broke up in flight, but even the fragments penetrated the 'ship'. The ranges at which the gun was used were very short: under 20 yards. A longer shot fired over 131 m lost about a sixth of its velocity but would still have penetrated the hull. The velocities achieved varied from 141–298 m/s, although the higher velocities using large charges of modern powder would not be representative of contemporary examples. The Royal Armouries' demi-culverin smashed through a target, equivalent to the hull of the *Mary Rose,* at velocities ranging from 130–199 m/s causing substantial damage.

The *Vasa* gun fired iron balls at transonic speeds that produced highly characteristic holes in the hull, with the splintering caused by the pressure wave ahead of it. The 5.5 inch diameter balls were unstoppable but the amount of damage they caused to the hull and the crew depended almost entirely on where they struck. Using modern techniques, employed by the military to assess damage to vehicle crews, it was found that contemporary wool clothing would have deflected 90 percent of small splinters, only large splinters would do much damage. Fred Hocker, the Director of Research at the *Vasa* Museum and responsible for the trials, considered that the ball created a large cloud of small splinters which were far less lethal than he

39 Nicholas Hall, 'Building and Firing a Replica Mary Rose Port-Piece', *Royal Armouries Yearbook*, Vol. 3 (1998), pp.57–66.

would have expected.[40] If on the other hand one of the large deck-supporting knees was hit, then the instant demolition of large timbers at 200 m/s would have certainly proved lethal to anyone nearby.

On other occasions a ball could pass through the hull a foot or two away from a sailor and leave him no doubt surprised and grateful but relatively unharmed. This perhaps helps to explain why so much ammunition could be expended at short range in sea fights without producing mass casualties: it was not principally enemy projectiles which killed the sailors but parts of their own ship. Sir John Hawkins experienced such a 'close call' during the sea fight at San Juan de Ulua. The gunner Job Hortop reported that:

> The Jesus was being wonderfully pierced with shot … our General [Hawkins] courageously cheered up his soldiers and gunners and called to Samuel, his page for a cup of beer, who brought it to him in a silver cup, and he drinking to all men willed the gunners to stand by their ordnance lustily like men. He had no sooner set the cup down out of his hand, but a demi-culverin shot struck away the cup and a cooper's plane that stood by the mainmast and ran out on the other side of the ship, which nothing dismayed our General, for he ceased not to encourage us saying 'Fear nothing! For God who hath preserved me from this shot, will deliver us from these traitors and villains.'[41]

Hawkins, whose experience of ordnance at sea was extensive, especially it seems at the receiving end, made this point relating to the relative effectiveness of different pieces of artillery:

> For although their [Spanish] Artillery were longer, waightier, and many more than ours, and in truth did pierce with greater violence; yet ours being of greater bore, and carrying a waightier and greater shot, was of more importance and of better effect for sinking and spoyling: For the smaller shot passeth through, and maketh but his whole [hole], and harmeth that which lyeth in his way; but the greater shaketh and shivereth all it meeteth, and with the splinters, or that which it encountreth, many times doth more hurt, then with his proper circumference: as is plainely seene in the battery by land, when the Saker, the Demy-Colverin, the Colverin, and demi-Canon, (being peeces that reach much further point blanke then the Cannon) are nothing of like importance for making the breach, as is the Cannon; for that this shot being ponderous pierceth with difficultie, yea worketh better effects, tormenting, shaking and overthrowing all; whereas the others, with their violence, pierce better, and make onely their hole, and so hide themselves in the Wooll [stone wall] or Rampire [earthen rampart].[42]

40 Fred Hocker, 'Ships, shot and splinters: the effect of 17th-century naval ordnance on ship structure', in J. Litwin (ed.), *Baltic and beyond: Change and Continuity in Shipbuilding*, Gdańsk 2015. Proceedings of the 14th International Symposium on Boat and Ship Archaeology, pp.221–228 (Gdansk, 2017)

41 Peter Padfield, *Guns at Sea* (London: Hugh Evelyn, 1979), p.39. From *The Rare Travels of Job Hortrop* (William Wright, 1591), Cotton MSS; also Haklyut, quoted by J. Williamson, *Sir John Hawkins* (Oxford: Oxford University Press, 1927)

42 Sir Richard Hawkins, *The Observations of Sir Richard Hawkins Knight, in his voyage into the South Sea. Anno Domini 1593.* (London: John Jaggard, 1622) p.148.

The impact of a ball on the reconstruction of the *Vasa* hull was impressive, producing a very neat hole. Similar results were achieved with the Royal Armouries culverin. What was more impressive was the damage done to the interior of the ship. As can be seen, the exterior of the hull did very little to impede the passage of the projectile. It would require several feet of seasoned oak to protect against a cannonball. (Photograph courtesy of Niklas Gustavsson, Combat Camera)

He makes the point that smaller diameter shot produce less collateral damage than larger calibre but lower velocity pieces. The other point he makes, about the injuries caused by splinters rather than by the ball itself, is also well expressed, when he writes that the projectile 'shaketh and shivereth all it meeteth'.

A recent series of tests conducted by a team of Israeli archaeologists investigating a wooden ship sunk in 1840 has produced some fascinating and apparently contradictory results.[43] Using a carefully scaled-down 'gun' and target, they found that high velocity steel balls traveling at 400–500 m/s punched a neat hole through the target and sped on their way. When much lower velocity shots were fired, at 200 and 150 m/s the entry hole was the same as the higher velocity shots but the generation of large splinters was much greater, especially in the case of the lowest velocity shot. At only 150 m/s the exit hole was the largest of any of the shots and the damage and splinter generation was the greatest. In the case of the 200 m/s shot some

43 T. Kahanov, Y. Me-Bar, D. Cvikel, A. Hillman (eds), 'Akko 1 shipwreck: the effect of cannon fire on the wooden hull', *Journal of Archaeological Science*, 39 (2012) pp.1993–2002.

These two images from the *Vasa* gun tests show the impact of two balls inside the hull. The balls were travelling at 340 m/s in both. In the upper photograph the ball has hit a 'thin' (45 cm) section of the hull. The cloud of small splinters, moving at a similar speed to the ball, although impressive is not that lethal. In the lower photograph the ball has hit one of the knees supporting the deck. The larger splinters are travelling at 180–200m/s. The ball in this photograph has already passed through the target. It is clearly the ship that kills the crew, not usually the projectile. (Photos courtesy of the Bofors Test Centre)

54 percent of the energy was absorbed by the target while in the case of the lower velocity, some 91 percent of its energy was absorbed, which explains the greater extent of the damage. It would be helpful to see tests conducted at full scale. The port-piece, firing a stone projectile at a modest 150 m/s, could have been a far more powerful weapon than previously thought.

Trollope in his analysis of English firepower at the time of the Armada was particularly impressed by the performance of stone firing guns:

> Damage caused by an 8inch perrier shot at the short ranges at which it will penetrate is in a class of its own. While splinters will fly, and there will be many more [than for smaller diameter iron shot] guns and carriages overturned. The shot has the capacity not just to punch holes, but also to shatter timbers. The shot whose diameter is a substantial portion of the width of any ships' plank, will not only penetrate but possibly start the planks at their fastenings or even carry away at section of the planking altogether. Between wind and water this must be a potential ship-sinker.[44]

He supports this assertion by referring to the transfer of perriers from garrison to sea service on the post-Armada new-build ships, where they were kept in service as late as 1623. It is also worth noting the much larger proportion of perriers in the Armada of 1588, about a quarter compared to a little over two percent in the English fleet.[45] Perhaps a matter of choice as much as necessity.

The projectile of iron or stone could produce a ragged hole in the hull of any wooden vessel, but would it prove fatal? The record of sinking ships through gunfire alone is not impressive. The first recorded sinking was at the battle of Zonchio in 1499, and well into the Napoleonic era although ships could be reduced to splintered wrecks above the water, they were rarely sunk by the direct effects of gunfire. At Zonchio three ships were sunk as a result of the catastrophic explosion of their magazines, as was the recently discovered *Mars*, sunk in action in 1564. The tactic employed across the centuries was to disable the ship and its crew not to sink it. The point of aim was 'between wind and water', not to hit below the waterline. One problem faced by shot when hitting the water was the very rapid loss of velocity that significantly diminished its ability to penetrate the hull.

Holes would be plugged with cones of soft wood, or with canvas and mattresses used externally to cover the damage. John Smith in his account of an imaginary seafight used: 'a plug lapped in Okum, and well farred in a tarpawling clout'.[46] In trials conducted in the mid nineteenth century in America a 32 lb shot with an initial velocity of 1,600 ft/s against seasoned white oak timbers produced some intriguing results. The target, hit at a range of 1,300 yards, was penetrated to a depth of 21 inches but:

---

44 Charles J. N. Trollope, 'The Guns of the Queen's Ships during the Spanish Armada', *Journal of the Ordnance Society*, 2002, Vol. 6 (1994), p.27.

45 I. A. A. Thompson, 'Spanish Armada Guns', *The Mariner's Mirror*, Vol. 61, (1975), p.362.

46 Smith, *A Sea Grammar*, p.78.

> The wood fibres at the edge of the tunnel had been pushed forward with the ball's passage and sprang back to their original position after the ball passed onwards. The diameter of the tunnel is therefore reduced to approximately ⅔ the diameter of the shot. The fibres in the centre of the tunnel are reduced to powder and flow out and around the ball. The passage of the ball creates a vacuum behind the ball which then pulls the fibre powder back in behind the ball and creates a 'dense pack' of powdered fibres.[47]

In effect the timber self-healed, significantly reducing the ingress of water.

Sinking a ship was a difficult task as was proved in 1588 where only one ship, the *Maria Juan*, was directly sunk by gunfire. Heavy guns could still cause horrendous damage to ships and their crews. The *Santa Maria* suffered 200 casualties through 'fight and sickness', and: 'was shot through four times, and one of the shots was between the wind and the water, whereof they thought she would have sunk, and most of her tackle [rigging] was spoiled with shot'.[48] At the heavy fighting at Gravelines the *San Sicilia* was battered:

> so heavily from their [English] guns that they completely shattered her ... [we had to] repair the damage from the many shots which the ship had received alow and aloft and from the prow to the stern, and below the waterline in places difficult to repair ... The ship remained so damaged and battered by cannonballs that she could move forward only with great difficulty.[49]

The *San Martin* was struck by over 200 balls and was only saved by a pair of divers who fitted oakum and lead patches to the hull. The *San Mateo*, surrounded by over a dozen enemy, became 'a thing of pity to see, riddled with shot like a sieve'. Worst of all was one poor ship seen to be 'all full of blood'.

What conclusions can be drawn from these accounts and recent ship-breaking trials about the effectiveness of late medieval shipborne artillery? Firstly, that large-calibre wrought iron chambered guns could certainly punch their stone projectiles through one side of a ship while the larger bronze pieces could send their shot right through any vessel. The damage that would be caused to the ship and the crew would depend very much upon which portion of the ship's structure it hit. The effective engagement range would at most have been at about 200 yards and probably far less. It was most unlikely that a ship would be sunk by gunfire or even that this was the intention. The target would not have been the ship but its crew. This is why the armament of medieval ships were predominantly small-calibre weapons such as the universal 'serpentines', mounted on swivels, especially on the fore and stern castles, where their withering fire could be concentrated on the enemy crew. Larger guns certainly could cause massive damage to the ship as well as the crew, as was demonstrated during the Armada, but it was still not in the interests of anyone to sink a valuable prize. The stress

47 Roger A. Haskins, 'Effect of Shot and Shell upon a Wooden Warship', *Newsletter of the Ordnance Society*, No. 135, October 2021, p.21.

48 Martin, *The Equipment and Fighting Potential of the Spanish Armada*, p.141.

49 Martin, *The Equipment and Fighting Potential of the Spanish Armada*, p.374.

on heavier armament and the reduction of smaller anti-personnel weapons was criticised by the more conservative English captains at the time of the Armada as a retrograde step.

The Range of Artillery and the Science of Ballistics

There was in the sixteenth century and still today an obsession with establishing and calculating the maximum range or *randon* of artillery pieces. Perhaps it would be better to enquire as to what target this piece could hit at what distance, under a list of fairly precise conditions. What is certainly not accurate are the figures so readily given by contemporary records for the maximum ranges of various guns. There is also a considerable difference between the range that a gun could achieve at its maximum elevation and the range at which it could effectively engage its target.

Guilmartin is rather dismissive of shooting at long range. He asserts that: 'it is probably accurate to say that most sixteenth-century gunners knew little (at least little that was correct) and cared less about accurate long-range shooting,The reason for this is not hard to find: sixteenth-century artillery was quite ineffective at long ranges'.[50] The combination of excessive windage (for smaller pieces) and the high induced drag for projectiles, allied to inconsistent powder and rule-of-thumb aiming, made shooting at anything significantly further than point-blank ranges a waste of ammunition.

Colin Martin in his thesis also draws upon the conclusion of John Guilmartin:

> The unreliability of sixteenth century range data was demonstrated by an analysis of the values given by Collado and by Diego Prado y Tovar, undertaken by Mr. J. W. Kochenderfer and his co-workers in the Firing Tables Branch, U. S. Army Ballistic Research Laboratory, Aberdeen Proving Ground, Maryland … who determined that the maximum ranges given by Collado and all the ranges given by Prado y Tovar which were subjected to analysis would have been attainable only with muzzle velocities in the neighbourhood of 6,000 feet per second, nearly five times the speed of sound and almost three times the muzzle velocities of modern small arms.
>
> Experimentally derived mid-nineteenth century data, most authoritatively given in the published work of Captain Thomas Jefferson Rodman, makes it plain that any muzzle velocity for black powder artillery be held in excess of 2,000 feet per second must suspect, particularly for the larger pieces … It seems clear that to the sixteenth century gunner, long ranges in general and maximum ranges in particular were a highly theoretical proposition. We should not therefore expect too much accuracy in what he tells us about them … Most of the so-called range figures which he left us are therefore little more than educated guesses which reflect his prejudices concerning the types of gun which he felt should have been able to shoot farthest and which he himself preferred when long-range fire was called for.[51]

50 Guilmartin, *Gunpowder & Galleys*, p.162.

51 Guilmartin, *Gunpowder & Galleys*, pp.279–281, in Martin, *The Equipment and Fighting Potential of the Spanish Armada*, pp.239–240.

When his home city of Brescia was ransacked by a French army in 1512, the young Nicolò Tartaglia was badly disfigured. He was attacked by a soldier in the church in which he had sought sanctuary and left for dead with deep cuts to his face and skull. It was only the devoted nursing of his mother that saved his life. The sword cuts were hidden by a full beard, but he was left with a permanent stammer. (Fig. 85 in Guttmann's *Monumenta*, 1584))

Part of the problem was that there was little understanding of the science of ballistics. Nicolò Tartaglia (1499/1500–1557) was the most important figure in the history of ballistics in this period. He had been brought up in poverty but emerged as a mathematical prodigy although prone to bad-tempered disputes with his colleagues. He published his enormously influential textbook *Nova Scientia* in 1537. It was the first attempt to explain the subject of ballistics specifically as a response to the needs of the new artillery.

The Aristotelian concepts of natural and violent motion were at this time universally accepted; these assumed that:

> The natural movement of equally heavy bodies is the movement they accomplish from a higher place to a lower one, perpendicularly and without any violence. Natural motion starts from zero and continuously increases. The violent movement of equally heavy bodies is the movement they accomplish with effort either upwards or downwards, to the right or the left, and is caused by a moving power. Violent motion starts fast and gradually diminishes.[52]

He also assumed that:

> Falling bodies fall at steady speed. Heavier things fall faster, the speed being proportional to the weight. The speed of fall of a given object depends inversely on the density of the medium it is falling through.

Thus, according to Aristotle, a projectile when launched by force followed a straight line and slowed until the violent motion stopped at which point natural motion took over forcing the projectile to fall. Tartaglia significantly modified this theory so that:

> the violent motion of an equally heavy body that does not follow the perpendicular to the horizon, never shows any perfectly straight part because of the gravity of that body which continuously pulls it toward the centre of the Earth. Nevertheless, that part [of the transit] that is not perceived as being curved is assumed to be straight, and that part that is evidently curved is assumed to be part of the circumference of a circle, as this [assumption] does not influence the argument.

52 Nicolò Tartaglia, *Three Bookes of Colloquies*, first book, 2r. p.89.

> Nevertheless, we shall suppose that part that is insensibly curved to be straight, and that which is evidently curved we shall suppose to be part of the circumference of a circle, as they do not sensibly differ.[53]

The Aristotelian model produced a very different trajectory to its actual parabolic one, a fact that was readily observable. Robert Norton pointed to the failure of reality to match this Aristotelian/Tartaglian model:

> Tartaglia in the fourth, fifth, and sixth Propositions of his second Booke of his Sciencia Nova (which I purpose one day to Translate, if God give life, and no other man prevent me) doth demonstrate, that the crooked Range of a shott made above the Leuell, maketh more then a Quadrant, and vpon the Levell a just Quadrant, and vnder the Levell, lesse then a Quadrant. But for all his subtill Demonstrations, it is not exactly so found by experience. For it is neither perfectly Circular elipsicall parabolicall, Hyperbolicall, nor Conicall, but meerely Helicall, or Helisphericall, according to the proper Levelling, mounting, or imbasing of the Peece assigned, which let now suffice.[54]

Hopefully this clarifies the matter in the reader's mind. Rather less elliptically Norton explains the constantly changing trajectory of the 'bullet' as he perceived it:

> But if any Peece be discharged vpon any Angle of Randon, albeit the violent motion contend to carry the Bullet directly by the line Diagonall; Yet the Perpendicular motion being not directly opposite, doth though unsensibly, even from the beginning by little and little drawe it from that direct and Diagonall course. And as the violent doth decay, so doth the naturall encrease: and of these two right lined motions, is made that mixt Curve Helicall Circuit of the Bullet.[55]

Norton was thus in agreement with Tartaglia that the effect of gravity on the projectile occurred as soon as it left the muzzle.

Bourne explained contemporary understanding when he drew and described the trajectory for shot: 'given violently by the blast of the powder'. There were four parts to the flight of a ball (see illustration overleaf):

A. The right line – perfectly straight
B. Flying of the shot upwards
C. The uppermost course for flying at the farthest distance
D. The circular falling or coming downwards which is the most circular of the trajectories.

Others argued that the flight of the projectile was in only three parts. The first was in a straight line, followed by a circular path and concluded in a vertical descent.

53 Tartaglia, *Nova Scientia* (np, 1558) second book, 11r,

54 Norton, *The Gunner*, p.12.

55 Norton, *The Gunner*, p.16.

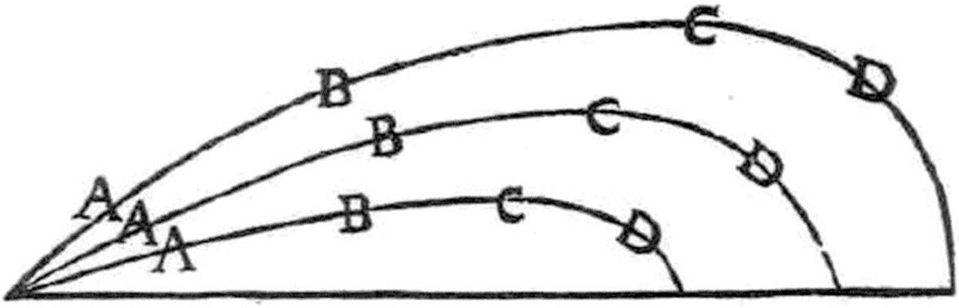

Bourne's understanding of the four parts to the flight of a cannonball. (William Bourne, *The Arte of Shooting in Great Ordnaunce*)

What part theoretical understanding played in the training or experience of a Tudor gunner it is difficult to say, Hall concludes that the: 'emphasis of the experts and of the artillery schools was on a sound education in straight forward field gunnery in which scientific ballistics played a very small part.'[56] The common complaint from Bourne and other afficionados was of the ignorance and incompetence of their fellow practitioners. Colin Martin provides a devastating summary of the arcane knowledge claimed by gunners:

> Most of this theoretical knowledge, however, we now know to have been unnecessarily complicated, usually inaccurate, often spurious, and generally of little practical help. Not to have easy access to or interest in such information [in 1588] may have been one of the English gunners' greatest strengths.[57]

This Aristotelian–Tartaglian view predominated throughout this period and well into the eighteenth century. Tartaglia did helpfully established that the greatest range for any projectile weapon could be achieved with an elevation of 45°.[58] He also concluded, rather less helpfully, that the maximum range of a gun would be 10 times its point-blank range:

> I found that the range of the above-mentioned shot elevated at 45 degrees above the horizon was 10 times the straight transit of a shot made parallel to the plane of the horizon, which is said by the bombardiers shooting at the blank point. On the basis of this evidence … using means of geometric and algebraic arguments, I found that a ball thrown along the mentioned 45 degrees above the horizon moves along a straight line which is about four times the straight line along which a ball moves when thrown parallel to the plane of the horizon, called by the bombardiers (as I said) shooting at the blank point.[59]

France considers that this conclusion was based on empirical data from practical trials rather than theoretical knowledge.

In the face of bewildering complexity presented by competing theories of the movement of projectiles, the simple gunner reverted to what he understood, simple rules of proportion. Thus, if the maximum range of the piece at 45° was helpfully 1,440 paces, then reducing the elevation by one degree would

56 A. R. Hall, *Ballistics in the Seventeenth Century* (Cambridge: Cambridge Uni. Press, 1952), p.2.

57 Martin, *The Equipment and Fighting Potential of the Spanish Armada*, p.376.

58 In fact 42°.

59 Tartaglia, *Nova Scientia*, epistle, third folio (unnumbered), verso. p.73.

A field piece elevated to 45 degrees by the simple expedient of digging a hole for its trail. Few gun captains would have been happy to emplace their precious piece in this undignified position. Whether this was attempted experimentally is unlikely, a similar conclusion could be drawn with less effort and danger with a humble crossbow and bolts. (Nicolò Tartaglia, *Nova Scientia*, epistle, second folio (unnumbered), 1558)

shorten the range by 1⁄45 or 32 paces. Lechuga without citing any evidence to support it, claimed that once the point-blank range was measured, elevation by every additional point (nine degrees) would result in an increase in range of 12 percent.[60] Lucar recommended a simple arithmetical system to establish the maximum range when the point-blank range was established. The point-blank range should be multiplied by 1,280 and the result divided by 320 (surely just multiplied by four). Unfortunately, he does not explain how he came to this calculation. Norton recognised that systems of proportion and arithmetical calculation simply did not reflect reality:

> But being mounted above 20 degrees, she [the gun] will shoote shorter & shorter, a litle at the mount of every degree to the best of the randon [range], according to the height & circular motio[n] of the bullet. If the peece be mounted to the best of the randon, the plummet will cut the 45 degree of the quadrant.[61]

In practice only mortars would be expected to have an elevation above 20 degrees for as Bourne himself remarked that: 'neither is there any service above the mounting of any piece of ordnance above 20 degrees except it be

60 Juan Luis García Hourcade, 'From Tartaglia to Lechuga. Artillery Engineers in the Renaissance', p.68, in Alicia Cámara Muñoz and Bernardo Revuelta Pol (eds), *Renaissance Engineers*, Fundacion Juanelo Turriano, 2013.

61 Smith, 'All Manner of Peeces', p.49.

a mortar piece'.[62] Later he asserts that: 'there is commonly no service to be done, above the mounting any piece of ordnance, more than ten degrees'.[63]

Smith asserted in a highly practical vein, that the only reliable way of finding the actual range of a gun was by systematic experimentation:

> Many Authors have taught how to make a table of randons, whereas some of them never shot in any peece of Ordinance in their lives. And for as much as I find their writing and reasons differing, I thinke it will be a very hard matter to make a perfect table of randons, except the same be tried and experimented with some peece of Ordinance in some convenient ground. I never heard nor reade of any that hath as yet fully put the same in practise.[64]

One experienced gunner, possessing both the interest and authority to carry it out was Luis Collado, Engineer of the Army of both Philip II and Philip III. He was also the author of the enormously influential *Practical Manual of Artillery*, a volume that went through 14 editions in three languages between 1586 and 1606. For his experiment he used a 3-pdr falcon and his results were as follows:

Table 3.

Point of quadrant	Degrees	Distance in paces/yards*
First graze	0	368/336
First	9	594/543
Second	18	794/726
Third	27	954/872
Fourth	36	1,010/923
Fifth	45	1,040/951
Sixth	54	1,053/963
Seventh	63	No distance given?
Eighth	72	Between second and third 874/799?
Ninth	81	Between second and first 694/634/?
Tenth	90	Shot fell back: 'quite near the gun'

* It is assumed that the pace referred to here was La Vara de Burgos, declared to be the official standard of length by Phillip II in 1568. It was 0.8359 metres or 32.909 inches long.

Clearly some error in loading or measurement occurred to give the longest range at 54° but the general pattern is clear. The table produces a fairly well-defined maximum range of a little under 1,000 yards and a practical maximum range, at the usual maximum elevation of 10 degrees, of

62 Bourne, *The Arte of Shooting*, p.27.
63 Bourne, *The Arte of Shooting*, p.35.
64 Smith, 'All Manner of Peeces', p.48.

about 550 yards. He concluded his experiments in wonderfully bombastic style condemning his illustrious predecessors:

> Silence therefore to Geronimo Rucelio and Nicholas Tartaglia and silence to all the other authors who, without foundation or experience whatsoever, state that knowing the range of the gun fired through the first point that of any other point will be known by calculating the differences between the shots by the rules of arithmetic. This they will never be able to prove.[65]

Galileo was at the end of this period conducting experiments and developing theories that would later become invaluable aids for the truly scientific gunner. In March 1637 he wrote, in some triumph:

> I wish to put an end to the treatise on projectiles and to send it as soon as possible to Signor Elzevir [his publisher]. I say put an end to because in revising and reordering it I have been discovering continually the most beautiful propositions in which this topic abounds; but I wish for now to close the treatise with a table which I have proved and calculated for artillery and mortar trajectories, showing their flights and with what proportion they increase and diminish according to the various degrees of elevation. The practice of this table will be useful to gunners, its theory of great delight to philosophers.[66]

It has been argued that he initially failed to include reference to air resistance in his calculations, but Rose has suggested that he did in fact fully understand the implications of this in predicting ranges. Unfortunately, his insights did not result in their enthusiastic application. The theories of Tartaglia and systems of proportion seem to have dominated military thinking for some time to come.

There are a few 'authentic' accounts of long-range fire. In the *Gunners Glasse* published in 1646, Eldred records that 'Queen Elizabeth's Pocket Pistol', a long 20-pdr, when fired at an elevation of two degrees hurled its projectile 1,200 yards. During the siege of Malta 1565 the cannons of Fort St Angelo bombarded the Ottoman forces massing for their assault on Fort St Elmo 1,000 yards away. In 1570–1571 at the siege of Famagusta a 60 lb culverin engaged an army in parade formation at a distance of over three miles, but this was again exceptional. It caused enough consternation for the event to be abandoned. Henry Hexham recorded one very long-range shot of six 'English' miles (nearer 10 by Google Earth). This was accomplished with a long-barrelled sling or culverin some 34 or 25 feet in length, shooting from Arnhem in Gelderland to Nijmegen.

Anyone concerned with accuracy always seeks to find out how well a gun and its crew could group its rounds, defined as to how close a number of rounds would fall within a specified space. Nathaniel Nye conducted some

65 Vicente Montojo, 'Primitive Ballistics, 1586–1610', *Journal of the Society for Army Historical Research*, Vol. 12, No. 48 ('Winter', 1933–1934), p.249.

66 Paul-Lawrence Rose, 'Galileo's Theory of Ballistics', *The British Journal for the History of Science*, Vol. 4, No. 2 (Dec. 1968), p.156.

invaluable tests to discover what could be achieved. The firing was deliberate and not under any particular pressure of time, or the strains of combat; the results were impressive:

> I have discharged a peece seven times in the space of 50 minutes, with the like weight of powder, shot, and elevation, and have found their ranges as following, the first shot was conveyed 416 paces, the second 436, the third 440, the fourth 432, the fifth 425, the sixth 410, the seventh 394.[67]

The gun showed remarkable consistency, which excluding the seventh shot, saw six rounds all dropping within 30 paces or 25 yards, at a mean range of 426.5 paces or 355 yards. Nye explained the variation between shots in terms of the perturbation of the air, the 'moyst' nature of a cold gun and then 'as the peece grows hotter, one shot wil not out-shoot the last, before it, but every time come shorter and shorter'.[68]

Effective Range in the Field and at Sea

The maximum range of a gun is very different from its effective engagement range, that is the distance that a target can be hit with significant damage done. In the field the engagement range of a target is determined by its visibility, speed of movement and the available stock of ammunition. Ammunition could not be wasted on a fleeting target when powder and shot were in short supply, as they often were. The effective range of artillery in the field was likely to be only some hundreds of yards. Norton provided some sensible guidance presented in the form of a question and answer.

> Question: How would you order things, to shoot at a Squadron of the Enemies Souldiers, and what Peeces would you choose?
>
> Answer: According to the distance I would choose a Peece that in a straight line can shoot home, if I may, be it with *Demy Culuering, Saker* or *Falcon,* and plant my Peece as neere parallel to the champion Plaine [field of battle] as I can, that the shot may range, and shoot at girdle height, unlesse the ground be stony, for then would I place my shot short of them, that grazing amongst the stones, that stones may doe them more spoile then the shot of it selfe can doe, by farre; but in no case would I shoot wide, or over them, for that were both losse and shame also.[69]

He is clearly looking for a 'flat shot' at point-blank range with the piece 'neere parallel', that is neither shooting up or down hill, a difficulty which would require some nice adjustment and judgement by the gunner. He recognised the danger of overshooting and sensibly recommended shooting short on stony ground to take advantage of the bouncing ball. A point-blank

67 Nye, *Gunnery*, p.32.
68 Nye, *Gunnery*, p.33.
69 Norton, *The Gunners Dialogue*, p.22.

range of a few hundred yards would seem the preferred range extending to perhaps 1,000 yards with the heavier pieces if the size of the target permitted.

In besieging a castle, it was essential to set up batteries as close as possible as the high induced drag on the spherical projectile led to a significant reduction in penetration at longer distances. In the examples he provides, Bourne sets the range from the target at some 200 yards, well within the range of even the smallest pieces of artillery. Thomas Smith recommended planting ordnance: 'within 200 or 240 paces of the object to be ouerthrowne'.[70] In 1590 John Sheriffe recommended that artillery was most effective for siege work within 80 paces of their target, a distance also recommended by Luis Collado. At the siege of Haddington the French commander, dissatisfied with the effect of his bombardment, moved a six-gun battery within 60 paces of the fort with more satisfactory results. At Montdidier in 1523 guns were brought up at night to within 40 feet of the walls and such was their effect that the town surrendered in four hours.

Cyprian Lucar was very clear as to the effective and ineffective range of siege artillery:

> To batter a wall of a Towne or fort, lay your Cannons if you can at the distance of 80 paces from the wall which shall be beaten downe, and in no wise (without constraint) more farther from that wall than 150 paces: for when Cannons doe lie 300 or 200 paces of from the wall which shal be battred, they are planted (as Luigui Collado writeth) in an inconvenient distance & vnmeete place to batter.[71]

In his *Discourse Dealing With Artillery and All That Is Required for It, With a Fortification* (1610), written by experienced engineer and gunner Cristóbal Lechuga, in answer to the question 'how far from the walls should the artillery be laid'? the answer was quite simply: 'the closer the better'. His qualification was that: 'as there are hazards when that close and the site does not allow it, it is advisable to adapt as best as you can and lay the artillery at the least possible distance'.[72]

At what distance could ships open fire at each other with a hope of hitting and causing damage? The recent trials of the *Mary Rose* and *Vasa* guns, provide evidence that their practical engagement range was not as different as their ballistic performance would suggest. In the case of the magnificent *Vasa* 24-pdr it was not the ability to do damage to a target even at 1,000 metres at question, but the ability to hit a ship at anything like that distance. The implication that a combination of the rolling of the ship, the movement of the target, uncertain speed of ignition, the 'loopy' trajectory and the difficulty of judging precise distance, meant that engaging a target at more than a few hundred yards would be a foolish waste of ammunition, unless a tactical opportunity was provided that might make such expenditure worthwhile. The stocks of ammunition that were provided were hardly generous, between 25 and 50 rounds per gun, in

---

70 A single pace measured 30 inches, a double pace of 60 inches was also used. Smith was probably using the single pace here. Smith, '*All Manner of Peeces*', p.67.

71 Lucar, *Appendix*, p.59.

72 Hourcade, 'From Tartaglia to Lechuga', p.67.

English ships throughout this period, not enough to allow wastage. The *Vasa* went to sea with barely 12 rounds per gun.

As an aside, it is interesting to note the very short range employed by both sides in the American Civil War by both sides and the long-range shooting employed by British troops when using what were often identical weapons, rifled muzzle-loading muskets. The 'loopy' trajectory of the Minie bullet made range judging and sight setting crucial for accurate shooting. Accurate judgement of range was a key element in the comprehensive British system of training, such training and practice was virtually unknown in the states. The American troops flat-shot at short range and suffered the horrible consequences. The author faces similar problems of distance judging when shooting his traditional longbow alongside modern archers, who have flat-shooting bows. Judging distance and elevation is a skill that comes from training and practice. Long-range shooting either on sea or land would be unlikely in this period.

Charles Trollope writing in 1994 about the Spanish Armada came to a similar conclusion to Fred Hocker based on similar calculations, although he was even more sanguine as to what was likely to be an effective range.

Howard Douglas in his book based on years of experiments concluded that for practical purposes a sailing ship rolled through 1 to 2 degrees a second.[73] He also noted that the delay between igniting the priming at the vent, should a quill or tube not be used [as in this period], and the departure of the shot after full ignition of the propelling charge was in the order of 1 second, seldom less and often more.[74] If therefore a gunner on one of the Queen's ships rolling upwards at one degree a second fired at a Spanish ship presenting a structural target 15 feet high, at the moment his gun pointed at the ship's waterline, he would only strike at the top edge if the target was less than 285 yards. To ensure an acceptable 50 percent hit ratio on the target this range must be reduced by half or more. If the rate of roll is two degrees per second, the range must be halved again to 75 yards. At 75 yards all guns capable of doing structural damage will be effective.[75] The 'water level range' (*waterpasdracht*)[76] or horizontal range of Dutch guns in the seventeenth century was considered to be only 75 to 100 metres at most.[77]

In Phillip II's oft-quoted advice to the Duke of Medina Sidonia suggests that the English tactic at sea was to use the long-range potential of their ordnance:

> Above all it must be borne in mind that the enemy's object will be to fight at long distance, in consequence of his advantage in artillery, and the large number of artificial fires with which he will be furnished. The aim of our men on the contrary, must be to bring him to close quarters and to grapple with him, and you will have to be very careful to have this carried out. For your information

73 Sir Howard Douglas, *A Treatise on Naval Gunnery* (London: John Murray, 1855).

74 In the *Vasa* trials, ignition of the main charge could take place up to three seconds after the priming was ignited.

75 Trollope, *The Guns of the Queen's Ships During the Spanish Armada*, p.26

76 Usually defined as point-blank, but looking along the line of metal so in fact pointing slightly up.

77 Puype, 'Guns and their Handling at Sea', p.17.

> a statement is sent to you describing the way in which the enemy employs his artillery in order to deliver his fire low and sink his opponent's ship; and you will take such precautions as you consider necessary in this respect.[78]

Contemporary English opinion was somewhat different, as Sir Richard Hawkins expressed it: 'How much the neere, so much the better'.[79] Sir William Monson made his position clear when he wrote: 'He that shooteth far off at sea had as good not shoot at all'.[80] During the heaviest fighting in the Armada William Wynter wrote to Sir Francis Walsingham that: 'when I was farthest off in discharging any of the pieces, I was not out of Shot of the harquebus, and most time within speech of one another'.[81] This would suggest a distance of less than 100 yards.

Spanish vessels were well equipped with modern artillery and their tactics certainly stressed the importance of their use at the beginning of an engagement. A treatise on fleet tactics of about 15 years before the Armada is quite explicit about how guns, large and small, should be used.

> Then shall the trumpet sound and at that signal all shall move in their aforesaid order; and as they come within range they shall commence to play with their artillery, taking care that the first shots do not miss … they shall always first to play the largest guns which are on the side or board towards the enemy, and likewise they shall move over from the other side those guns which have wheeled carriages to run on the upper part of the deck and poop. And then, when nearer, they should use the smaller ones, and by no means should they fire them at first, foe afar off they will do no hurt, and besides the enemy will know there is a dearth of good artillery and will take better heart to abide an attack. And after having come to close quarters then they ought to play the lighter artillery. And so soon as they come on board to grapple all kinds of other arms should be used … more particularly first missiles, such as harpoons [gads or topdarts] and stones, handguns and crossbows, and then the fireballs aforesaid as well as from the tops as from the castle, and at the same time caltrops, linstocks(?), stinkballs, grenades and scorpions for the sails and rigging.[82]

Clearly the intention was for the guns to precede the decisive close-quarter action not themselves to decide the outcome. To take the weather gage was judged important to allow the gunpowder smoke to blind the enemy, not because it gave a ship a serious tactical advantage when it came to manoeuvring. This was an indication of how the Spanish perceived a sea battle, which was as a land battle conducted on water.

---

78 Stephen Usherwood, *The Great Enterprise: The History of the Spanish Armada as Revealed in Contemporary Documents* (London: Bell and Hyman, 1978), p.68.

79 N. A. M. Rodger, 'The Development of Broadside Gunnery, 1450–1650', *The Mariner's Mirror*, Vol. 82, No. 3, p.308.

80 *The Observations of Sir Richard Hawkins*, ed. James A. Williamson (London: Argonaut Press, 1933), Monson, *Naval Tracts IV*, p.43 and Rodger, 'Broadside Gunnery', p.309.

81 Usherwood, *The Great Enterprise*, p.110.

82 Padfield, *Guns at Sea*, p.32; Alonso de Chavez, courtier to King Charles V, in *Espejo de Navegantes*, *c*. 1530

The Last Voyage of Drake and Hawkins makes it clear that closing within musket shot and then boarding was the preferred tactic and when that was not possible the outcome was likely to be unsatisfactory:

> Our enimie playenge upon us with theyr ordinance made our gunnors fall to it ere we were at musket shot & never no nerer could I bringe them though I had no hope to take any of them but by boordinge, here we popt away powder and shot away to no purpose for most of our gunnors would hardly have striken Paules steeple had it stoode there.[83]

Criticism of English gunnery of the period may appear somewhat harsh but was repeated by a close observer of the defeat of the Armada. An English master gunner named William Thomas. who was at Flushing, wrote the following to Lord Burghley in September of that momentous year:

> if it had pleased God that her Majesty's ships had been manned with a full supply of good gunners according to the forces they carry ... it would have been the woefullest time or enterprise that ever the Spaniards took in hand ... What can be said but our sins was the cause that so much powder and shot spent, and so long time in fight, and, in comparison thereof, so little harm?[84]

There was even an understandable disdain for the developing obsession with race-built galleons with heavy armament. Ships with a high stern and fore castles well provided with small pieces had proved their worth after bow chasers and a single broadside had done their part. Sir William Monson, an Armada veteran represented the views of the old guard when he wrote:

> Such a ship that hath neither fore-castle ... nor any other manner of defence, hath no fowlers, which are pieces of great importance, after a ship is boarded and entered, or lieth board and board; for the main ordnance stands her in little stead ... but a murderer or fowler being shot out of their own ship, laden with dice shot, will scour the deck of an enemy.[85]

The debate concerning the comparative armament of the English and Spanish vessels in the Armada campaign has generated much valuable data.[86] The discussion focuses on the supposedly superior long-range armament or heavier 'throw-weight' of English ships, proposed by Lewis and Thompson respectively. As we have seen there was very little difference if any in velocity and engagement range between cannons, culverins and perriers. This was especially the case as the engagements took place at very short range. Thompson records that: 'Repeatedly over the nine days of the battle, units of the two fleets came not merely within culverin range, but within musket

83 *The Last Voyage of Drake and Hawkins*, ed. Kenneth Andrews (Hakluyt Soc. 2nd. S. 142, 1972), pp.104–5, in Rodger, 'Broadside Gunnery', p.309.

84 Martin, *The Equipment and Fighting Potential of the Spanish Armada*, p.374.

85 Monson, *Naval Tracts IV*, p.92.

86 Please note the invaluable work of Michael Lewis, Colin Martin and Ian Thompson in this field.

range, within half-musket range, and finally within hailing range, even within pike range'.[87] What seems to have mattered more in determining casualties and damage, was not so much what hit the ship but where the ship was hit. Thus, the more rounds that hit a target the greater the likelihood of causing serious damage. The Spanish shot around 100,000 balls in the various engagements. They killed one sea captain and a few score sailors and failed to cause significant damage to a single English ship a singularly unimpressive performance. Of much greater importance in explaining the failure of the Spanish, were the tactics employed by both sides, the confidence and competence of the crews and especially of their captains and commanders.

John Reynolds the Proof Master after testing some drakes in 1627, concluded that that: 'for shooting at eleven or twelve score [yards] which is considered by most seamen to be far enough at sea, these pieces have done such execution … as more is not required'.[88] Colin Martin in his thesis on the Armada concludes that: 'it is probably true to say that in sixteenth-century sea fighting very little damage was ever caused by gunfire at ranges beyond 500 yards, and that the chances of regularly hitting even a large target such as a ship beyond 200 yards were fairly remote'.[89] The conclusion from contemporary and experimental evidence would suggest that a maximum realistic range for opening fire with any hope of doing serious damage was little more than 200 yards and to be certain of causing havoc, the range should be reduced to well under 100.

It was certainly the case that even if the effective range remained very short a heavier armament for warships was increasingly preferred. In 1618 a Commission of Reform in England concluded that:

> Experience teacheth how sea fights in these days seldom come to boarding, or to great execution of bows, arrows, small shot and the sword, but are chiefly performed by the great artillery breaking down masts, yards, tearing, raking and bilging the ships, wherein the great advantage of His Majesty's Navy must be carefully maintained by appointing such a proportion of ordnance to each ship as the vessel will bear.[90]

This was an opinion shared by the Swedish navy in the 1620s; the driving force behind their adoption of a heavy gun armament was the Swedish king Gustav II and the result was the *Vasa*.

One question the author had not thought of asking was whether the range of a gun firing from land to sea was more or less than one firing from sea to land? Diego Ufano in an *Artillery Treatise and How to Use Artillery* (1613) includes a dialogue between a captain and a general on just this knotty question. His general propounds the theory that shooting from the sea has less range because of the unsatisfactory behaviour of the ship when

87 Thompson, 'Spanish Armada Guns', p.356.

88 R. M. Towes and P. McCree, 'A Note on Drakes', *Journal of the Ordnance Society*, Vol. 6 (1994), p.39.

89 Martin, *The Equipment and Fighting Potential of the Spanish Armada*, pp.246–7.

90 Robertson, *Evolution of Naval Armament*, p.21.

the gun recoils. The captain disagrees and describes the whole business as an: 'extremely mysterious phenomena'. He goes on to explain that:

> the force with which the vessel fires the bullet landward makes the bullet seek shelter and natural rest, and the force with which the shot is fired from land to sea naturally has to fight against the two elements, namely, water and wind, because the force is such, I tell you, that leaves the land that the damp and the wind force stirs it up and shakes all over the place and it has been demonstrated that at low tide the shot from land will be longer than at high tide and the shot fired from sea to land arrives with greater ease.[91]

It seems doubtful that the reader will be any more convinced than the author with this explanation.

Rate of Fire

If engagement ranges were short, the rate of fire that could be achieved was of critical importance. There is evidence for the limitations on the usage of artillery pieces in siege warfare; in 1513 at the siege of Tournai the culverins were limited to 36 shots per day, and the bombard to only five. There is unfortunately little hard evidence for the rate that could be achieved in the heat of battle in the field.

One key element in determining the rate of fire is the number and training of the crew. What is so surprising when looking at contemporary illustrations of artillery is the sparseness of their crew: in the case of an illustration of the Carberry hill the guns are fired in the absence of any crew at all![92] Large English pieces in Henry VIII's reign were provided with a gunner and an assistant-gunner or *matross* who received a daily pay of 16d and 8d respectively. This size of crew is obviously inadequate: a minimum of three[93] are required to carry out the simple processes of loading and gun laying. In the field, Napoleonic guns would have five serving on the gun and perhaps double that number preparing ammunition and assisting in manhandling the piece.[94] The contemporary illustrations do not show large gun crews, but it is inconceivable that weapons of such value should be undermanned, and more than likely pioneers or surplus infantrymen would be told off to provide the necessary muscle power.

At the siege of Boulogne in 1544 the battery of the *Lord Admyral* consisted of three cannons, three demi-cannons, three culverins, two demi-culverins and three sakers and was manned with the assistance of 500 pioneers. The Scottish artillery train that accompanied James IV to England in 1513 with five cannons and 12 culverins of various sorts had a train of 500 to accompany it, or almost 30 men per gun; surely some would have assisted

---

91 Hourcade, 'From Tartaglia to Lechuga', p.71.

92 National Army Museum engraving.

93 The crew of the author's falconet is three, and they are very busy when serving the gun.

94 Otto von Pivka, *Armies of the Napoleonic Era* (Newton Abbot: David & Charles, 1979)

in their manning? The numbers of crew needed to serve a variety of guns for French artillery in 1639 were as follows:[95]

Table 4. Crew Numbers for French Artillery in 1639

Gun	Gunner	Assistant	Pioneers
Canon	2	3	30
Grande Culverin	2	2	24
Bastard Culverin	2		12
Minion	2		6
Falcon	1		4
Falconet	1		4

These figures represent, under what might be considered ideal conditions, a reasonable apportionment of manpower for each weapon. The rate of fire suggested by the same author would be one round every six minutes for the heaviest and three minutes for the lightest pieces, commendably fast for the large and disgracefully slow for the small. Lucar, in his detailed review of what rate of fire could be achieved, cites the following:

> An ordinarie demy Cannon duely fortefied with mettall, will serve to batter, and may safely for offensive and defensive service be 108 times charged and discharged within the space of 5 houres in one day, and it ought to have three Gunners and fifteene Assistantes or Labourers.[96]

He considers that cannons could only manage 30 shots in a day while culverins could manage 60 at most to ensure their safe operation. To fire more rounds would be to risk at the very least muzzle droop and at the worst to put the gun and its crew in peril. Lucar was very clear about how a heavily used gun was to be treated:

> After a peece with manie shootes is made very hot, it changeth his colour, & shooteth weaklie, and then to save it from breaking, you ought to coole it within with a spunge wet in cold water, or in two parts of cold water, & one part of vineger, or in lie mingled with a little water: and lay all ouer the peece, or at the least from the touchhole to the mouth, sheepes skinnes with long wooll on them dipped in the said cold water, or in 2 parts of cold water, & 1 part of vineger, or in lie and a little water, which lie stoppeth the powres [pores?] of the mettall in the peece, and causeth it to resist heate.[97]

At the English siege of Boulogne in 1544, guns of demi-cannon and above were allocated 360 rounds but sakers 100 and falcons even fewer with only 62 rounds per gun. The very large numbers of rounds for the heavier guns

95 See Appendix IX in this work.
96 Lucar, *Appendix*, p.48.
97 Lucar, *Appendix*, p.59.

were required to sustain a prolonged bombardment.[98] Guns in a siege could be expected to provide sustained fire for as long as their ammunition lasted.

Guilmartin wisely points to the danger of making assumptions about the way in which technology can be used:

> Cannon went to sea in significant numbers only in the sixteenth century; by the middle of the eighteenth century, the manufacture and employment of naval artillery was rationalised, standardised, and organised along lines which seem familiar to us today. As a result, we tend to analyse ships and guns of the sixteenth and seventeenth centuries in terms of the technology, tactics, organisation, and attitudes of the eighteenth and nineteenth. Even the cannon work against themselves: as early as the 1520s bronze cannon were being cast which were similar in overall appearance to cannon cast for the next three centuries. Since there was little change in appearance, we assume that there was little change in substance. This implicit assumption is false.[99]

The development and employment of artillery at sea is more complicated and nuanced than it might appear. The obsession with long-range gunfire and rate of fire which dominated the thinking of the nineteenth and twentieth centuries cannot readily be applied to this earlier period. The *Mary Rose* was not a Tudor version of HMS *Victory;* its armament, tactical philosophy and strategic purpose was very different. The armament of the *Mary Rose* underwent many changes during her various commissions with a heavier armament being accommodated and cast bronze guns taking pride of place on her gun deck between 1513 and 1545. The *Vasa* showed great similarity to the *Victory*, with its clear attempt to produce a common and very heavy armament whose purpose was to pummel an opponent into submission.

Much of the detailed evidence for firing rates comes from the Armada of 1588 and primarily from Spanish sources. The assumption that has been made was that the design of the Spanish sea carriages, the preponderance of soldiers rather than sailors as gun crews and the preferred tactics of boarding an enemy ship, rather than employing gunfire to disable her, led to a disregard if not disdain for using artillery at sea. This is supported by some of the available evidence provided by the Spanish themselves. The Captain of Medina Sidonia's flagship Alonso de Vanegas reported that the English fired 2,000 rounds to 750 from Spanish ships, this despite the fact that Medina Sidonia himself declared that, 'The guns on our flagship were served well and rapidly'.[100] Even so the flagship with 48 guns fired only 120 rounds in one hour, a very low rate indeed. The 22-gun *Trinidad* fired only 94 rounds over three days of fighting and the 20 guns of the *Santa Barbara* fired but 264 rounds over four days with two thirds shot on one day alone. The *Almiranta Francisco*, with 21 guns fired 241 rounds in all, an average of 11.5 per gun, with her two heaviest pieces, both full

98 Letters and Papers, Foreign and Domestic, No. 272 RO 13, 'The proporcyon of ordnaunce'.

99 J. F. Guilmartin, *The Guns of the Santíssimo Sacramento,* no page number, <https://www.angelfire.com/ga4/guilmartin.com/Santissimo.html>

100 Usherwood, *The Great Enterprise*, p.128. Letter from Medina Sidonia to Philip II, 11/21 August 1588

cannons having shot an average of only 11 rounds. For a fleet fighting for its life the quantity of ammunition fired seems absurdly, nay embarrassingly low.

The simple explanation would be that their reloading rate was very slow as a consequence of the self-imposed problems that they faced. This is supported by the 1589 account of van Linschoten, a Dutch passenger aboard the 1600-ton *Santa Cruz*, a Portuguese carrack which met an English privateer: 'When we shot off a piece … we had at least an hour's work to lade it in again, whereby we had so great a noise and cry in the ship as if we had all been cast away'.[101] This chaotic and sclerotic effort must seemingly be contrasted with the performance of hyperactive English seadogs.

This view was supported by the report of a Spanish officer when fighting the *Revenge* in 1588, which was 'Letting fly every way from both her broadsides, soo that she seemed to repeat her fire as rapidly as any harquebusier'.[102] This picture is not however supported by the record of Sir William winter that the *Vanguard* at the height of the fighting on 8 August 1588, fired 500 rounds from her 32 guns over nine hours. This was a rate of 1¾ rounds per gun per hour. The allocation of ammunition for English guns was also to modern eyes laughably small. The Anthony Roll and other records, surveys and expeditions from 1545, 1572, 1585 and 1596 all show that almost all ships were provided with no more than 25 rounds per gun.

It is a mistake to visualise sea battles in the sixteenth century as reduced versions of Trafalgar or the Battle of the Nile, with ships on parallel course engaging in an artillery duel where speed of reloading was a crucial factor in determining victory. Throughout most of this period guns in the bow and stern were deemed to be the most useful, especially when facing galleys rather than carracks and galleons. To 'give him [the enemy] the prow', to open the action with the most powerful bow pieces was the intent of any proficient sea captain and then to use broadside and stern pieces. After this he could reload at comparative leisure before entering the fray once more. This tactic was described by John Smith in his *Sea Grammar*: 'begin with your bow peeces, proceed with your broad side, & let her fall off with the wind, to give her also your full chase, your weather broad side, and bring her round that the sterne may also discharge'.[103] Dr Hocker makes this important point when considering what broadside guns could actually be brought to bear on a target:

> Up through the 1620s, the ports themselves were built so that they did not all point directly abeam, but were bowed, so that those nearer the bow were angled forward and those near the stern angled aft. This made it difficult to aim all of the guns at a single target. Instead, each gun fought as an individual. This can also help to explain the low rate of fire.[104]

The use of the broadside was not new. As early as 1540 the English ship *Barbara* encountered a Portuguese vessel off the coast of Brazil which: 'Shote

101 Martin, *The Equipment and Fighting Potential of the Spanish Armada*, p.372.
102 Rodger, *Broadside Gunnery*, p.310; Corbett, Drake II, p.260, in Rodger.
103 Smith, *A Sea Grammar*, p.60.
104 Correspondence with Dr Hocker, 26 July 2021.

off his hole syde of ordynaunce at us. And then he caste aboute on thothersyde of us also, and shotte al his syde at us agyne to thentente to have soncke us'.[105]

In the case of the Armada the numbers of rounds fired as referred to by Alonso de Vanegas, could be explained by the relative position of the ships and their ability to bring guns to bear. The Spanish were after all in a close formation, pursued by the English ships engaging in a very slow chase; only at Gravelines was there what might be called a sea battle. In the absence of turret mounting, guns could only fire when the ship manoeuvred into a position where they could bear on the target. The key to successful gunnery was not the gunner so much as the helmsman. It is though clear that the Spanish faced greater problems than the English whose race-built galleons allowed them much greater opportunities to bring all their guns to bear. The Spanish heavy guns were certainly underused, as archaeological and documentary sources attest. Colin Martin summarises the dire position that the Spanish faced:

> The probability is, therefore, that once close action was joined most Spanish ships only managed to get off their previously prepared salvo, after which sustained gunfire ceased ... it seems certain that the big ship-killing guns were never able to apply a continuous close-range cannonade against the enemy, although this was the only way in which they might have achieved success.[106]

Spanish ships could allocate one gunner and six soldiers to each main gun aboard the Armada ships. Allocation of sailors and gunners in English vessels was established at three men for every five tons, of which one man in seven was expected to be a gunner. This would be enough to allocate at least one gunner to every piece of demi-culverin size and above. The crewing for guns as large as a demi-cannon was put at two to three men in 1588, rising to four by the 1630s, but the additional men were sailors rather than gunners. This of course led to a shortage of gun crew when the ship was manoeuvring, as explained by this account of an action in 1628:

> In the beginning of the fight I had all my gunnes in, and all my sailes out (for otherwise I could not have reached them), so that I suffered much for want of men before I could fit the sails and bring the gonnes to their due bearing, otherwise they should have had many more shottes out of my shippe.[107]

Thus, the Spanish had more men at their guns, but many were soldiers and inexperienced in handling artillery in a new and dangerous environment. The English had few men available to serve if engaged in manoeuvring their ship, but they were certainly used to serving together and had much handier pieces to load and fire. Neither would have been able to produce a Nelsonian display of firepower. In the *Vasa* trials a crew of seven could load and fire their piece

105 'Voyage of the Barbara to Brazil: Anno 1540', ed. R G Marsden, in *The Naval Miscellany II*, ed. J. K. Laughton (NRS Vol. 40, 1912, 3–66, p.31), in Rodger, 'Broadside Gunnery', p.307.

106 Martin, *The Equipment and Fighting Potential of the Spanish Armada*, pp.370–371.

107 *Journal of a voyage into the Mediterranean by Sir Kenelm Digby*, A. D. 1628, ed. J. Bruce, Camden Society, Vol. 96 (London: Camden Society, 1868) p.12, in Rodger, 'Broadside Gunnery', p.313.

every 90 seconds, a real achievement, while a crew of four could achieve one round every five minutes or better.[108] This would suggest that a small crew would be perfectly acceptable in action, as long as they were well trained.

There has been a tacit assumption that at the beginning of this period guns would have been secured to the hull. When the majority of the large calibre guns on heavy fixed wooden beds would have been breech-loading and with limited traverse this would not have greatly mattered. If truck carriages and modified land carriages, with muzzle loading pieces mounted, were allowed to recoil inboard, this would have assisted reloading. It has been assumed that the practice of fixing the guns to the side of the hull would damage the ship. Which may have been the case in lightly built merchant vessels equipped with heavy pieces. On the other hand, Dr Hocker has pointed out that although the 24-pdr *Vasa* guns were allowed to recoil 1.5 m, the gun 'does not slow down significantly, in that space, so that the shock load on the ship structure is about the same as if it were tightly bound [to the hull]'.[109] Dr Hocker suggests 'that the development from fixed to recoiling guns was not a strictly chronological process, but that there were a number of factors affecting whether guns were held tight to the side or allowed to roll backwards'.[110]

The evidence from the *Vasa* would suggest that the 24-pdr guns on the main gun deck would have recoiled inboard far enough to be reloaded from within the hull. On the upper deck however, the space did not permit the guns to be brought inboard, which would have required them to be loaded by the crew balancing on the chainwales.[111] In the *Vasa* trials Dr Hocker commented that this practice was not perhaps as tricky as it may seem, as: 'The chainwales (channels) are actually quite broad, providing a very stable platform for working the guns, if not very protected. The only pair of upper deck gunports not over the chainwales does have enough room behind the gun to allow even the long 3-pounders to be brought inboard for loading'.[112]

The sixteenth century had begun with a revolution in naval warfare, the gun port. This together with carvel construction meant that more heavy guns could be mounted than ever before and ships benefited from the 'new' artillery which appeared to fill these new gun ports. The tactics and equipment available were changing and developing over time, so that by the end of the period the *Vasa* would emerge with a predominantly big-gun armament, loaded and fired in a similar manner and with the same effect as the ships at Trafalgar. In 1625 Lord Wimbledon the commander of the English fleet established a new set of instructions. The crews were now expected to man the tackles used to run the guns out, that is after they had been allowed to recoil inboard and then reloaded, increasing the rate of fire. The number in a gun crew was therefore determined by the weight of the piece.

---

108 Correspondence with Dr Hocker, 26 July 2021.

109 Correspondence with Dr Hocker, 11 July 2021.

110 Correspondence with Dr Hocker, 11 July2021.

111 Correspondence with Dr Hocker, 11 July 2021. The chainwale, pronounced channel, consisted of a number of iron rings to which the deadeyes supporting the shrouds were attached. This was sometimes provided with a ledge or shelf.

112 Correspondence with Dr Hocker, 26 July 2021.

This chapter concludes unashamedly with the most wonderful description of derring-do by the extraordinary Captain John Smith (1580–1631) of Pocahontas fame, sometime Governor of Virginia and Admiral of New England. He describes a fictional sea battle between a Spanish and English vessel. The combat is inconclusive but provides an excellent description of how a vessel was fought and includes details of damage repair, as well as highlighting the mariner's natural terror of fire:

> Master how stands the chase? Right on head I say; Well we shall reatch him by and by; What's all ready, Yea, yea, every man to his charge, dowse your top-saile to salute him for the Sea, hale him with a noise of trumpets; Whence is your ship? Of Spaine; Whence is yours? Of England; Are you a Merchant, or a man of War? We are of the Sea; He waves vs to lee-ward with his drawne sword, cals amaine for the King of Spaine, and springs his loufe,[113] give him a chase peece with your broad side, and run a good berth ahead of him; Done, done, We have the wind of him, and he tackes about, tacke you about also and keepe your loufe, be yare at the helme, edge in with him, give him a volley of small shot, also your prow and broad side as before, and keepe your loufe; Hee payes vs shot for shot; Well, wee shall require him; What are you ready againe, Yea, yea. Try him once more as before, Done, done; Keepe your loufe and loge your ordnance againe; Is all ready? Yea, yea; edge in in with him againe, begin with your bow peeces, proceed with your broad side, & let her fall off with the wind, to give her also your full chase, your weather broad side, and bring her round that the sterne may also discharge, and your tackes close aboord againe; Done, done, the wind veeres, the Sea goes too high to boord her, and wee are shot thorow and thorow, and betweene wind and water. Try the pump, beare vp the helme, Master let vs breathe and refresh a little, and sling a man over boord to stop the leakes; that is, to trusse him vp about the middle in a peece of canvas, and a rope to keepe him from sinking, and his armes at liberty, with a malet in the one hand, & a plug lapped in Okum, and well farred in a tarpawling clout in the other, which he will quickly beat into the hole or holes the bullets made; What cheere mates, is all well? All well, all well, all well; Then make ready to beare vp with him againe, and withall your great and small shot charge him, and in the smoke boord him thwart the hawse, on the bow, mid ships, or rather then faile, on his quarter, or make fast your graplings if you can to his close fights and sheare off. Captaine we are fowle on each other, and the ship is on fire, cut any thing to get cleare, and smother the fire with wet cloathes. In such a case they will presently be such friends, as to help one the other all they can to get cleare, lest they both should burne together and sinke; and if they be generous, the fire quenched, drinke kindely one to another; heave their cans over boord, and then begin againe as before.[114]

113 A nautical order to keep the ship's head to the wind, thus to stay clear of a lee shore or some other quarter. A sailing vessel that could point higher to windward and hold its speed better than another was said to stand apart or to sail a-luff, a term that later became aloof with a very different meaning. <https://see-the-sea. org/nautical/naut-body.htm> (accessed 24 September 2021).

114 Smith, *A Sea Grammar*, pp.60–61.

4

Designing the Gun

> All that can be said of gun design is that it was largely a matter of trial, error, and personal idiosyncrasy in which a range of successful forms emerged, and less successful ones were abandoned, in an almost Darwinian process of evolution.[1]

This is the conclusion that Colin Martin came to when writing his thesis on Spanish guns from Armada wrecks. His somewhat exasperated tone is quite understandable, and it is a feeling the author shares when trying to make sense of the bewildering variety of *versos, bases, couleverine moyennes, passavolante, springards, bombards* and *cannon de batir* that appear in contemporary accounts and modern literature. The author is pleased to say that what is perplexing for us was infuriating for those unfortunate enough to command and crew such weapons. In 1587 Cyprian Lucar wrote of the confusion of types that the gunner faced:

> But through the intollerable fault of carelesse or vnskilfull gunfounders all our great peeces of one name are not of one length, nor of one waight nor of one heigth in their mouthes & therfore the gunners bookes & tables which do shew that al our great peeces of one name are of an equal length, & of an equal waight, & of an equall heigth in their mouthes are erronious.[2]

There follows an attempt to provide a simple system to categorise and explain the variety of ordnance available in this long period, but why was there such a variety of ordnance in shape, size, material and mounting?

One of the principal reasons that artillery of the same period was so varied was that ordnance supply was not the monopoly of government. It was only around the middle of the sixteenth century that there were attempts by the state to standardise weapons for administrative convenience. Artillery could be purchased by the nobility and gentry, by towns, merchants and ship owners and to whatever specification they wished. They could be given as gifts as were the 19 bronze cannon and 1,023 cannonballs bought by Henry VIII on behalf

1 Martin, *The Equipment and Fighting Potential of the Spanish Armada*, p.255.
2 Lucar, *Three Bookes of Colloquies*, Appendix, p.33.

of the Hospitaller order. A sea captain might well make a purchase of new or second-hand wrought iron ordnance, cheap to purchase and if not state-of-the-art, more than enough to scare off pirates and other ne'er-do-wells. In 1518 John Marshem the Mayor of Norwich purchased at least one robinet from Flanders. In 1544 the City of Norwich purchased six falconets from John Owen which were probably used against and captured by Kett's rebels in 1549. Thomas Gresham provided one of his merchant ships that came to grief in the 1570s, with both cast and wrought iron guns from his own foundry.

If purchasers were a very varied bunch so were those who made the guns. Gunfounders were highly skilled craftsmen and although they may have sometimes been by 'appointment' this does not mean that they were simply servants of a state institution. Some were closely associated with a regal or imperial employer. Hans Poppenruyter for example was chosen as the Royal Gunfounder by Charles V in 1520 and was the principal supplier of guns to Henry VIII at the beginning of his reign. Dynasties of gunfounders such as the Ghiberti, Gioardi and Alberghetti were associated with city states, in their cases, Florence, Genoa and Venice. Others travelled to a specific court or foundry sometimes casting guns, bells or other objects of virtue.

Guns and bronze guns in particular could also have a very long life of many decades or even centuries and even if of 'obsolete' design could still prove useful. In addition, there were inevitably a large number of foreign pieces acquired by purchase or capture to be found in ships, arsenals and fortifications. This included, for example, many captured Scottish pieces found in the Tower of London in 1547.

The 'New' Artillery 1475–1500

Francisco Lopez-Martin has followed the development of the design of the cannon from its fourteenth-century inception to the nineteenth century.[3] He notes the use of bronze ordnance throughout the period but highlights the predominance of the wrought iron gun in inspiring the design of bronze pieces in the late medieval period. Thus, reinforcement bands can be seen on bronze guns of the late fifteenth century, mimicking the rings and hoops surrounding the staves of wrought iron guns. Bronze guns were cast with larger diameter reinforced sections in front of the breech, copying the peculiar stepped appearance of bombards such as *Mons Meg* and *Dulle Griet*. In the case of wrought iron guns this feature was an essential part of their design, providing structural strength and safety at the key junction of breech and barrel but performing no role whatsoever in a bronze cast piece. The casting of bronze cannon exploiting the qualities of that material, rather than mimicking the shape of wrought iron pieces, was to appear in the last decades of the fifteenth century.

3 Francisco Javier López-Martín, 'Historical and technological evolution of artillery from its earliest widespread use until the emergence of mass production techniques'. A thesis presented in the Sir John Cass Dept. of Art, Media & Design, for the degree of PhD. London Metropolitan University. June 2007,

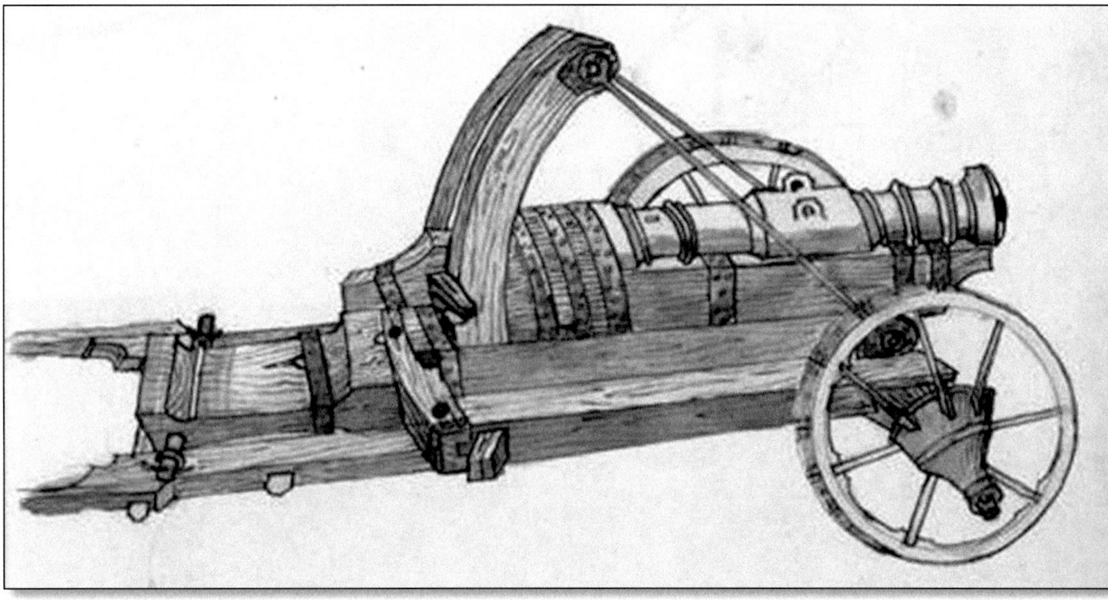

Above: A fine example of a bronze gun with a smooth barrel with cast rings, as well as the characteristic swollen midriff, all redolent of wrought iron construction. The barrel is fitted into a substantial wooden bed that pivots on the carriage beneath. The very heavy carriage is supported by surprisingly spindly wheels. The elevation arc does not appear to have any peg holes but there does appear to be a wooden wedge in place, perhaps performing that role. It is fitted with lifting lugs, a feature of this 'new' artillery. (Heidelberg, Mönch 1496 cpg 126_0059, courtesy of the University of Heidelberg Library)

Below: The contrast of this gun with contemporary bronze pieces is obvious and important. The barrel is of a constant diameter and would probably have fired an iron ball at comparatively high velocity. The trunnions would transmit the recoil into the massive wooden bed in which the gun is set. The whole impression is of a light, manoeuvrable and hard-hitting piece. (*Feuerwerk- und Büchsenmeisterbuch* – Shelfmark, Fol. Cgm 356 pp. 63–64. From the end of the fifteenth century. Courtesy of the Bavarian State Library)

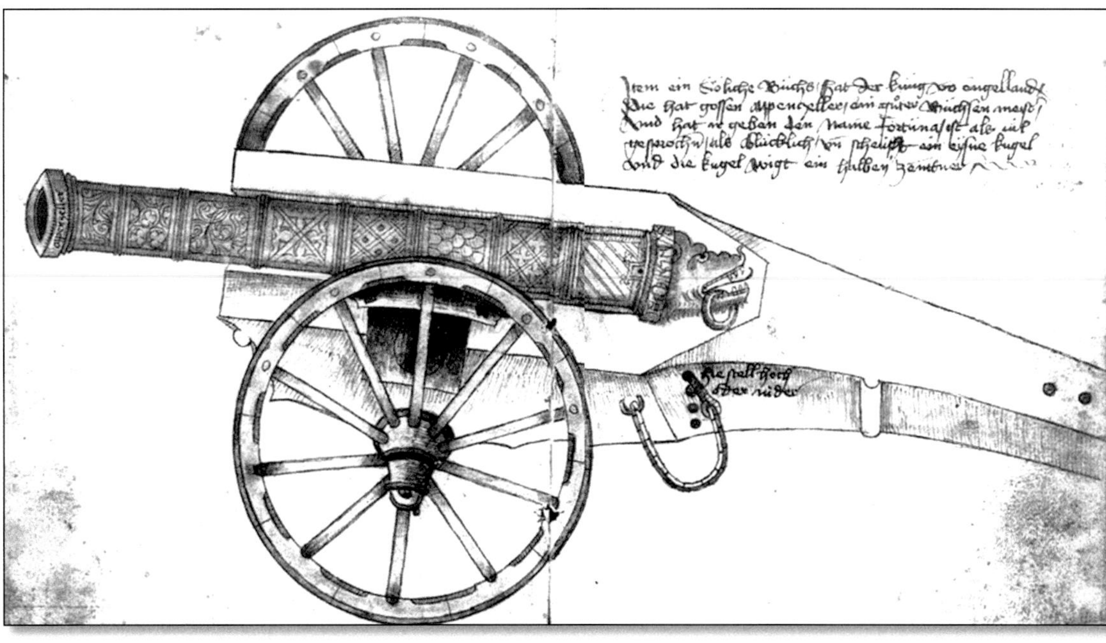

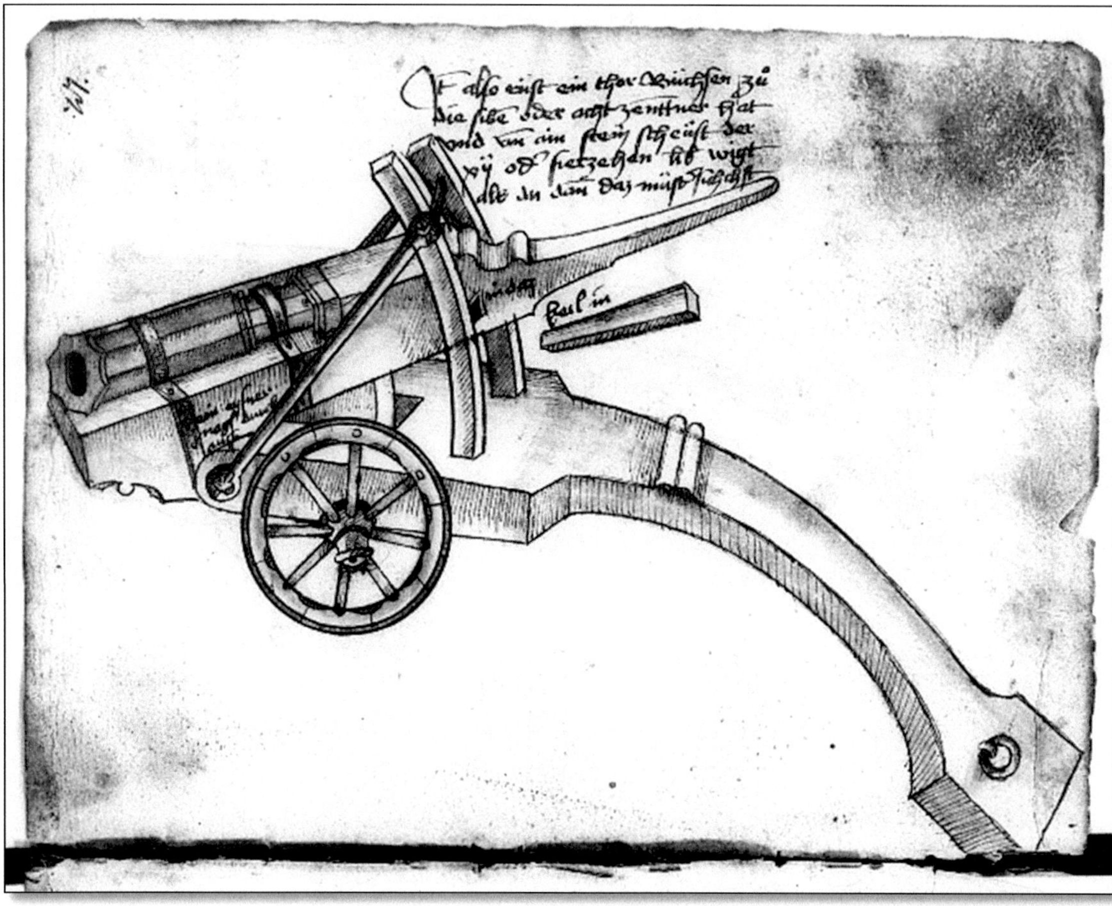

Above: This example of a 'traditional' contemporary field gun can be seen in the same book as the previous image. The barrel is faceted, it has a thickened middle section and is seated in the upper wooden mount. The breech end is flat so that it can impart the recoil fully to the mount. In the absence of trunnions, the gun mount is pivoted at the front and uses an elevation arc adjusted with a wedge, rather than a hole and peg system. (*Feuerwerk- und Büchsenmeisterbuch* – Shelfmark, Fol. Cgm 356 p. End of the fifteenth century. (Courtesy of the Bavarian State Library)

Facing page: This famous engraving by Albrecht Dürer of a cannon with attendant landsknecht was finished in 1518. The gun is of transitional form, with the swollen midriff and elevation arc from the earlier period and the trunnions (very ill-placed), lifting lugs, carriage and limber from the later. It may have been drawn from life, although if it is, less spindly wheels must have been used. In the first decades of the sixteenth century there must have been many such hybrid guns in service. (Courtesy of the Cleveland Museum of Art – Open Access).

López-Martín identifies the 1480s as the key period in development as bronze guns begin to develop their own design philosophy. He highlights the bronze pieces on display in Neuchatel which can be precisely dated to 1488. These were also identified as important by Robert Smith from the Royal Armouries in his article on late medieval artillery.[4] With multi-faceted barrels, as were many of the late medieval bronze pieces, the Neuchatel guns are in many respects very simple in design, with the metal thickest at the breech and thinnest at the muzzle. They are cast with trunnions, and conclude with a convex cone cascabel and a fitting for a vent cover. Perhaps the most important feature is the bore which was of a constant diameter. They are of the form that, would in its fundamentals, be seen in use for almost 400 years.

The artillery that emerged in the 1480s was to provide the models for both the Emperor Maximilian and King Charles VIII. Large bombards carried on carts and emplaced on the ground continued to remain part of the arsenal as did many wrought iron pieces, but the 'new artillery' would soon come to predominate. These were cast bronze pieces where the metal at the breech was thickest and with a barrel that tapered towards the muzzle. The barrel was fitted with trunnions and mounted on a resilient but manageable carriage, firing iron shot. The cascabel no longer needed to be flat as it was no longer fixed in a wooden bed which would absorb the recoil of the gun.

4 Smith, 'All Manner of Peeces', pp.130–138.

The recoil was now transmitted through the trunnions into the trail. The cascabel now provided an opportunity for decoration and what decoration there was no longer mimicked the prevailing but outmoded Gothic style but would copy the exuberance of the Renaissance. The barrel would now be fitted with lifting lugs that would develop into the form generally referred to as 'dolphins'.

The artillery train of Charles VIII in Tours in 1488 prior to the descent on Italy, was described by a spy to Ferdinand of Spain and indicates that old and new were now operating side-by-side.[5] The informer pointed out that many of the pieces 'can fire from the gun carriages while they are pulled by the horses'. This might suggest the ability to conduct a medieval drive-by shooting but we all know what he means. Of the 97 pieces some 16 were large and medium, iron and stone-throwing bombards, there were also 24 *canones* shooting iron balls of 20 cm (8 in) 'directly from their carts', 12 smaller cannons firing from their carts but with stone balls and four carriage mounted serpentines firing iron shot.[6] The balance of weapons was made up of carriage-mounted *culebrinas* of small calibre and shooting lead shot.

The table below is an impressive but rather confusing and, in some cases, extremely dubious collection of data. The length of gun barrels could be readily ascertained but the performance of the guns was merely conjecture or perhaps tavern boasting recorded as gospel truth. It was clearly a collection of guns in transition, combining stone firing pieces as well as what were almost certainly bronze guns firing iron and lead shot of the new design, fitted to gun carriages.

Table 5.

Gun	No.	Mounted	Lgth (cm)	Wt (kg)	Cal. (cm)	Wt of shot (kg)	Shot	Rate of fire per diem	Range (km)	Penetration (metres of wall)
Lombardas	8		330	4,600	40		Iron	12	10	8.4
Lombardas	8		270			55.2	Stone			5
Canones	24	Carriages	200	1,380	20	36.8	Iron	28		5
Canones	12	Carriages	200	1,150		27.6	Stone			3.3
Serpentines	4	Carriages	360		20	25.7	Iron		2	5
Culebrinas	41	Carriages	340/ 240/ 210	1.104/ 644/ 552/ 338/			Lead			

One gunfounder who was clearly caught up in this period of rapid change was Bonnacorso Ghiberti in Florence. He proudly recorded: 'On September 22, 1498, I have tried, and I have handed to the Dieci a French-fashioned, bronze cannon with lion heads on the trunnions and classical ornaments on

5 López-Martín, 'Historical and Technological Evolution of Artillery', pp.323–324.

6 López-Martín, 'Historical and Technological Evolution of Artillery', p.324.

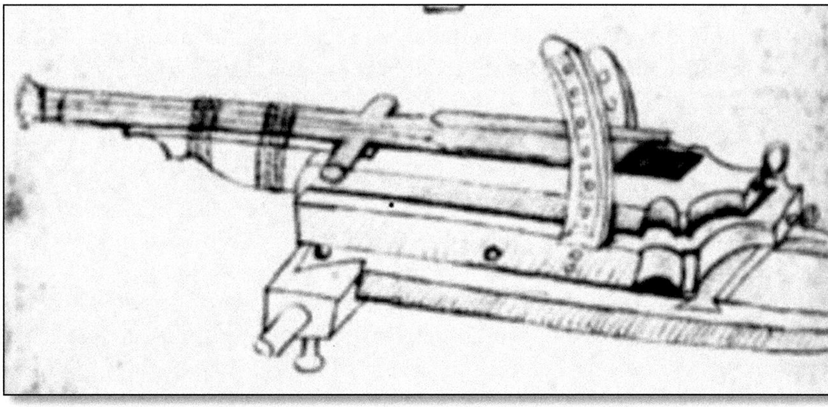

A curious drawing from the *Zibaldone* of Ghiberti. The barrel is very similar to the faceted Norwich Great Hospital gun but for reasons unknown it has been tied down to, or rather into, a wooden bed which is equipped with an elevating arc. Is this a piece of whimsy or a refusal to accept the latest innovation? (Banco Rari 228, in BNF, 82v. With the permission of Ministero dei Beni e delle Attività Culturali e del Turismo, Biblioteca Nazionale Centrale di Firenze)

the muzzle'.[7] This contrasts with his drawing from the 1480s where the guns are mounted in wooden beds pivoted on a trail and where he experiments with elevating screw jacks, folding trails and ratchets, all systems made irrelevant by the trunnion. He does illustrate trunnions on a mortar and even has one gun with trunnions tied to a wooden bed pivoting on a trail.[8] He was clearly aware of the trunnion but seemed initially unconvinced as to its utility.

While some transitioned to the 'new' artillery, what might be considered 'old' or obsolete designs continued to be produced. An excellent example is the *Pluckeroseken* cast in 1509 and lost with the sinking of the *Engelen* in 1518. It was a magnificent breech-loading bronze bombard with a swollen midriff, without trunnions and seated in a wooden bed. If bronze guns continued to be cast in old forms, wrought iron guns also continued in use and in popularity. Large wrought iron guns were to remain the predominant armament in Tudor warships such as the *Mary Rose* until well into the mid sixteenth century, although they were clearly being supplemented and replaced by cast bronze guns in the decades before. The numerous wrought iron breech-loaders had an extraordinarily long lifespan. Ruth Brown when considering merchant ship armament explained that:

> There was still a place for specific types of wrought iron ordnance: slings or murderers – small breechloaders with chambers which could be made by the anchor-smiths at ports. Throughout the seventeenth century merchant ships often carried a mixed armament of small calibre cast iron muzzle-loading guns and wrought iron breechloading swivels.[9]

The *Bronze Bell* which sank off Tal-y-Bont in the early decades of the eighteenth century was found to be armed, at least in part, with small wrought iron guns. Manufacture of large wrought iron guns, on the other hand, was by the mid sixteenth century clearly in decline. The last order in the Empire

7 Ansani, 'The Life of a Renaissance Gunmaker', p.769.

8 Banco Rari Folio 89 v.

9 Ruth Rhynas Brown, 'Guns for merchant shipping', *Journal of the Ordnance Society*, Vol. 22 (2010), p.26.

for such guns was given to the workshop of Sebald Pögl in Innsbruck in 1532, although existing large stocks of wrought iron guns clearly continued in use for decades after.

The transition from a predominantly wrought iron to cast bronze and cast iron guns in the Tudor navy has been meticulously recorded by Robert Smith. The data is summarised in the table below.[10] The author's categorisation into heavy, medium and light guns here is only a general indication as there was such a variety of ordnance over this period, even so the trend is still quite apparent. The urgent purchases of large numbers of cast bronze guns by Henry VIII soon after his accession would suggest that at the beginning of the period England was somewhat behind the European trend at the beginning of the century.

Table 6.

	Wrought iron			Cast bronze			Cast iron		
Date	Heavy	Medium	Light	Heavy	Medium	Light	Heavy	Medium	Light
1514	63	246	489	22		49			2
1546	231	201	1,383	83	167	6			
1576*	94	94	761	75	227	10		54	2
1603				338	574		13	96	

* At this time 37 percent of wrought iron ordnance was held in store, compared to only 15 percent of the bronze pieces being held there.

If we exclude the light pieces of artillery that would be discarded over the decades as tactics developed, then the percentage of wrought iron to cast pieces as the total changed from 94 percent in 1514 to 71 percent in 1546 and by 1576 only 35 percent were wrought iron. At the end of Elizabeth I's reign in 1603 all the guns were cast pieces with 89 percent being bronze.

Barrels, Cascabels, Dolphins and Trunnions

The 'new' guns took on important changes in their details. The shape of the barrels, cascabels, dolphins and trunnions, as well as the style of decoration all were subject to modification and a form of evolution until preferred arrangements became generally adopted. The exterior of the late-fifteenth century bronze gun barrels was often faceted and sometimes had an attractive barley twist or sugarstick form. The faceted style was continued into the first decades of the sixteenth century. Faceted pieces were manufactured in France from the reigns of Charles VIII to Henri II and they were also cast in Naples

10 Robert D, Smith, 'Ships and Guns of the Tudor Navy 1495–1603', in *Ships and Guns: The Sea Ordnance in Venice and in Europe Between the 15th and the 17th Century*, ed. Carlo Beltrame and Renato Ridella (Oxbow Books, 2011) pp.109–112.

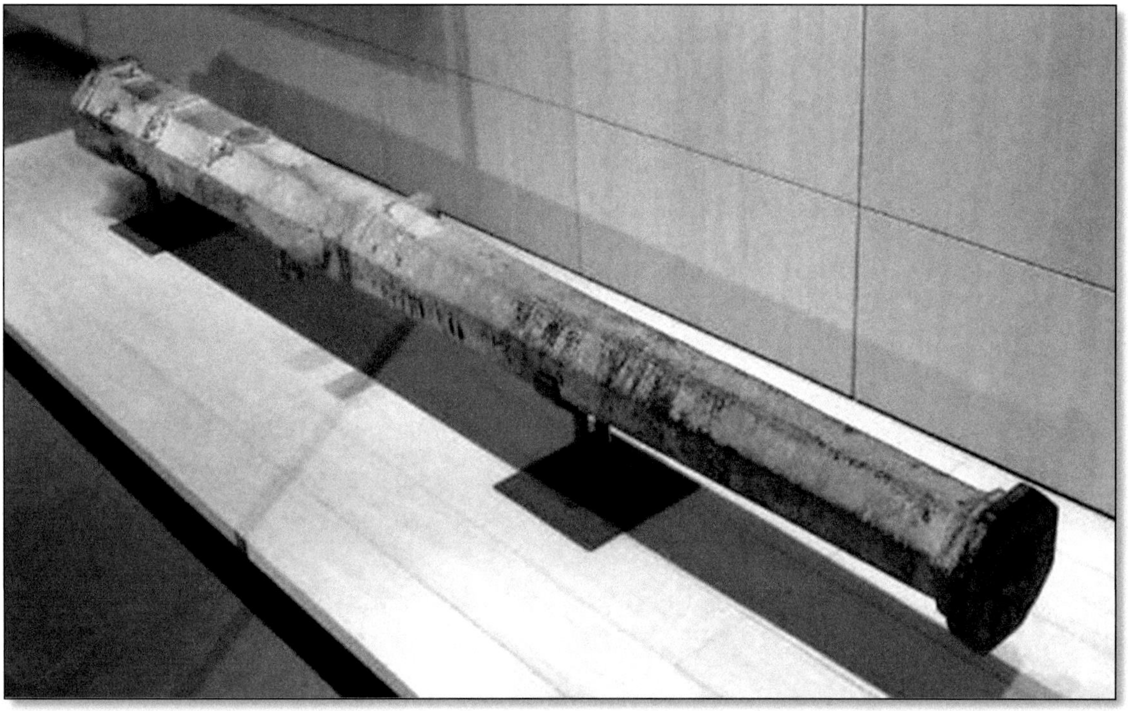

This demi-culverin was probably cast by Alessandro Gioardi in Italy for the Order of St John, on the orders of King Henry VIII. It is a fine long-barrelled faceted piece useful on land or at sea. It bears the arms of Phillipe Villiers de l'Isle Adam, Grand Master 1521–1534, and the coat of arms of England, rather crudely applied. The vent has been enlarged from use, then reamed and screwed to fit a replacement bronze or possibly iron pierced screw bolt. Overall length: 109.25 in (2.775 m); length muzzle to touch hole: 100 in (2.55 m); diameter of bore 3.15 in (80 mm); weight of projectile 3¾ lb (1.7 kg); diameter of projectile 3 in (76 mm) – the normal windage for Italian pieces was 1⁄20 of the bore diameter. (Author's photograph)

and Genoa with the Gioardi family producing several surviving examples.[11] There is an extant octagonal gun for James V of Scotland, cast possibly by Borthwick, which shows the influence of French 'stranger smelters'. The facets, like the barleytwists added nothing to the strength of the gun and were more difficult to produce as a pattern than a simple round form. They also gave less opportunity to the founder to apply the flamboyant decoration that so appealed to their royal and aristocratic patrons. It is not surprising that they were almost completely superseded by simpler round barrels by mid century. The gunfounder Hans Pegnitzer, who died in 1509, was succeeded by his son Endres (d. 1549/1554?) who now replaced the faceted designs of his father with round Renaissance cannon type.

The Italian gunfounders seem to have taken a lead in the manufacture of round barrels starting in the last years of the fifteenth century, with examples from the foundries of the Alberghetti, Francisci, Petrus and Giovanni Battista dell Tolle. It was soon taken up across Europe. Leonardo da Vinci's

11 Renato Gianni Ridella, 'Two 16th century Papal esmerils in the Cleveland Museum of Art, Ohio, and some notes on bronze pieces of ordnance with a polygonal section', *Journal of the Ordnance Society*, Vol. 19 (2017), pp.5–38.

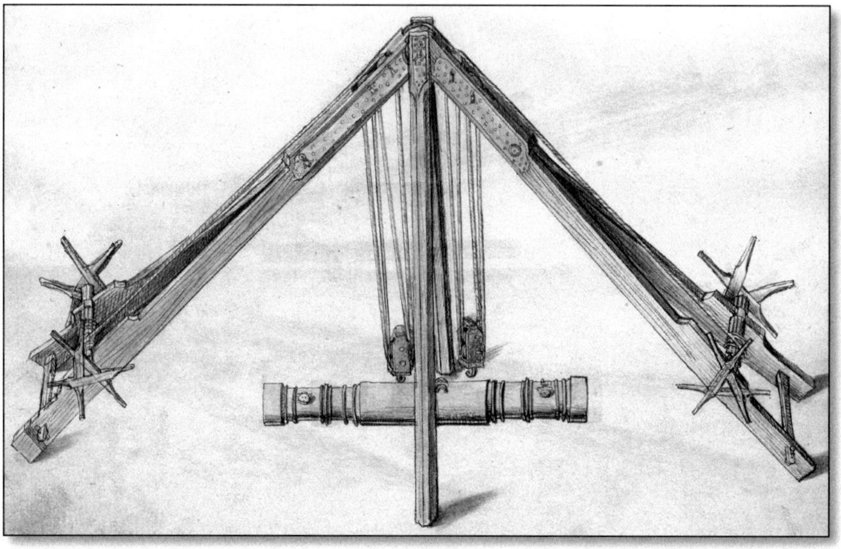

An essential accompaniment to any artillery train was the means of mounting and dismounting barrels. This is a most sophisticated windlass system connected to a pair of block and pulleys. The barrel is very much a conventional late medieval form, but it is clearly equipped with lifting lugs. These would develop into an essential feature of barrels: the dolphin. (*Feuerwerksbuch*, Martin Merz – Shelfmark, Fol. Cgm 599 p. 71. Courtesy of the Bavarian State Library)

drawings of French cannons completed 1513–1515 shows one gun entirely round, a form clearly now in vogue in France. One of the definitively 'new' guns which still exists is the Mulhausen gun which can be dated to 1515. It is entirely round with a long slender body and divided into two parts and provided with two semi-circular parallel lugs with one more at the breech end. A further excellent example of the new style is the *Greif* (griffin) cast by Master Simon for the Archbishop of Trier in about 1522. It is round in form with the thicker breech end and has a slender moulding around the middle before a slimmer chase. The decoration is of renaissance putti and the lugs are curious half-snakes/half-dolphins.

As a consequence of the trunnion, the rear of the breech no longer had to be set into a wooden bed to absorb the recoil. The Cascabel could now become a decorative as well as functional part of the barrel. It could be used as a lifting point or used for belaying a breeching rope but even so it provided tremendous decorative opportunities for the creative gunfounder. The Neuchatel and Heidelberg style guns were both cast with a convex cone cascabel, fitted with a lifting ring. Hans Poppenruyter used a highly characteristic curlicued tail cascabel, on a rather lovely bronze falconet with a faceted breech and round chase, now to be found in the Hotel de Ville (Dinant).[12] The cascabel and muzzle designs were often a variation of the mouldings found at the base of columns. Not especially a Renaissance form, found as it was in medieval buildings both ecclesiastical and secular. It was both attractive and easy to produce as a mould, either of turned wood or by using a stickle board on clay. Other guns of the late fifteenth and very early sixteenth centuries were fitted with a 'tiller' or rectangular projection, not set centrally, this would have been inserted into the elevating arc associated with

12 Nico Brinck, 'Three cannon salvaged from the Meuse in Dinant, Belgium', *Journal of the Ordnance Society*, Vol. 28 (2021), pp.45–53.

Left: A bronze saker cast by Francesco Arcano in 1529 for Henry VIII. This barrel has plain turned mouldings and a band of foliage including Tudor roses and fleur-de-lys by the base-ring. It demonstrates an appreciation of the Renaissance style of decoration; this was an age when weapons and armour displayed their owners' appreciation of the contemporary aesthetic. The vent is hardly worn at all in comparison to other guns in the collection. (Royal Armouries xix.165; author's photograph)

Left: Cast between 1530 and 1540 by Peter Baud and at present in the Tower collection of the Royal Armouries (XIX.17). The Tudor rose is surmounted by a crown supported by two very sweet little putti, holding for no discernible reason some ears of wheat and a branch bearing fruit. It is suggested that the work was done by one of the Florentine craftsmen working in England, Giovanni da Maiano or Benedetto da Rovezzano who worked with Peter Baude on other projects. The rectangular panel is inscribed in clear italic and not gothic script, HENRICVS OCTAVVS DEI GRACIA ANGLIE ET FRANCIE REX FIDEI DEFENSOR DNS HIBERIE. Henry VIII, by the Grace of God King of England and France, Lord of Ireland. (Courtesy of the Royal Armouries)

earlier carriages. A good example of this type is the falconet found today in the Great Hospital in Norwich, a piece which should be much better known.

One of the important features of the 'new artillery' was the appearance of lifting lugs positioned midpoint on the barrel, to allow it to be lifted on and off easily. These lugs emerge in the Austro-German area in the late fifteenth, early sixteenth century, with the first example dateable to 1473.[13] The lugs vary from simple open handles to geometrical shapes or paired heads of snakes but are not commonplace until the second decade of the century. After this they take many forms: paired snakes, dog heads, and hands become common. The dolphin appears as the predominant form, combining efficacy with aesthetic appeal in the 1520s, especially after they were taken up as a feature by Gregor Löffler after 1530. In Spain lugs were not used at all until the late 1530s, possibly then influenced by Löffler's use of them for Imperial weapons.

Renaissance style would now dominate the decoration used, sometimes it seems to excess. The fluted Corinthian column could now be reproduced on the chase of the barrel. Peter Baud's impressive three-barrelled breech-loading canon was cast in the form of a Tuscan column.

In the dedicatory cartouche the script would now be in stylish and legible Italic rather than ghastly Gothic. The rediscovery of Nero's Golden Palace

The lion's mouth cascabel of this gun is almost as impressive as the gunner's beribboned codpiece. There seems to be an excess of felloes and strake plates on the wheels and the proportions of the gun seem somewhat exaggerated. Style and swagger were essential features of the sixteenth-century soldier and his weapon. The shot box on the right has a gabled lid which is very sensibly lockable. Flat lids encouraged the weary to rest, inadvisable if you were to sit on gunpowder. The linstock is an interesting example of the type, long and fitted with the serpentine mount for match cord of a form usually associated with arquebuses. (*The Gunner* by Franz Brun, 1559. Courtesy of the Cleveland Museum of Art – Open Access)

13 *Feuerwerksbuch*, Martin Merz – Bayerische Staatsbibliothek München, Shelfmark, Fol. Cgm 599 p.71.

beneath Rome led to a passion for complex mixture of animal, human and plant forms. Known as 'grotesque' from the Italian word for cave 'grotta', this style came to predominate. The motifs could include faces with exaggerated features, acanthus leaves, flames and rays, sometime in the form of long friezes all the way round the barrel. *Putti* or naked cherubs would now find themselves astride a gun barrel rather than disporting themselves among the clouds. Even the humble trunnions could be found displaying the busts of, no doubt bewildered, Roman emperors. Pearwood and boxwood moulding boxes allowed details such on trunnions and cascabels to be copied and reused as in the case of the dolphins of Gregor Löffler, ensuring speedier production and widespread adoption.

Designing the Barrel

One of the first and most important decisions when designing a gun was to determine the length of the barrel. The assumption that was commonly made was that the longer the barrel the greater would be the range. The contemporary opinion was expressed by Norton, for:

> The longer the chase of the piece the stronger the stroke ... whereby it cometh to pass that long Culverings carry further than great Cannons although with less powder: yet the force is better entertained by their greater length and better fortification to endure the full charge of powder.[14]

Smith agreed, citing the following result:

> Two peeces in all respects equall, save onely that the one is something longer than the other, discharged with one like quantity in powder and bullet, shall make severall grazes, according to the length of the cilinder of the peece, the longer shall outshoote the shorter.[15]

Norton attempted to explain what occurred to make the shot fly further, attributing it to the creation of a vacuum within the barrel. He also recognised that a stronger piece meant that a larger charge could be used:

> The generation of this fire being made in an instant, the rarity ariveth at once, the violence prest to chase out the shot, the flame flyeth making it issue out, and Ayre to come in to prevent emptinesse, and all as it were in an instant. Therefore the longer the Chase of the Peece is (being fortified and loaded accordingly) the more effectuall shall the Action be, and the stroake the more violent, whereby it commeth to passe that long Culuerings carry further then great Canons although with lesse powder: yet the force is better entertained by their greater length and better fortification to endure the full charge of powder.[16]

14 Norton, *The Gunner*, p.238.
15 Smith, 'All Manner of Peeces', p.99.
16 Norton, *The Gunner*, p.8.

This self-evident truth was unfortunately not always supported by practical experience as in this account, again from Norton:

> Luigo Coluado in his 21 Chapter mentioneth, that an extraordinary Culvering of 48 Dyametres of her bore in length,[17] being loaded with a Charge of powder vsually allowed to Culverings (proportioned to the waight of her Shott) did shoote fewer paces, then afterwards when 8 dyametres of her length was cut off from her Mouth to her Breech wards. And then also 6 dyametres more being likewise cut off, and in like sort, shee being loaded and discharged at the third time, convayed her Shot farther then either at the first or second.[18]

Henry Hexham reported the results of a trial conducted by Maurice of Orange which also questioned the current assumption relating barrel length to range. Unfortunately, he did not provide an explanation or consider the major implications:

> Yet this [longer barrel equating to longer range] doth not alwaies hold true for Count Maurice Prince of Orange of blessed memorie, tryed once a half Canon vpon the Strand at Scheveling, giving it the due charge of powder, and caused the Canonier to levell in vpon its highest range, at 6,000 paces, to trye this conclusion, to see how farre it would carry, when it was shorter, caussed some two foote and a half of yt to be sawne off, made it be loaded againe, shoot it off, and found that it carried it bullet as farre as it had done before.[19]

Bourne noted that pieces destined for naval service were shorter than those normally cast for land use but without a significant effect on performance:

> They [the 'Queenes Nauie'] have devised to make their Ordnaunce shorter than the accustomed manner, and so by that meanes they are lighter than the peeces before time made, and yet as serviceable as the longer in some points, shooting that weight in pouder, and ye shotte that the heavier doth, in all poyntes as the other: for that mettall that is taken from the length of the peece, hurteth not the fortifiyng of the peece.[20]

He is however of the belief that: 'the troth is, that the longer peece doth shoote further than the shorter', although he admits to some contention on the subject.[21]

This was an opinion held by some historians until recently. Peter Padfield explains that the privateers' enthusiasm for 'long' guns was: 'because the longer the gun, the longer the explosive gasses have to work on the shot, thus the greater the muzzle velocity and the greater the *effective* range'.[22]

17 What might be described as a 'basilisk'.

18 Norton, *The Gunner*, p.102.

19 Hexham, *The Third Part of the Principles of the Art Military*, p.3.

20 Bourne, *The Arte of Shooting*, p.13.

21 Bourne, *The Arte of Shooting*, p.24.

22 Padfield, *Guns at Sea*, p.35.

Colin Martin has pointed to the difference between the performance of modern propellant and gunpowder which help explain why contemporary understanding of gunpowder was so unhelpful:

> With modern propellants, the increase of temperature and pressure within the chamber serves to increase the burning rate and hence the rate of evolution of propellant gases as the projectile moves along the barrel. Careful shaping of the propellant grains further enhances this desirable effect. Black powder, on the other hand, burns in an entirely different manner. The rate of decomposition is essentially independent of temperature and, above a quickly reached threshold of about 350 pounds per square inch, of pressure. Thereafter the rate of evolution of propellant gases within a black powder cannon is in all practical respects constant. As the ball accelerates down the barrel a point will therefore be reached beyond which the projectile will expand the volume behind it faster than the decomposition of the propellant can keep it pressurised. Extending the barrel beyond this point will thus result in a reduction of the muzzle velocity and range of the gun.[23]

Guilmartin argues that although the highest possible velocity could be achieved with a barrel of 25 calibres, increasing muzzle length over 12 calibres would lead only to a slight increase in muzzle velocity (12 percent), and that over 18 calibres there would be no measurable increase. The rapid deceleration of the inefficient projectile would mean that there was no discernible difference in velocity irrespective of the length of the barrel. He concludes that: 'the maximum ranges of all 16th Century cannon firing cast iron projectiles were therefore essentially the same regardless of cannon length'.[24]

Irrespective of the evidence it was assumed that a longer barrel resulted in a longer range. There was a real advantage for a long barrel which was cast breech down, as this would ensure a greater 'head' of metal, increasing the density of bronze in the breech. The gun would therefore be stronger which was the principal objective of the gunfounder and encouraged the use of a larger charge. A long barrel would also provide a longer sight base and therefore potentially greater accuracy. The trend was for such long-barrelled culverin types to be of 25 calibres or more, while cannon types had barrels of up to 25 calibres in length.

Once the type of gun was chosen then the actual design was based upon a system of proportions based around the diameter of the projectile. Systems of proportion were already used when bell-founding as recorded by Bonnacorso Ghiberti. In the case of the bell, it should be 13 parts wide to 14 parts high and thickness was calculated in terms of parts of the width varying from 1/20 to 1/28. One of the earliest descriptions of the rule of proportions for the 'New artillery' came from a study of the French artillery that so impressed the Italians in 1494. Ghiberti the great Florentine gunfounder helpfully noted down the specifications of the French pieces in the *Zibaldone*:

23 Martin, *The Equipment and Fighting Potential of the Spanish Armada*, pp.240–241.
24 Guilmartin, *Gunpowder & Galleys*, p.149.

> Another chortaldo 25 or passavolante will weigh about six thousand libbre, if it has a length of six braccia, 26 and if it fires eighty libbre of lead … French gunmakers are accustomed to cast the breech of their passavolanti three shot thick, that is, one for empty space and two for bronze, that is, every side of the chamber as much thick as the gap. This is their own way to craft guns with a shot weight of ten libbre of lead, 27 or less. And for the pieces with a shot weight of thirty, or forty, or fifty libbre of lead, they made these sides two and half shot thick [1.25 times the diameter of the shot in metal on both sides of the bore].[28]

Ghiberti identified and described here the key feature of the new artillery: it was a simple tube with the thickness of metal at the breech about the same diameter or slightly larger than the projectiles it fired.

Biringuccio, some 40 years later, described a similar system of proportions, based on the diameter of the ball (which was 1/20 less than the diameter of the bore for Italian guns). The thickness at the breech of cannons should be three quarters the diameter of the ball and at the muzzle before the mouldings or 'cornice' one third.[29] He recommended as a 'precaution and embellishment', the use of reinforcement rings where the powder ends as well as before the trunnions. Culverins and demi-culverins should have more metal at the breech, being the same thickness as the full diameter of the ball. He highlighted the apparent rationalisation of the wide range of weapons such as *springards*, *cerbottane*, *caccia-cornacie* with sakers, falcons and falconets.[30] These new weapons were built with the same proportions of the culverin with breech thickness the full diameter of their shot and at the muzzle a half diameter.

Bourne writing a generation or more later considered how much metal should be in the piece to ensure safety and came to pretty much the same conclusions. At the breech where the touch hole is placed: 'the metal be in thickness as high as the shot round about the sides of the piece, and somewhat thicke, and if that the metal be not in thickness as much as the height of the shot, then it is too slenderly metalled'.[31] The ever-helpful and informative Norton, several decades later, provides us with a list of the thickness of metal required at three points on the barrel, at the vent, the trunnions and the 'necke', which is the point on the muzzle before the swell of metal provided by mouldings. The proportions vary according to the type of gun being cast. The measurements are in the diameter of the bore:[32]

---

25 Known variously across Europe as a curtell, curtow, cortaldi, kurtowe, all roughly equivalent to a culverin.

26 One Braccio = 12 palmi = 583.5 mm or 1 foot 11 inches.

27 Iron shot weighed using lead weights.

28 Ansani, 'The Life of a Renaissance Gunmaker', p.766. Ricordanze di Bonaccorso di Vettorio di Lorenzo Ghiberti, in ASII, 13230, 19r. The reference is in Banco Rari 228, in BNF, 87v–88r: 'Uno chortaldo overo passavolante…'

29 Biringuccio, *Pirotechnia*, p.225.

30 Firing 12 lb, 6 lb and 4 lb shot respectively

31 Bourne, *The Arte of Shooting*, p.62.

32 Norton, *The Gunners Dialogue*, p.9.

Table 7.

Type	Touchhole	Trunnions	Necke
Cannon double	1	11/16	7/16
Cannon ordinary	7/8	10/16	6/16
Cannon lessened	3/4	9/16	5/16
Culverins	1 1/8	15/16	9/16
Culverins lesser	7/8	10/16	6/16

The need for a degree of standardisation was met by the use of patterns, such as those held in the Tower in 1547, or by drawings showing the diameter of the barrel at key points. For Genoese guns of 1550, the thickness of metal just behind the mouldings at the 'necke' should be ⅜ the diameter of ball, just before trunnions 6/8 and at the vent 7·5/8, very similar to the figures above.[33] In Collado's *Treatise* of 1586 he illustrates this with a diagram of carefully calculated concentric circles. The largest circle was at the *culata* (breech near the vent), then the orechioni (trunnions), the *cuollo* (muzzle neck) and finally the *bocha* (bore).[34]

Pieces conforming to the standard for that type of gun would be described as 'legitimate' or 'common'. If more metal was used it would be described as 'fortified' or 'double fortified'. Two late-sixteenth century guns from the *Neustra Señora* wreck were cast for the defence of Lima with more metal than was used even for what were considered reinforced demi-culverins. They were certainly too heavy for regular sea service but their greater wight of metal and long barrels would have been of use if engaged against enemy warships in a prolonged fight. Their barrels would have heated more slowly and be less prone to bursting.[35]

If less metal was used, the barrel would be described as 'lessened'. Common pieces would be of the length specified for their type, if they were longer barrelled then they would be called 'extraordinary' and if shorter they would be 'bastard'. An example of lessened pieces are the falcons of St Peter's Parish and Castle Cornet on the Island of Jersey. These two guns had 'lessened' breeches with metal 84 percent and 80percent of the diameter of their bores.

The principal modification of the barrel came with the addition of the 'reinforces'. The reinforce is itself an interesting concept, as it was clear that the design of the 'new' artillery already acknowledged the importance of having the thickest metal at the breech. The first development in the design of the barrel was to increase the diameter of the breech section. This 'reinforce' was intended to strengthen this critical area, and extended from the vent to the furthest portion of the charge and shot. The creation of a second reinforce, to

33 Renato G. Ridella, 'Dorino II Gioardi: "A 16th century Genoese Gunfounder"'. *Journal of the Ordnance Society*, Vol. 16 (2004), p.32.

34 Renato G. Ridella, 'Examining two bronze cannon recovered from the wreck 54 of the frigate Neustra Señora de la Mercedes', *Journal of the Ordnance Society*, Vol. 28 (2021), p.59.

35 Renato G. Ridella, 'Examining two bronze cannon recovered from the wreck 54 of the frigate Neustra Señora de la Mercedes', *Journal of the Ordnance Society*, Vol. 28 (2021).

produce the recognisable 'stepped' appearance, has been attributed to the use of more powerful corned powder. This in itself would not have significantly affected the pressure generated in the barrel if the corned powder had a reduced saltpetre concentration.

Ridella has drawn some interesting conclusions from his rigorous analysis of artillery design, especially pieces cast in Italy:

> Some authors state that this solution [the double reinforce], introduced in German and Flemish bronze pieces of ordnance early in the sixteenth ·century, is due to the need to strengthen the barrel owing to the adoption of the corned powder. I think, on the contrary, that it developed for ornamental reasons as the presence of two reinforces does not modify greatly the structure of a piece in comparison with a single reinforce one. For this purpose it should have been necessary to have more evident steps in the external shape as can be seen in pieces from the end of the nineteenth century. Besides, we know that corned powder was probably first introduced in Flanders around 1450, half a century before modern muzzle loaders were cast. In all the Italian states the single reinforce remained in use through the sixteenth century until the 1680s, for example in Venice, while Genoese bronze cannons maintained this shape into the 18th century. The true strengthening was obtained by increasing the thickness of the barrel walls in the three crucial points (vent, trunnions and muzzle neck) causing a consequent increase in the weight of pieces.[36]

He argues that the second reinforce was a German–Flemish innovation associated with Gregor Löffler who as Master of the Imperial foundry introduced it to Spain. The reinforces were in some cases only nominal. The barrel's reduction in diameter from breech to neck was often continuous and the reinforce rings were solely decorative and did not indicate a step down to a thinner portion.

Trollop when considering the design of English cast iron guns from 1543 to 1660, showed that the thickness of metal at the breech remained the same and the reduction in length of the first reinforce and the introduction of the second which reduced the amount of metal and saved weight, was due not to stronger powder but to more reliable serpentine powder. Heavier barrels were introduced into naval service after 1589 perhaps to take advantage of corned powder with their higher proportion of saltpetre. The vent circumference remained the same but the reinforce now extended 2⁄7 of the gun's length. The new culverin therefore weighed 37 cwt and was 8½ ft in length as compared to the 33 cwt and nine feet of the earlier. As a consequence, the trunnions were now positioned forward to 3⁄7 of the length to ensure a good balance. This change demonstrates the incremental nature of the improvements taking place.

One of the most important innovations was the trunnion, the 'axle' around which the gun pivots. This was probably invented in the Duchy of Burgundy, although Venice has also claimed this honour and it has been attributed to the famous French Bureau brothers. Carman refers to an entry

36 Renato Gianni Ridella and Antonina Palazzolo, 'A bronze falcon from Federico Musarra's workshop found off Genoa in 1959 and a brief survey of ordnance production in Sicily through the 16th century', *Journal of the Ordnance Society*, Volume 21 (2009), p.5.

in the archives of Lille referring to the making of trunnions for a serpentine in 1465.[37] The evidence for Burgundy comes from a small number of guns lost to the Swiss in 1476 and guns with trunnions are illustrated in the Swiss Chronicle of Diebold Schilling of 1478–1483.[38] It seems certain that the trunnion was certainly in use by the last quarter of the sixteenth century and probably before.

Previously the recoil had been transmitted through the bed of the carriage by the flat breech end of the piece, or if breech loaded by its flat end of the separate chamber wedged into the barrel. The trunnions placed a new strain on the carriage at two precise points, which could split the wood of the cheeks or trunk in which they were seated. Early gun carriage can be seen to address this problem by using iron strapwork reinforcing this area, as well as iron plates clenching the wood together. Tim Jones was responsible for the extensive strapwork that surrounded and clenched the forepart of the author's carriage, cleverly mating it with the capsquares that held the trunnions in place. It was clear that it was all designed to ensure that the carriage was protected from the recoil of the gun, and not to protect it from damage in transit.

The dimensions of the trunnion were readily calculated, again by proportion. They should be: 'Dyametre one Calibre of her proper bore in thicknes, and also one in length, onely lessening 1/30 of a Calibre, tapering by little and little towards their outward end of them'.[39] This profile created problems for the foundry casting the author's gun, as the junction between the trunnion and the barrel was a possible weak point where there could have been cracking. This required slightly more metal to be used in casting and then removed using a cold chisel and hard graft. The taper ensured that the barrel centralised itself when being moved, rather than slipping from side to side. Smith recommended that two thirds of the trunnion be set into the carriage although the evidence for earlier carriages suggest that it was close to fully set. It was important that the barrel did not 'bounce' either when travelling or fired, so that a close fit was needed, something that Tom Davies spent a lot of time and effort achieving on the author's carriage. The trunnion enabled the gun to be elevated or depressed more easily and permitted greater accuracy when aiming. The importance of a close-fit for the trunnions was very noticeable when Tom and Tim refitted the trunnions and capsquares.

The positioning of the trunnions on the barrel was absolutely crucial. Biringuccio explained the consequences of getting it wrong:

> for if he puts them too far back the muzzle hits the ground, lifting the breech up in the air, and it can be said that they are useless guns because they cannot be fired to strike where desired. If the trunnions are put too far forward the gun loses itself in the bed between the wheels so that it cannot be loaded well, for it is so heavy at the loophole that one man by himself cannot handle it without one or two helpers. It is a reprehensible thing if the master allows it to fall into either of these errors.[40]

37 W. Y. Carman, *A History of Firearms* (London: Routledge Kegan Paul, 1955), pp.34–35.

38 Diebold Schilling, Amtliche Berner Chronik, Vol. 3, p.347.

39 Norton, *The Gunner*, p.73.

40 Biringuccio, *Pirotechnia*, p.257.

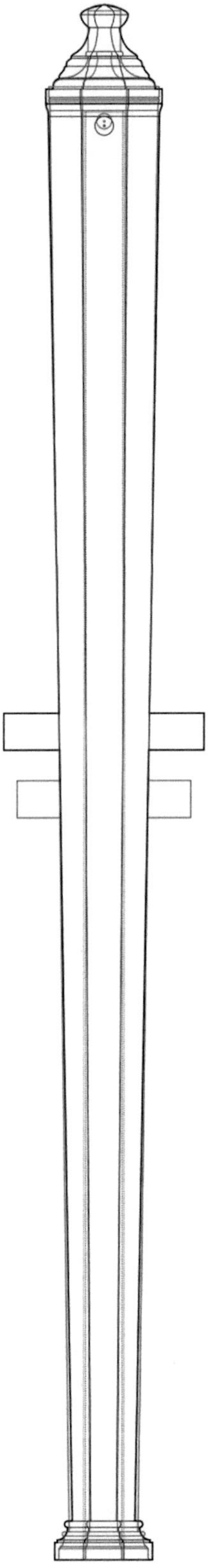

Clearly this was a Goldilocks conundrum that as always would be resolved with reference to a rule of proportions. Biringuccio recommended that they should be placed ⅖ of the distance along the barrel from the breech end. Smith recommended the following:

> Measure the length of the bore of the peece, from the mouth to the breech, devide that measure by 7, and multiply the summe that commeth in the quotient by 3, the product will shew you how many inches or other measure the trunions ought to stand from the end of the lowest part of the concavity of the sayd peece at the breech.[41]

Using his calculations, and taking the author's falconet based upon examples from two falcons of about 1500–1520 as an example, his calculations would place the trunnions 716 mm from the end of the breech, whereas in fact they are 733 mm. Happily Norton concurs with Smith in general principle. If the gun was a perrier type with a sub-calibre chamber and relatively thin barrel walls the trunnions were placed a little nearer the breech to ensure good balance.

Left: This is an overlay of the two barrels the author and Thomas Davies studied when considering the design of their own falconet. They are both falcons held in the collection of the Royal Armouries at the Tower of London (XIX–14) and Fort Nelson (XIX-15). Although there are differences in decoration and the trunnion position has shifted the general proportions are very similar. The Fort Nelson falcon was probably cast by Gregorio I Gioardi in around 1500 while the Tower example was probably a Flemish product of around 1520. They are both long-barrelled types with a calibre of approximately 32–34. (Illustration courtesy of Thomas Davies). For more information on these pieces see Renato Gianni Ridella, 'Two 16th century Papal esmerils in the Cleveland Museum of Art, Ohio, and some notes on bronze pieces of ordnance with a polygonal section', *Journal of the Ordnance Society*, Vol. 19 (2017).

The height of the trunnion on the barrel is a very rough indication of its date.In the first few decades of the sixteenth century the trunnions were placed on the centreline, by mid-century they were set slightly lower and by the end of the century they were placed between ⅗ and ¾ of the diameter of the gun from the top. Cyprian Lucar in 1588 asserted that the gunfounder should: 'set the trunnions so as ⅔ partes of the circumference of the Peece in that place where the trunnions shall be set, may bee seene aboue vppon'.[42]

A Proposed Typology for Wrought and Cast Guns

> That it is possible to say of the guns that we call antique as well as of those that are modern to us today that I have never found a uniform size in any kind that is seen. In short it seems to me that in every age men have proceeded to make them best for their purposes according to the wishes of whoever has them made or of the masters who make them.[43]

41 Smith, 'All Manner of Peeces', p.72.

42 Lucar, *Appendix*, p.36.

43 Biringuccio, *Pirotechnia*, p.223.

If there is a subject likely to bring a historian to the point of despair it would be an attempt to create a simple typology of ordnance in this period. There is a mass of conflicting and often apparently contradictory information and assertion provided by contemporary sources and other historians. Smith and DeVries in their scholarly study of Burgundian artillery make a heroic attempt at establishing the different types of ordnance. To attempt a similar study across Europe and the whole of this period is a project that could absorb several lifetimes and would not greatly add to understanding. Late medieval artillery could best be understood by its purpose rather than by name, names which as Smith and DeVries point out were often used for only a relatively short period of time or designated pieces which varied considerably in size and which overlapped in dimensions with pieces with a different name.

Smith and DeVries wisely relied upon the remarkable survival of Burgundian pieces and access to the extensive documentary material for their study of Burgundian artillery. The present author will use the remarkable assemblage of ordnance from the *Mary Rose* as the basis for a simple wrought iron typology, to which must be added the medieval super-gun, the bombard.

The port-piece: a heavy, short-range gun, firing stone shot.

Wrought Iron Gun Typology

The *Mary Rose* excavation has provided historians with a remarkable assemblage of guns that can, for the first time, be related to the considerable documentary evidence. There are many problems when trying to provide a definitive typology. The first is the wide variety between individual but similar pieces. These were the product of separate craft workshops and manufactured over many years, although they could clearly be identified as belonging to a class or identifiable type of gun. The poor condition of many of the wrought iron guns or sometimes portions of guns has made precise measurement at best difficult; even so it is possible to distinguish between classes of guns.

These were the principal armament of ships and fortresses, especially the Device forts, alongside bronze pieces. They are large bore breech-loaders with a relatively short barrel, firing stone shot and mounted on heavy two-wheeled carriages. The hull or barrel was made of a single layer of staves overlaid by rings and hoops. There were some guns of up to 10 inches (250 mm) diameter bore but usually less; those in the Tower armoury are recorded as averaging 6½ hundredweight for the barrel (330 kg), and about 300 lb (136 kg) for the chamber. The barrel was about three times the length of the chamber. The chamber length was approximately three times the bore of the barrel.

Summary specifications for the type:

Bore diameter 6–8 inches (150–200 mm)
Shot weight 4.38 kg (9.71 lb) at 144 mm to 9.9 kg (21.83 lb) at 193 mm
Barrel length about 15 bores
Overall length 2.5–3 m (9–10 ft)
Length mounted 4 m (13 ft)
Weight mounted 2. 5 tons/tonnes

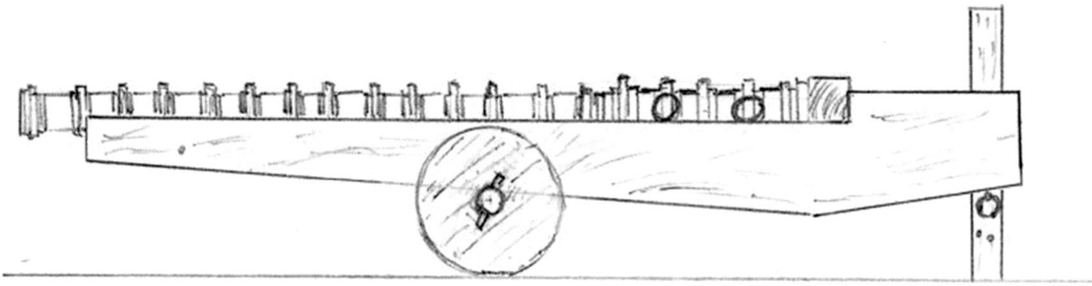

The sling – a lighter, longer-barrelled piece firing iron shot. (Author's drawing)

The Sling

The sling comes in a bewildering variety which includes 'single', 'demi' and 'quarter'. Its name derives for the German for serpent which fits its very large calibre (bore diameter to barrel length measurement). Although in appearance it seems simply to be a reduced-size version of the port-piece, it is a very different beast. The key to understanding it is to appreciate that it fired iron and not stone shot. It had barrels that were as long or longer than those of port-pieces, with a calibre of 20–30. Closer examination also showed that the barrel is proportionately much thicker especially at the collar, the critical point between breech and barrel. It is clear that the purpose of the sling was to shoot its iron shot at higher velocity and over a longer range. The breech is especially long as a proportion of the barrel and especially strongly built; even so they seem to have been prone to failure. All of this would suggest that the powder loadings were higher than for the port-piece or fowler and that the purpose was to produce a long-range piece.

In Alexzandra Hildred's *Weapons of Warre*[44] it is suggested that the slings were the iron equivalent of the 'new' culverin. This assumption is based on the probability that they shared the same ammunition, as where they are recorded as being present there is no reference to shot specifically for them. They were mounted on wooden beds, as were port-pieces, and some at least were fitted with wheels.

Proposed sizes of slings by muzzle diameter and weight of shot:

Sling 95–103 mm, 7.5 lb/3.4 kg
Demi-sling 83–88 mm, 4¾ lb/2.2 kg
Quarter sling 60–65 mm, 1½ lb/0.7 kg

44 Alexzandra Hildred, *Weapons of Warre: The armaments of the Mary Rose*, Vol. 3 (Exeter: Short Run Press Ltd, 2011)

Fowlers – Smaller Stone Firing Pieces

The fowlers were quite extensively used on land and sea and appear to have been a smaller version of the port-piece, lacking the strengthening and iron projectiles of the sling. With a bore of between four and five inches and a relatively short barrel of about a metre, it was again mounted on a wooden bed and was provided with wheels.

An early sixteenth-century wrought iron swivel gun. Simple, and comparatively cheap to make, it would have been a valuable addition to both civilian and military vessels. Barrel length 34.5 in (8.77 cm), overall length 64 in (1.63 cm), weight 172 lb (78 kg), bore 3 in (76.2 mm). (Royal Armouries XIX.3; author's photograph)

Bases and Swivel Guns – Anti-Personnel Pieces

There were bronze, wrought and cast iron guns of this type in use. If of wrought iron, they could be constructed on the stave principle or made from a single sheet of iron which could be bent and hammer welded around a mandrel. The breech was usually made from a single piece of wrought iron, there being no great need for a more complex construction. They sometimes had a surprisingly long barrel of around 30 calibres and a bore from ½–2½ inches in diameter. They were always provided with separate breeches equipped with handles that could then be slotted into place and held with a wedge, hammered firmly in. One of the common warnings to gunners was not to stand where a poorly inserted wedge could fly out when the piece was fired. Those fitted with bronze barrels were generally referred to as being *Portingall* or Portuguese. There were also examples of hybrid construction with a bronze barrel but wrought iron breech holder and breech.

Robert Smith has helpfully produced a typology of such weapons, considering their method of construction in detail. In practice they all performed a single purpose irrespective of the materials and methods used. They were primarily if not solely anti-personnel weapons, mounted on swivels to allow them to sweep the deck of an enemy ship. They could use shot of iron 'dice' or lead, or in an emergency shoot whatever came readily to hand. Those issued by the Tower between 1563 and 1589 weighed 168 lb (76.2 kg), while their chambers weighed a further 8 lb (3.6 kg) each. A representative piece would have a bore of about two inches, a barrel length of around 40 inches and a breech length of rather less than 12 inches.

The longevity of wrought iron swivel guns was extraordinary, extending well into the seventeenth century. They were probably one of the most commonly available and cheap pieces, suitable for the self-defence of merchantmen as well as a valuable short-range weaponry of warships. There were 892 bases recorded in the Anthony Roll out of a total of 1814 iron guns which is an indication of their prevalence and perceived utility. The 'shrympes' or war carts that accompanied Henry VIII to the siege of Boulogne in 1544 were equipped with very long-barrelled bases, which were later dismounted and used in static defences.

The *Pumhart von Steyr* is one of the most famous medieval bombard types. It weighs approximately 10 tonnes and has a sub-calibre powder chamber surmounted by a massive barrel tapering from 76 to 88 cm. It would have fired a 690 kg stone projectile, from a barrel one a and a half times its diameter. Forged between 1448 and 1452 probably as a gift for emperor Friedrich III by the famous Schachner family of iron smiths in Upper Styria. It is difficult to say how far it was in any sense a practical weapon even then. (Author's photograph)

The Bombard

For a century from 1375 the bombard was the essential accompaniment to any siege, replacing the trebuchet as heavy artillery. The size of the projectile mattered more than anything else as the velocity was comparatively slow, using a low proportion of propellant weight to projectile. The powder was contained in a thick-walled sub-calibre chamber able to withstand a higher pressure than the thin walled relatively short barrel. The massive bombards such as *Mons Meg* and *Dulle Griet* that came from the same Flemish workshop in the mid fifteenth century had significantly longer barrels with a calibre of about five.

The bombard could be a wrought iron piece but by the mid fifteenth century was more and more likely to be of cast bronze. Rate of fire varied but would average a handful of shots in a day. Very large bronze pieces were usually constructed in two parts for ease of movement with the parts joined with a screw thread. They were comparatively low velocity weapons compared to the new cast bronze ordnance but still performed a useful role in a siege. They were designed on the basis of simple proportions. In 1497 Bonaccorso Ghiberti used 12,000 libbre of metal to cast what was even then an obsolescent piece a bombard, with the following proportions: the barrel was seven and a half times the diameter of the shot and a sixth of the diameter thickness, while the metal at the breech was a quarter of the diameter of the projectile and at the rear a third.[45]

45 Ansani, 'The Life of a Renaissance Gunmaker', p.766.

Dulle Griet or 'Mad Margaret' in Ghent, Belgium. The bigger sister of *Mons Meg* and the Basel Bombard: all were built in the workshop of Jean Cambier for the Duke of Burgundy and represent different versions of the same basic design. The barrels are 7⁄10 of the total length, and the internal length of the barrel is five to five and a half times the bore. They also share the same distinctive barrel shape. These were not the absurd aberrations one might have thought but impressive and effective contributions to the siege train of any medieval autocrat. (Karelj, public domain, via Wikimedia Commons)

Cast Gun Typology

> The development of bronze ordnance was a far more complex process than has been hitherto suspected … it may have overlapped the development of wrought iron construction and … it varied considerably from place to place, driven largely by economic considerations.[46]

It is clear that bronze guns sometimes shared the design characteristics of their wrought iron counterparts until the last decades of the fifteenth century. The period between 1480 and 1530 did however see the emergence of the archetypal cast barrel that would continue in more or less the same form for over three centuries. The guns that emerged, developed into four broad classes that performed quite specific purposes and superseded the various medieval bronze and wrought iron guns that preceded them, sometimes inheriting their names as well as their roles.

It was clear in the last decades of the fifteenth century that the emerging bronze 'new' artillery gave considerable advantages to those who possessed it. This point was made quite explicitly in 1498 when the magistracy of the Republic of Genoa received a submission from the Council of Elders which declared:

> we think it has been known for some time that … the French, have introduced heavy bronze ordnance aboard their ships. So, although this ordnance is placed in small vessels, shooting from a long range, it is enough to defeat every big ship. For this reason, nowadays, our sailing is dangerous even if our ships are of a large size, which everyone knows, and they are very well equipped with iron bombards.[47]

46 J. F. Guilmartin, 'The Cannon of the *Batavia* and the *Sacramento*: Early Modern Cannon Founding Reconsidered', *IJNA* 11, 1982, no. 2: pp.133–144.

47 Ridella, 'Dorino II Gioardi', p.39.

The response of the Genoese Office of the Sea was to order all vessels over 480 tons to equip themselves with two bronze cannons weighing in total 2,382 kg in addition to the numerous wrought iron pieces they already carried. What these new pieces replaced were the long barrelled wrought iron guns shooting iron balls, called the 'sling' or 'passavolante'.

The classes of predominantly wrought iron artillery would be replaced by new classes of cast bronze pieces. The heavy artillery of the late medieval period represented by *Mons Meg* in Scotland and its sisters was supplemented and then replaced by the cannon. Longer range fire would be provided by the culverin shooting iron shot. Short range wall and ship smashing guns firing stone shot, the bronze perrier or pedrero, would supplement the port-piece and fowler. The many varieties of anti-personnel weapons often swivel-mounted and breech-loaded could be wrought iron or bronze or with a bronze barrel and wrought iron breech and chambers.

The classification of cast guns cannot readily be achieved using the weight of their projectile as the defining factor. The modern distinctions made between artillery pieces, where weight of shell and purpose are closely correlated and where the basic principles of design are the same do not match contemporary practice. A heavy culverin could fire a much heavier shot than a lighter cannon and a cannon-perrier could have a larger bore than a cannon but shoot lighter shot. How they principally differed was in their breech shape and calibre, that is the length of the gun tube not the diameter of its barrel.

Cannons were on the whole large-bore weapons with barrels of about 15–25 calibres, shooting iron shot and with the thickness of metal at the breech rather less than the diameter of its shot. Culverins provided the medium and field guns had barrels considerably longer than a cannon at over and sometimes well over 25 calibres. Again these shot iron balls with the breech thickness the same diameter of its shot. Cyprian Lucar, drawing on a range of English and Continental authorities, came to some definitive conclusions as to what barrel length for a variety of current, obsolescent and foreign types should be in calibres.

Falconet 34
Faucon, minion, pssavolante, saker, moiane [culverin moyenne] and culvering, 32
Quarter cannon 28
Demie cannon 21 and 25
Double cannon 18
Basiliske 24[48]

The culverin was a term that could be used to describe a type of long-barrelled guns with a breech thickness equal to the diameter of the bore. Thus, the falconet and falcon were of culverin design, although they fired projectiles only a fraction of the weight of a 'true' culverin.

The cannon-perrier although it appeared very similar to a cannon was unique in its design. It had a sub-calibre breech, previously seen in medieval

48 Lucar, *Appendix*, p.34

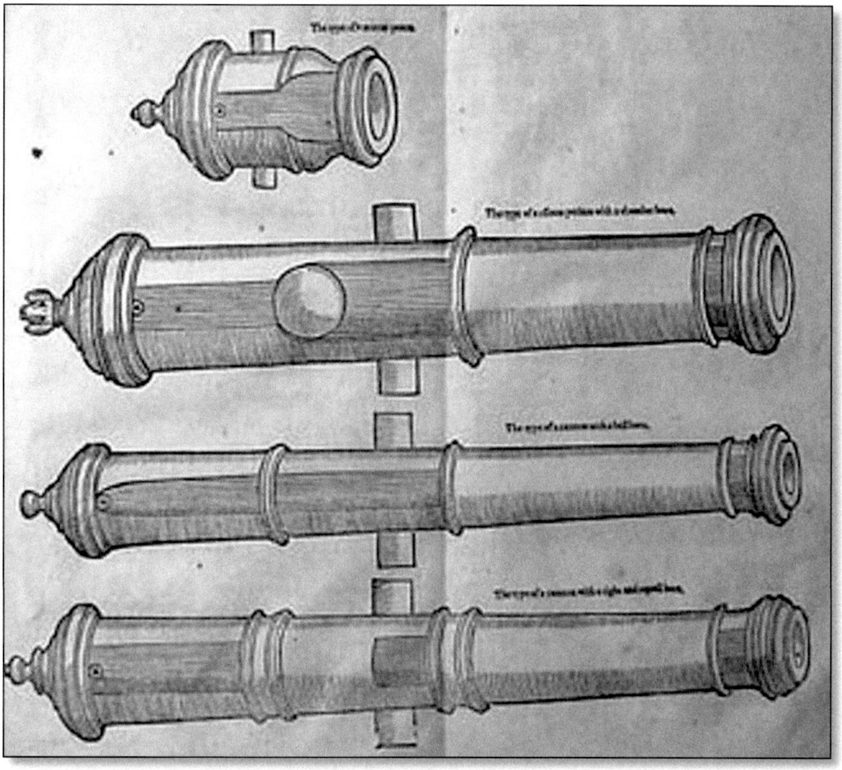

The principal types of cast guns, as illustrated in Tartaglia. They are from top to bottom: mortar, perrier, culverin and cannon. Note the sub-calibre chamber in the mortar and perrier as well as the unusual conical chamber in the culverin. (Nicolò Tartaglia, *Three Bookes of Colloquies* (1588), pp. 34–35)

pieces, with a shorter and lighter barrel and fired stone shot. The cannon perrier was very economical in metal and powder, the sub-calibre chamber held a small charge suited to a relatively light stone projectile. This design produced relatively low barrel pressures allowing for a thinner, lighter barrel and cheaper gun. This characteristic of high breech and low barrel pressure was particularly apparent during the Middelaldercentret trials of their bronze gun. The stone balls had to be hand carved and the price of labour in most of Europe made stone more expensive than iron, except in Portugal and the East.

One weapon that shared some of the perrier's features was the mortar. The use of high angle fire and indirect fire to discomfort the enemy was well understood in the Middle Ages. The trebuchet provided direct fire against walls and indirect fire against the castle or city within. Biringuccio declared unequivocally that mortars: 'are not esteemed by us moderns'.[49] This is a strange conclusion to draw as high angle fire using large diameter projectiles appears to have been an aspect of siege warfare throughout this period. Smith pointed to the special role that mortars could play when other artillery could not be brought to bear:

> Morter peeces were invented onely to annoy the enemy, when other Ordinance cannot be vsed against them, as being charged with stone to beate down the houses of the enemy, or to fal amongst men being assembled together, or charged with balles of wild-fire to burne the enemies ships, houses, or corne.

49 Biringuccio, *Pirotechnia*, p.227.

A well-prepared mortar position. The ammunition is stored very safely in case of a premature explosion in the barrel. The sophisticated trench system was probably a response to the threat posed by counter-battery fire, deadly at the short range at which the mortar was emplaced. One of the crew may be ringing a bell ordering the lucky bombardier to fire, either way he has placed himself safely out of the way, presumably he was the officer in charge. (Walther Litzelmann: *Vortrab zu der Arckalay – Büchsenmeisterbuch* – Shelfmark, Fol. Cgm 909 p. 287, courtesy of the Bavarian State Library)

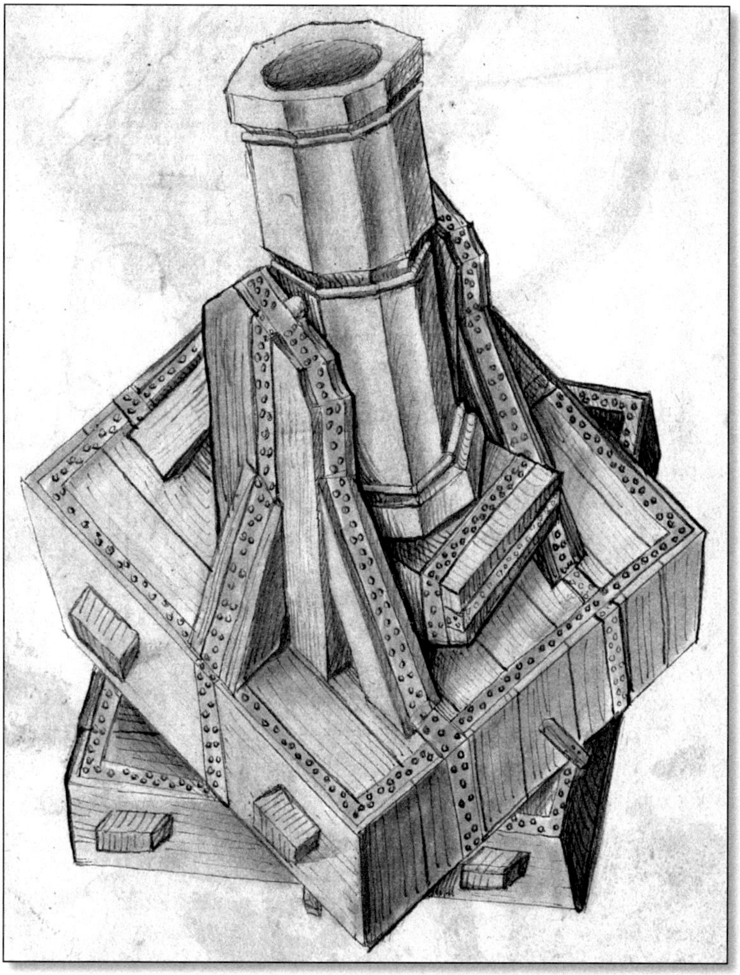

Left: an early example of a cast bronze faceted barrel mortar, equipped with trunnions, set on a complex mounting. This permitted the barrel to be traversed as well as elevated. How necessary or practical such a mounting would be is questionable. The importance of a stable base to ensure accuracy was important, the swivelling system would not have helped. Range could have been adjusted by altering the weight of the propelling charge rather than by changing elevation. (*Feuerwerksbuch*, Martin Merz v2, shelfmark, fol. 0004546000016, courtesy of the Bavarian State Library)

Below: an illustration from Bourne provided by an artist who had clearly neither seen a mortar, nor understood how a quadrant worked. He has put the ruler in the right place, though. It is not always safe to assume that a contemporary illustration is an accurate representation of what it claims to show. (William Bourne, *The Arte of Shooting in Great Ordnaunce*

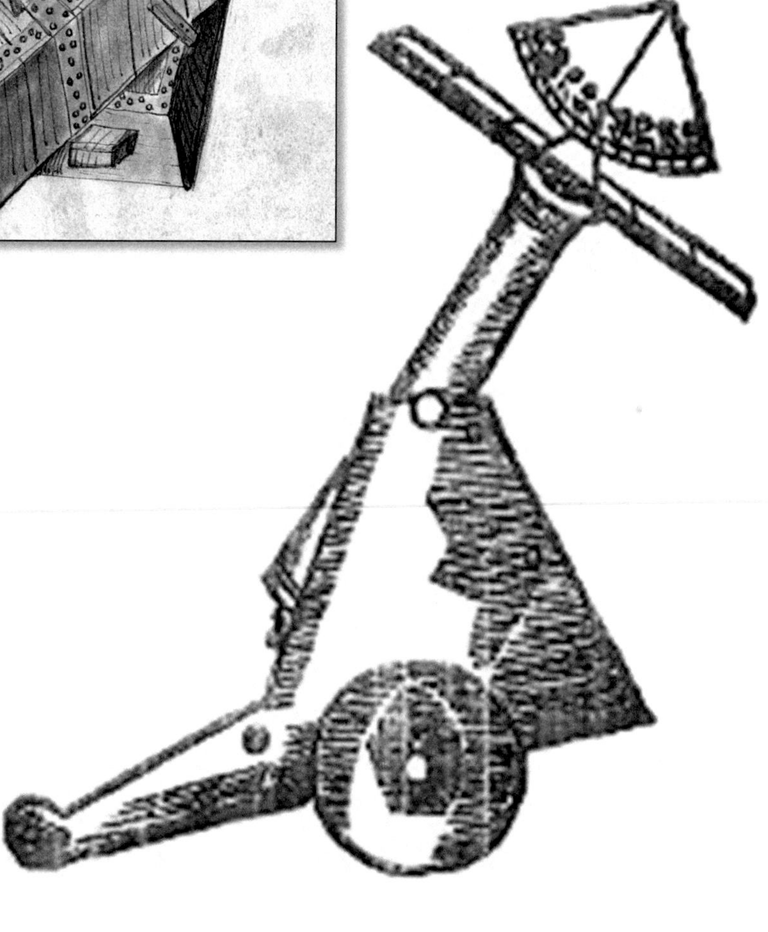

The manufacture of large diameter cast hollow balls filled with a bursting charge and complex pyrotechnic projectiles, gave to the mortar a capability that was not readily shared with conventional artillery. Explosive projectiles required a fuse that could be set alight either when the mortar was fired or by lighting the fuse just prior to being fired. Neither method was entirely satisfactory or safe. This was amply demonstrated in 1639 by Dutch merchants to the Japanese Tokugawa Shogun:

> The first shot fell too short, yet was observed to fall into a deep marshy hollow wherein rice was planted, between seventeen and eighteen feet deep, and consequently in their opinion it was either lost, or could not possibly take effect, albeit it proved to be the contrary, for shortly afterwards it burst with such violence that all the mud, slime and filth was hurled so high into the air, that all who saw it were astonished, and particularly the regents who could not show enough amazement. At the second shot the bomb exploded in the mortar, whereby the gunner's face was severely burned and all the rest us were wounded more or less.

In all 11 rounds were shot, not one of which hit the target, with several exploding in the barrel.

Proposed Typologies

Adrian Caruana in his study of Tudor artillery declares that: 'Tudor (or for that matter Renaissance) artillery is generally considered to be beyond the limits of practical understanding,' but continues more hopefully, 'it is possible, through comparative analysis and diligent study, to reduce it to a perfectly logical system.'[50] By a careful analysis of bore circumference or compass, Caruana has divided Tudor artillery into two main groups, with the cannon royal and the robinet both being outliers. The interval between the bore circumference of the heavier guns was three inches and that between the lighter guns two inches. The highest common factor was the unit of ¾ of an inch, while the design unit in length and diameter was ¼ of an inch.

Table 8. The basis of English gun design *c.* 1540: the compass (circumference) of shot

Gun	Compass (inches)	Intervals of inches	Intervals of ¾ in
Cannon Royal	24	2¼	3
Cannon	21¾	3	4
Demi-cannon	18¾	3	4
Culverin	15¾	3	4
Demi-culverin	12¾	3	4
Saker	10½	2¼	3
Minion	9	2	2
Falcon	7½	2	2
Falconet	6	2	2
Robinet	5¼	1	1

50 Adrian B. Caruana, *Tudor Artillery 1485–1603* (Bloomfield: Museum Restoration Service of Canada, 1992), p.3.

The mathematical sequences illustrated here are to be found in guns prior to the death of Henry VIII and probably represent a move towards centralised control and greater uniformity.

Charles Trollope has produced an analysis of gun design for English pieces from 1543–1660.[51] The key features were circumference of the piece at the vent, the position of the trunnions and the length of the first reinforce. Trollop has defined the following characteristics of English guns from 1543. The circumference of the breech would be 10 times the diameter of the bore. By dividing the circumference by π would give a radius of 3.2. With the diameter of the bore being one, then the thickness of metal on either side would be 1.1 times the diameter of the bore, very much the same thickness as proposed elsewhere. The trunnions were positioned at or just beyond two fifths of the gun's length and a decorative ring was cast at the half-length point. The muzzle had simple muzzle rings and not the swells of later pieces.

After 1587 guns increased in weight to take advantage of more powerful corned powder and the trunnion was moved to 3⁄7 of the length to compensate for the increase weight at the breech. There was a clear difference in barrel length between longer, land-service guns and their shorter sea-going equivalents and naval pieces had swelling muzzles as opposed to muzzle rings.

Standardisation

The frustration, nay desperation, of Cyprian Lucar in 1587 with the lack of standardisation has already been referred to. In 1592 a similarly frustrated Luis Collado noted that there were over 200 different sizes of loading tools for artillery pieces in the Castle of Milan. This was a number that he reckoned could be reduced to only 11 if the weapons had been standardised. The problem was bound to occur as weapons were procured piecemeal from any number of independently minded gunfounders. Obsolete guns and 'foreign' pieces acquired through purchase or capture would add to the problems.

Ammunition stocks were always comparatively small, and in peacetime in arsenals and fortifications each gun could be supplied with ammunition suitable for its own use. If guns were to be issued for field service or naval use then problems soon became apparent and may be one reason why the issue of ammunition was so niggardly. Shot sent from the Tower of London to the army sent to suppress the Western rebellion in 1549 was 'as fit as a shoe for a man's hand'.[52] The solution was clearly to standardise the weapons and their ammunition.

Gaspart Bureau the French Master of Ordnance 1444–1469 attempted to standardise weapons in an effort that continued in the reigns of Charles VIII and Louis XII. The adoption of the *Calibres de France* by the early sixteenth century was significant progress, and this was reinforced in the reign of Francis I with the establishment of the six calibres of artillery, the details of which were recorded in 1537 and confirmed by Henri II in 1552:

51 Charles Trollope, 'The design and evolution of English cast iron guns from 1543 to 1660', *Journal of the Ordnance Society*, Vol. 14 (2002), pp.51–65.

52 Frances Rose-Troup, *The Western Rebellion of 1549: An account of the insurrections in Devonshire and Cornwall against religious innovations in the reign of Edward VI* (London: Smith & Elder & co., 1913), p.254.

Table 9.

Name	Length ft and in	Bore in inches and lignes/mm	Weight in French lb	Weight of shot in lb and oz
Canon	9 ft to 9 ft 6in	6–3 /169	5,200–5,400	33 to 33–4
Grande Couleuvrine	9 ft to 10 ft	4–11/132	4,000–4,200	15–2 to 15–4
Couleuvrine Batarde	9 ft	3–10/103	2,500	7–2 to 7–3
Couleuvrine Moyenne	8 ft	2–9/74	1,200–1,250	2
Faucon	6 ft 4 in to 6 ft 10 in	2–4/63	700–760	1–1
Fauconneau	6 ft 4 in	1–10/49	410–420	0–14

The Founder Royal of Spain, the famous Remigy de Halut, established the general characteristics of Spanish ordnance such as barrel length, bore diameter, metal thickness and total weight, after 1534. Charles V established five patterns for artillery in 1544 and there is evidence in England for similar progress at this time. In the 1547 inventory of the Tower there are several references to: 'little gonnes of brasse ... being patrons [patterns] for Cannons,' with other patterns for a demi-cannon and a culverin.[53] Adrian Caruana concluded that the calibres for new royal artillery were established either in 1537 with the issue of the Artillery Charter, or possibly in 1543 with the foundation of the Ordnance Office.[54] Another attempt was made to achieve a degree of standardisation for Spanish forces by Philip II in 1587. He ordered his master of artillery to ensure that the pieces to be cast for the Armada of 1588 conformed to the dimensions, proportions and bronze composition of those cast by *Gregorio* (Gregor Löffler), thereby establishing a pattern for future guns.

Sir William Wynter as Surveyor and then Master of ordnance of the Navy was able to introduce a much-needed period of standardisation from 1569–1589 by the 'Proportion of 1569'. There were to be only six calibres and defining characteristic was to be the weight of shot of 32, 18, 9, 5¼, 4 and 2¼ pounds. It is debatable whether any such policy was ever implemented, let alone successfully concluded. Whatever success may have been won in the short term was bound to be compromised by the longevity of cast ordnance and the acquisition by purchase and capture of 'foreign' weapons of different patterns and periods.

Lightweight and Experimental Weapons

Gunners and commanders have always sought guns that were not only more powerful and with a longer range but guns that were lighter and more manageable. Examples of light guns on field carriages are to be found in

53 David Starkey, *The Inventory of King Henry VIII (the transcript)* (London: Harvey Miller Publishers, 1998), p.104. No 3829.

54 Adrian Caruana, *Tudor Artillery 1485–1603* (Bloomfield: Museum Restoration Service of Canada, 1992), p.7.

the illustration of Martin Merz from the 1470s and the 1496 *Kriegsbuch* of Philipp Mönch. Both illustrated light field carriages, drawn by a single horse, with short single and sometimes multiple barrels. The lightest guns for field use were the falcon and falconet while the robinet was little more than a scaled-up arquebus, equivalent to the *arquebus-a-croc*. The falconet weighed 500 pounds or more and with its carriage could be moved by a crew of three or four. It was long-barrelled and of the culverin design with a bore of about two inches shooting an iron or lead ball weighing about a pound. The falcon weighed significantly more making it something of a handful to move about a battlefield without horses or a lot of sturdy pioneers. It is notable that all gun carriages were well provided with hooks to which ropes could be provided for the man-handling of the guns. These were still only capable of firing small-calibre shot and what was needed was a significant means of lightening larger guns to provide more firepower.

The apparent, at least partial, solution to this problem of weight, was the 'drake', which appeared about 1620. The principal change in design was to use a tapered chamber, something that had been recorded and rejected by Biringuccio in 1540. This would use a reduced powder charge and therefore the metal around the breech could be lessened. The barrel was also shortened to between 13 and 15 calibres. This resulted in a much lighter piece. Its origins are less obscure than those of the leather guns and seem to have been the work of a father and son team. Hans Heinrich and David Friedrich von Sigroth, originated from Germany but came to serve Gustavus Adolphus in 1619. The death of the father led to a slowing of development and the Swedish King encouraged the development of the light, but flawed, leather gun.

The abandonment of the leather gun by the Swedes led to a return to the Sigroth's design and the result was the regimental 3-pdr. This was a cast bronze gun weighing 120 to 140 kg, which could be drawn by one or two horses and crewed by two or three men. Some 42 of these would serve the Swedes well at Breitenfeld in 1631. The gun that gained common acceptance was not the Sigroth piece, but similar guns cast in the United Provinces: the 3-pdr *Kleine draak* and 6-pdr *draak*, from which their common English name 'drake' derived. These were first used in 1622 during the Spanish siege of Bergen op Zoom. Ten such pieces were purchased from the Dutch by England in 1625 for the expedition against Cadiz. The gunfounder John Browne cast the first bronze and iron drakes in England in 1626 and they were to prove popular in the English Civil Wars. One of the disadvantages associated with them was the violence of their recoil. John Reynolds the Proof Master wrote in 1627 that, 'Their reverse [recoil] is so much that I doubt they may hazard the breaking of their tacklings and ringbolts and so deliver their shot uncertainly which must be helped by adding new metal'.[55] Despite these problems Browne was encouraged to cast more examples in 'iron of the refined kind', probably to avoid the cost of bronze pieces. Experiments were made in standard cast iron as well as clearly superior 'refined' iron:

55 Towes, '*A Note on Drakes*', p.39.

Table 10.

Gun*	Length	Bore (inches)	Ordinary iron weight	Refined metal weight	Powder weight
Whole culverin	8 ft 6 in	5¼	40 cwt	33 cwt 3 qr	
Demi-Culverin	8 ft	4½	29 cwt	22 cwt	
Sakers	8 ft	3½	22 cwt	18 cwt	
Drakes of iron					
Demi-Cannon	7 ft 8 in	6¼		12 cwt 2 qr	9
Culverin	7ft 7 in	5¼		14 cwt 2 qr	6
Demi-culverin	5 ft 2½ in	4¼		8 cwt 3 qr	4

There were other experiments with different alloys to reduce weight. A 1523 report from Italy refers to a copper alloy made in an area of Saxony called Islever near Herfort. The metal was easy to work but not fragile either. It led to a reduction in weight of 40 percent, although the guns themselves were also shorter.[56] One unusual metal suited for cannon casting according to Lucar was lead, of which great quantities were available.[57] He recommended that the thickness of metal at the breech should be 1½ times the diameter of the shot and ⅔ at the muzzle. He also very sensibly warned against overheating such a piece!

Drakes were seen as valuable replacements for older pieces of bronze ordnance such as sakers, fowlers and port-pieces which were: 'of small force to pierce the side of ship with Stone, bar or base shot, but that drakes are nimbler and more proper for their uses'.[58] One of the early examples dated to 1631 in the Boston Guildhall Museum had a barrel which weighed 123 kg and was 15 calibres in length. The drake was adopted by Spanish and Austrian forces who had experienced the effectiveness of these pieces at the hands of their enemies, the Dutch and the Swedish.

Perhaps the most intriguing of the new 'light guns' were two examples discovered in the wreck of the *Batavia*, a Dutch vessel that set off in 1628 on behalf of the Dutch East Indies Company. The majority of the guns found on the wreck were conventional cast guns, with 21 bronze and five iron guns constituting the majority of her armament. There were two guns which defied analysis until one was sectioned in order to discover its secrets. It was a new and improved version of the old wrought iron hoop and stave gun but with an exterior which looked almost exactly like a conventional cast piece.

The gun was 2.28 m long and had a bore of 147 mm. If cast of bronze it would have shot a ball weighing 24 pounds. Jeremy Green in his detailed analysis of the Batavia wreck provides this description of its construction:

> It seems, therefore, that the gun was constructed in the following manner. A copper sheet forming the bore was set onto a mandrel. Onto this were mounted

56 Howard L. Blackmore, 'The take-down gun', *Journal of the Ordnance Society*, Vol. 2 (1990), p.1.

57 Lucar, *Appendix*, p.37.

58 Towes, 'A Note on Drakes', p.41.

> six flat, iron staves, held in place with the wrought iron bands or hoops [30 wrought iron bands ranging in thickness from 16 to 56 mm] which were probably heat-shrunk in place. The chamber was fitted into the end of the piece and the staves hammered over to lock it into place. Iron pins were mounted around the trunnions, and the trunnions fitted on in an (at present) unknown way. The outer copper sheathing was moulded around the gun, using the pins to locate and centre the ironwork. Presumably, the whole was covered in clay, up-ended (breech up) and heated. The molten solder was then poured into the top until the interior was filled up. The cascabel-disk was then soldered into place, and the clay and mandrel removed.[59]

One peculiar feature of its construction was that the vent passed through the cascabel from the conical chamber. The use of a separately constructed conical chamber would suggest similarities with the drake, where a reduced charge and barrel length was used for light and shorter-range work. It may be assumed that the light construction of the piece, weighing half that of an equivalent bronze gun, would have precluded it from firing a full-charge and full-weight projectile. Two similar guns are extant: one was in the Rotunda collection and the other found underwater on the Little Bahama Bank. It seems unlikely that such guns were ever widely used. Patents from 1627 and 1633 by Jan de Rycker and Bartlet Cornelis Smidt refer to composite guns made with a wrought iron barrel and provided with a copper sheath. The piece was only four feet long and weighed 230 lb but claims were made for full-size guns shooting iron shot up to 24 pounds in weight.

A type that was widely used at least for a short period was the leather gun. This was something of a misnomer as only the exterior was sheathed with leather. There are examples of two guns made from actual leather, made of numerous rolled sheets, with trunnions, cascabels and reinforcing rings. These are to be found in the National Museum in Kathmandu, Nepal; they were captured in 1792 and apparently never fired. The real 'leather gun' was a Swedish innovation and was in fact made from copper sheet and iron wire. The thin sheet of copper was fitted round a mandrel and the two edges lead soldered. The copper tube was now wound round with iron wire and provided with a tapered breech. The ends of the tube were reinforced with metal pieces, probably shrunk into place. The iron bound tube was then sealed with hemp and plaster while wooden lathes and metal strips formed the trunnions. An external cover of thick leather was then shrunk on. Two examples of Swedish guns are to be found in the German Historical Museum where they rather conservatively date them to between 1620 and 1650. They are both light pieces, shooting projectiles of about a pound in weight with a length of about 2 m and a bore of 5.4 cm and 6 cm in diameter.

The origins and originator of the design are not clear. It is suggested that Philip Eberhard constructed the first example in 1622 in Zurich and that Colonel Melchior Warmprandt after visiting him returned to Sweden with the idea. They

59 Jeremy N. Green, *The Loss of the Verenigde Oostindische Compagnie retourschip BATAVIA, Western Australia 1629: An excavation report and catalogue of artefacts*, Western Australian Maritime Museum, Fremantle, Western Australia BAR International Series 489 (1989), pp. 46–47.

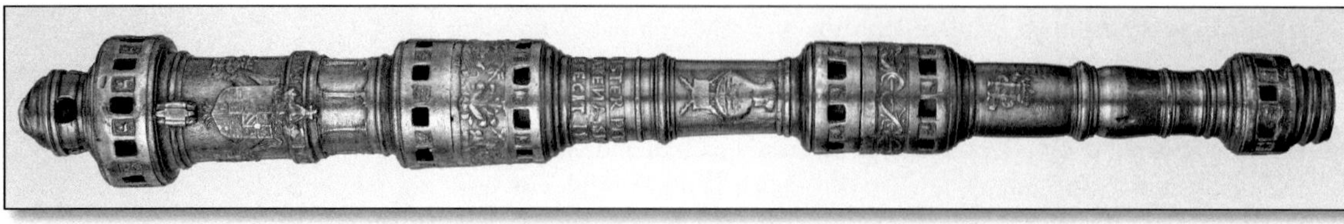

This unique surviving piece was made in 1523 by Pietro d'Arena. It is a model of a gun that would probably be over 15 feet (4.6 m) long when full size. It consisted of four sections that screw together (one is missing), which allowed it to be disassembled for transportation. Cast into its decoration are a portrait of Charles V, his coat of arms, and other personal insignia, suggesting the model was created as a gift for the Emperor. (Courtesy of the Metropolitan Museum of Art, Open Access)

were deployed on active service in East Prussia against Poland by Swedish forces between 1627 and 1629. Whoever designed the guns, they proved of considerable use although presenting major and ultimately insuperable problems. The guns were light and handy and gave Gustavus Adolphus a significant advantage. They provided superior firepower to assist his infantry at short range. One apparent innovation was in its ammunition, where the 18 0.8 in balls used by the gun, were attached to the cartridge in a single fixed round, significantly reducing loading times. Leather guns were rapidly replaced in Swedish service after 1629 as they had a tendency to overheat quickly in action, potentially disastrous for the gunners reloading them.

It has been said that Robert Scott and Ludvig Ripp presented leather guns for trial purposes at Skeppsholmen naval base in 1628. Scott after a short sojourn in Denmark, where he tried to interest Christian IV in the gun, returned to England where his experiments continued at Vauxhall. On his death the work was continued by his nephew James Weymss. Scotland favoured the use of leather guns. The service by Scots under Gustavus was perhaps how the technology and tactics found their way there. The design was somewhat different; the inner core of such guns was made from iron, not copper and they tended to be shorter and lighter and usually found in pairs or even as quadruple mountings. The use of the leather gun in England and Scotland during the English Civil Wars is well beyond the remit of this book and the author recommends Stephen Bull's excellent book *The Furie of the Ordnance* for anyone interested in this period.[60]

If a gun could not easily be made lighter then perhaps it could be divided into several pieces that could later be assembled. In 1610 Thomas Pitt, a gunfounder and 'a rare workeman' produced a bronze or copper model of a gun that would suit any piece from demi-culverin or less, that he claimed had no single part weighing more than 200 (pounds). The pieces would be cast in sections and Pitt boasted that damaged sections could be replaced, if need be. There was no necessity for gins, block carriages or limbers as the parts could easily be carried in wagons with much greater economy in horses and wagons. Above all it meant that ordnance could be borne over mountain and marsh which were impassable to conventional artillery trains. The problem of weighty field carriages was also solved by designing 'flat-pack' carriages that could be assembled on site from parts. A falconet made of nine pieces was constructed and proved by Pitt in 1612 but after this effort no more progress was made.

60 D. Stevenson and D. Caldwell (1977), 'Leather Guns and Other Light Artillery in Mid-Seventeenth Century Scotland', *Proceedings of the Society of Antiquaries of Scotland*, 108, pp.300–317. Retrieved from <http://journals.socantscot.org/index.php/psas/article/view/8975>.

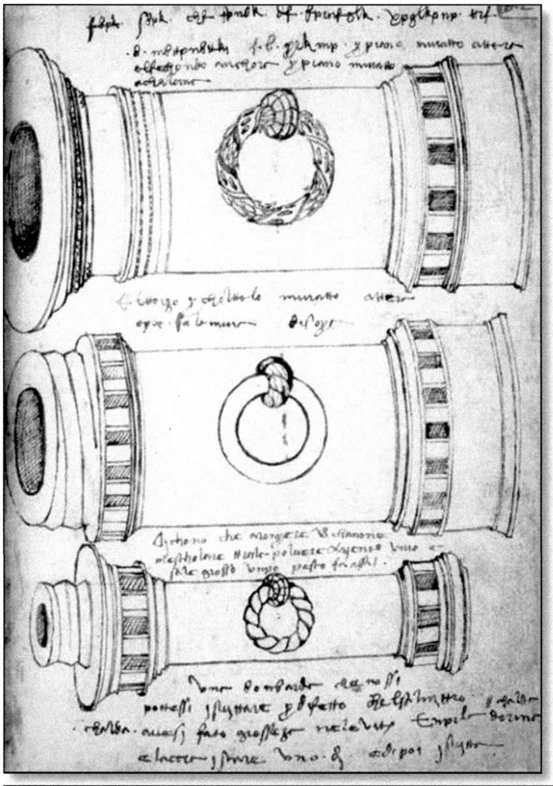

Left: one large bombard cast by Lorenzo di Giovanni and, Ludovico di Guglielmo in May 1496 weighed 17,000 libbre and very unusually was made in three parts, two for the barrel and a separate breech. Although perhaps deemed obsolescent it was still a viable castle crusher, using up a stock of massive stone balls. (Banco Rari 228, in BNF, 82v. With the permission of Ministero dei Beni e delle Attività Culturali e del Turismo, Biblioteca Nazionale Centrale di Firenze)

Below and bottom: Breech-loading had so many obvious advantages but also problems. These were gas obturation (escape) and the extreme pressures on a removable breech. Both solutions above were workmanlike, using a removable breech held in place with a wedge but neither went into production. It would only be in the mid nineteenth century that engineering and metallurgical advances would permit breech-loading pieces to be manufactured, and even then, the early products of Krupp and Armstrong were plagued with problems. (Walther Litzelmann: *Vortrab zu der Arckalay – Büchsenmeisterbuch*. Shelfmark, Fol. Cgm 909 pp. 299–300. Courtesy of the Bavarian State Library)

In the midst of a busy scene a 'wooden' bombard is being loaded at the siege of Boulogne in 1544. It is difficult to say whether the super-firing piece is an actual gun or a large pyrotechnic device. At least 100,000 round shot were expended during the siege. It is also difficult to see what this bizarre gun could contribute significantly to the terrors experienced by the battered burghers of Boulogne. Detail from the Cowdray engraving. (Dr Dominic Fontana, with kind permission)

Biringuccio referred to a gun where the barrel could be screwed together. Mid century, Bartolomeo Campi was rumoured to have produced a 'take-it-to-pieces' (*Cannone con sua fornimente di pesi*) gun on behalf of King Henry II of France.[61] Whatever advantages may have been claimed for such pieces, the disadvantages were also apparent, principally cost and reliability, and they do not seem to have progressed into production.

Other oddities deserve our attention although they were not in practice real guns. At the siege of Boulogne by Henry VIII in 1544, in addition to his large conventional batteries, were two large wooden barrels which were fitted with superimposed 'real' guns. Perhaps the guns mounted on them were used to 'fire' large clouds of pyrotechnic smoke to simulate a massive charge, who knows? These were returned to the Tower after the siege which suggests that they were of some, probably novelty, value. After the main siege the outlying castle at Hardilot refused to surrender, relying as it did upon the difficulty of bringing artillery up to bombard it. Lord Poynings had two wooden cannons built which he: 'artificially coloured like brasse'. Brought up at night their unexpected and unwelcome appearance led to the immediate surrender of the rather less than heroic garrison.

61 Blackmore, 'The take-down gun', pp.1–4.

5

Guns from Forge and Foundry

The vast majority of guns in this period were either hammered from wrought iron in a forge, or cast from bronze or iron in a foundry. Whatever material was used, the skills required were considerable and the logistical and technical problems faced many. Weapons, then as now, were at the cutting edge of technological development and the effort and expense devoted to them was exceptional. The artillery train of an army was calculated to cost about a third of total expenditure and at sea a single ship could dispose of more firepower than even a large field army could boast. The forges, foundries and furnaces spewed out an increasing volume of deadly and often exquisitely made pieces.

Wrought Iron Guns

Wrought iron is a wonderful material. By heating iron bearing ore with charcoal, the product a spongy bloom of slag and iron was then vigorously hammered producing a malleable form that could be worked easily. This was wrought iron. The quantity of iron produced as a single bloom was initially small, only perhaps fifteen kilogrammes but the introduction of the Catalan furnace and water powered bellows led to much greater quantities of higher quality being produced.

Wrought iron varies greatly in quality, dependent upon such factors as carbon and phosphorus content and grain. It is noticeable when studying original guns, or in the manufacture of reproductions, that the smiths chose the best material for areas of the gun most prone to stress and less good where structural strength mattered less. Contemporary descriptions make it clear that there was an unofficial grading system, sometimes defined by country of origin, with Swedish and Spanish iron especially prized, or by how many times the iron had been worked which was expected to improve its quality.

Wrought iron artillery were assembled from their component parts using a stave and hoop construction redolent of wooden barrel making, although in this period barrels were clenched together with hazel. Rather than iron hoops. The staves were assembled around a wooden former and the rings and hoops heated and slipped over them. There is no evidence that the staves were welded, in fact this was a practical impossibility. The close fitting of the staves is a testimony to the skills of the blacksmith. The cooling process led

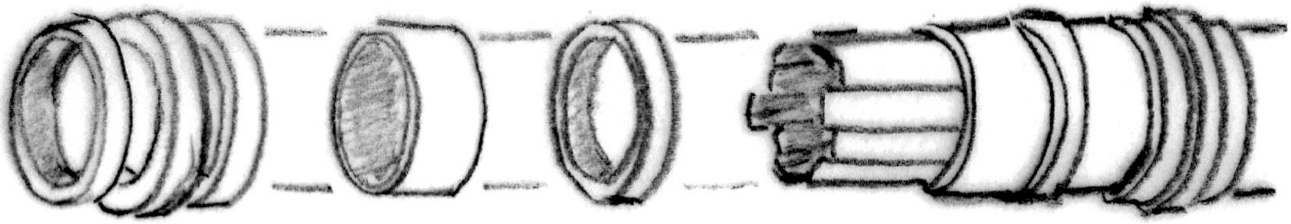

Above: this image demonstrates the system of hoops and staves used by medieval blacksmiths to construct artillery of all sizes. The author has not discovered when and why rings or hoops were used, and their positions seem to vary from piece to piece; he has asked a working blacksmith and he could not give an answer either. When an old piece such as this is left to rust its parts separate very much like this, revealing its construction. When in good condition and covered in layers of preservative paint they almost appear like a single forging hiding their sophisticated construction. (Author's drawing)

Below: The wrought iron gun has often been seen as fundamentally 'inaccurate, largely ineffective and often unsafe'. In practice it was none of these things. The balance that was achieved between the materials used and the stresses created by the stone shot with a moderate powder loading, meant that it was a fearsome weapon for its time. It could not match the power or glamour of the new cast bronze guns, but in its way it was just as much a technical marvel.

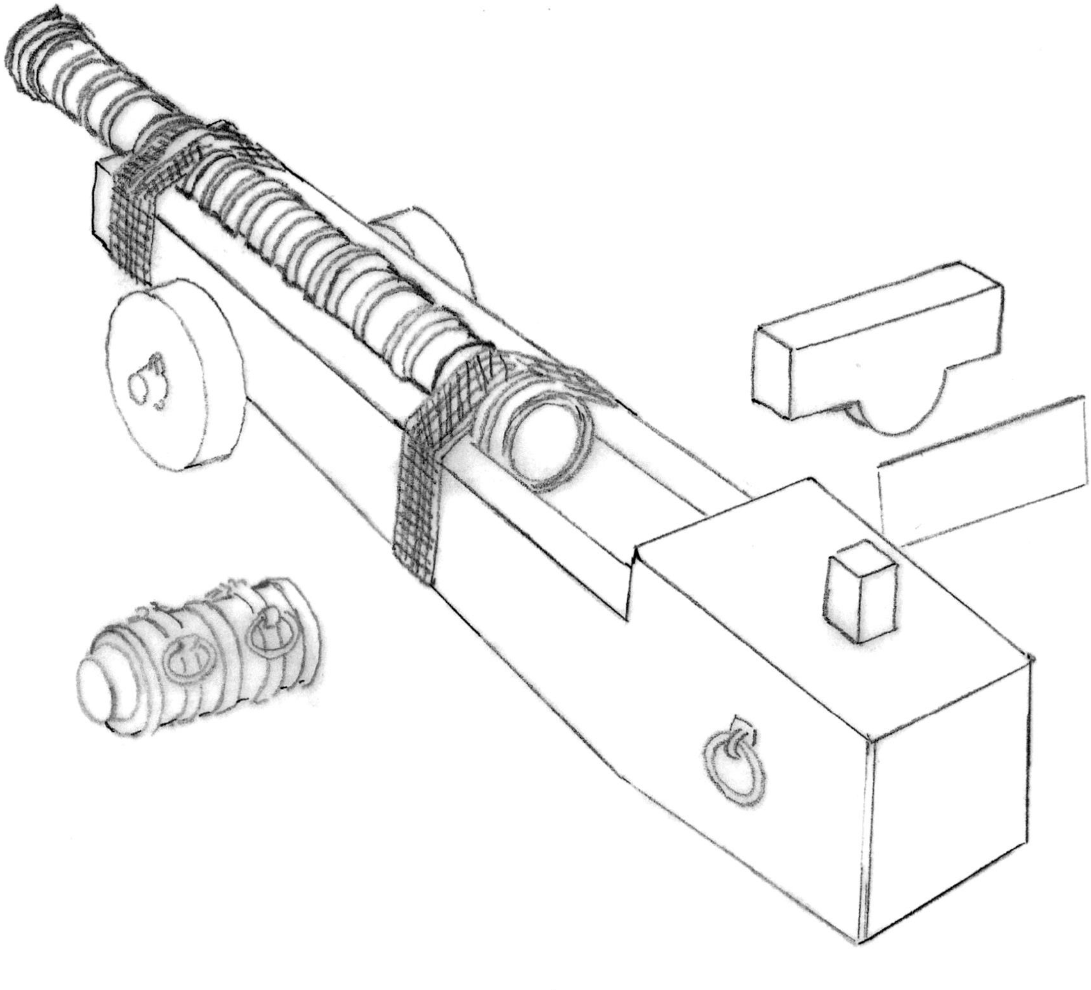

to the shrinkage that held the staves together and resisted the pressure of the propellant charge. Although sounding simple the precision required in their manufacture was exceptional, as the shrinkage was a matter of a few percent and required careful measurement of every part.

This technique meant that it was relatively easy to make the barrel or 'hull' as it was sometimes referred to. The barrel would be provided with a separate removable breech or more likely breeches. This system would create problems if a muzzle loading gun was intended and the rear of the barrel plugged. The largest guns, generally referred to as bombards, could not be constructed with a separate breech; it would simply be too difficult to make and unwieldy to use. The breech could be made from a single welded wrought iron block or of reinforced stave and hoop construction. This would then be permanently attached to the barrel with a tongue and groove arrangement as in the case of the bombards.

A separate breech system provided considerable advantages over a muzzle loading gun, particularly at sea or in the confines of a castle's tower. Firstly, a number of breeches, at least two, were provided for each piece which could then provide a high rate of fire without heating the individual breech too much. The breech could be loaded easily, especially in enclosed spaces, and long loading tools were unnecessary. Thirdly, the gun did not need to be withdrawn into the limited space found in a ship or castle tower for reloading, adding to both speed and convenience. Finally, the chambers could be sealed from damp and kept loaded ready for action at a moment's notice.

The Making of a Reconstruction

With great imagination and at not inconsiderable expense the Royal Armouries together with the Mary Rose Trust committed themselves to the making of a port-piece. This would be an accurate replica of an excavated (example MR81A3001). Much was to be learnt from both the manufacture of the piece and the later firing trials. It was decided to use traditional techniques and materials as far as possible. A firm of respected blacksmiths, Chris Topps and Co., was chosen and puddled wrought iron was used. Charcoal-smelted wrought iron was too rare and expensive a material and its use deemed impractical. Tests on medieval wrought iron have shown that it was well up to the standard of Victorian puddled wrought iron. In 1861 tests were conducted on two wrought iron guns dated to the reign of Henry VI, it was found that they had a tensile strength of 55,528 lb per square inch, not dissimilar from the best puddled wrought iron produced 400 years later.

A wooden mandrel was constructed upon which were set the nine staves that made up the bore. The front of the staves was thickened and turned over to hold the hoops and rings in place. The staves were curved to fit the mandrel, with their edges chamfered in order to produce a smooth interior surface. As with the original pieces there was no gap between the staves, which had suggested to some observers that the staves were welded together. An aiming ring was the first to be placed around the barrel followed by a single layer of bands and alternating single and triple hoops. Each ring or hoop had to be 3

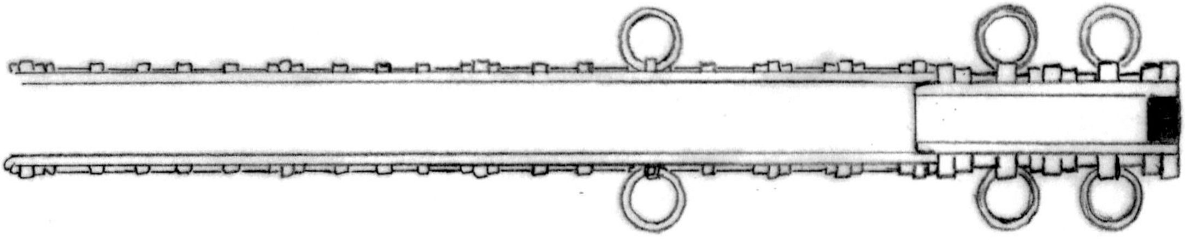

Above: this image shows the difference between the thickness of a sling's barrel and breech and the importance of achieving a tight fit between the two. The Middelaldercentret trials of an early cannon have shown the considerable difference between the chamber and barrel pressures achieved in early artillery. The obvious weak spot would be where the two joined, which explains the damage to the *Mary Rose* port-piece during trials and of course the fatal injury done to *Mons Meg*. (Author's drawing)

This massive single block of elm was needed to accommodate the large wrought iron barrel. The bed had to match the shape of the hoops and rings of the barrel, as when the gun was fired the barrel was forced forward, while the breech was forced backwards. The bed therefore had to be both deep and substantial. The cut-out portion underneath the barrel and breech would have reduced the overall weight but may have been to allow water to drain away rather than rusting the in situ barrel. Elm was a preferred wood for any portion of a ship that was likely to become waterlogged. (Author's drawing)

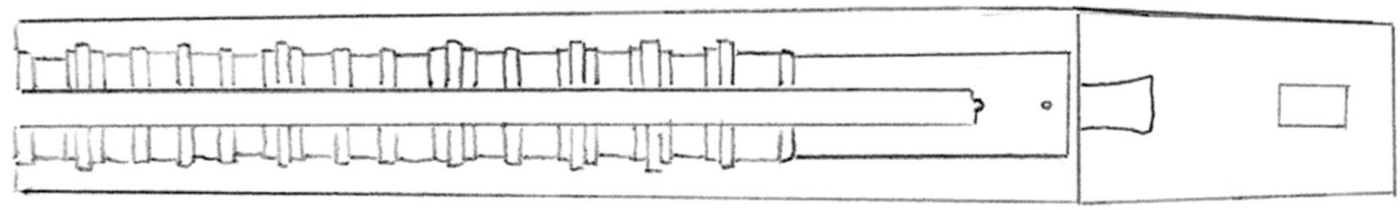

mm wider when heated than the diameter of the bore, to ensure than when shrinkage took place it would grip securely. In addition, each piece had to be cooled separately before another was added. In total 34 hoops and 17 bands were fitted. The bore was tapered by ½ in from muzzle to chamber end, as in the original. The overall length of the barrel was 3 m and the internal bore 7¾ in (175 mm), the minimum thickness of metal in the barrel was 1 inch. Once completed the core was burnt out and the staves cut the to their final length.

The chamber was made of two layers of staves with the upper covering the gaps in the lower, the two layers totalled 1⅜ inches (35 mm). The interior was 100 mm in diameter at the breech plug and 115 mm at the mouth. It was fitted with 13 bands and hoops adding at least 1½ inches to the thickness of metal. The whole of the breech was then heated, and the muzzle end welded and the breech plug inserted and welded. This was perhaps the largest task when it came to the size and quantity of material and took over 40 hours with sometimes three men working at once. The final chamber was some 22½ inches (890 mm) long. The project took some 240 hours of hard and hot work even using some modern tools to speed the process.

The carriage was made from a magnificent piece of green elm some 13 feet (4 m) in length and 20 inches square. The axle was made of ash, as were the spokes of the wheel while the fellies and hub were again of elm. The total weight of the gun on its carriage was 1.1 tons/tonnes. Rather unusually, although as was the custom, the wheels were not shod with iron and the spokes were fitted in pairs to their fellies. The bed of the carriage was carefully finished so that the hoops and staves of the completed gun would fit snugly. The barrel was also held in place by strong ropework at the end by the breech and towards the end of the wooden carriage.

The workmanship was excellent and certainly matched any of the examples excavated. The conclusion that could be drawn was that the manufacture of such weapons was well within the competence of contemporary blacksmiths and carpenters.

Casting Guns – Patterns and Moulds

The gunfounder, if Biringuccio is to be believed and if he was at all like Biringuccio himself, was loquacious, opinionated, prone to boast and bombast and a consummate craftsman. Working without the benefit of modern scientific understanding and equipment, much was achieved through a process of trial and error, accepted custom and a constant search for improvement. In his *Pirotechnia* Biringuccio provided the first and best description of the complete process until a French exposition by Saint Remy in 1697, which in itself differed only slightly. Robert Norton in 1628 provides a summary of the process but adds little to Biringuccio. The only significant change that took place in the entire process over the centuries was casting the piece as one, with the bore now being drilled out. This was begun early in the eighteenth century and became common in France by the mid part and in England by the latter part of that century.

The author has spent much labour both understanding and then editing his account as although the process is linear his account is not and often doubles back on itself. He begins with a long disquisition on the trials and tribulation on anyone engaging in metal casting, for: 'this art holds the mind of the artificer in suspense and fear regarding its outcome and keeps his spirit disturbed and almost continually anxious. For this reason, they are called fanatics and despised as fools. But with all this, it is a profitable and skilful art and in large part delightful.'[1] He considers whether success is the product of luck or judgement but concluded that failure:'shows a lack of skill or too little diligence in the art, and I am of the opinion that every error proceeds from these and not from fortune.'[2] A desire to speed on the process or skimp with materials was the cause of most failure and a potential cause of a disaster with not just a spoiled casting but molten death showered on those present.

He considered that the gunfounder should be a master of all the crafts and skills required in the process. He should be draughtsman and sculptor, carpenter and smith, brick-layer and mould-maker, and conversant with all the tools of the trade including lathes, 'saws, gravers and files'. He identifies five distinct phases in the casting of a cannon:

> The first is to make the moulds … the second is to bake them well; the third is to arrange and seal them up well in the pit; the fourth (the most important and the soul of the whole) is to melt well; and the fifth is to put enough material into your furnace so that you fill your moulds to overflowing.[3]

1 Biringuccio, *Pirotechnia*, p.214.
2 Biringuccio, *Pirotechnia*, p.215.
3 Biringuccio, *Pirotechnia*, p.260.

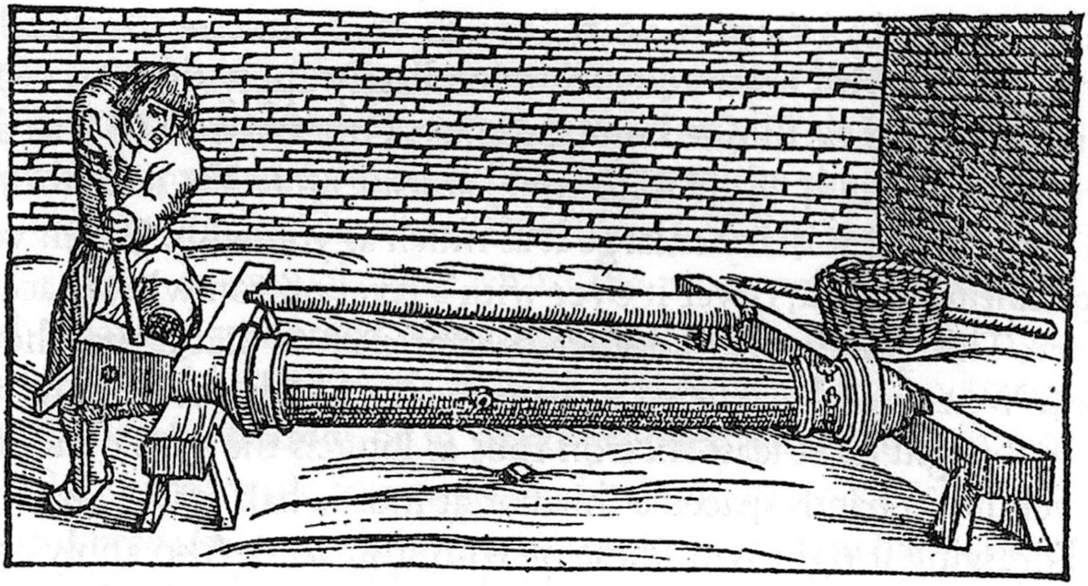

Making a pattern, from Biringuccio's *Pirotechnia*. This cannon takes the form of a fluted column, a design that gave no greater strength to the gun, but which pleased the prevailing enthusiasm for all things classical. The bronze cannon could be as much an *objet d'art* as a *machine à tuer*.

Characteristically he forgets the first, which is to make a model or pattern from which the moulds were made.

The full-scale model of the gun would be constructed in three main parts. The largest was the open-ended barrel, to which would be attached a breech, closing it. A core-tube would then be inserted into the empty barrel, which when the gun was cast would be removed (with difficulty) to produce the empty bore of the gun.

He recommended that the barrel mould should be a spindle, made from a single piece of fir, long, dry and well seasoned. It should be long enough for a feeding head to be attached and to rest on solidly made trestles upon which it can be turned on a lathe or carved to the required shape. If quality fir was not available then chestnut, pine, cyprus or dryoak could be used and spliced together in the same way as a mast. If the gun was too big or the wood too small then it could have rope wound tightly round it, to which layers of clay could be added. He described no particular type of clay although he suggested that the clay used for making tiles was probably suitable. He strongly recommended that the clay should be well pugged, meaning it was free of air, pebbles and other detritus and of a uniform consistency. If the gun was round, a strickle[4] board could be used when the pattern was turned on the trestles, to produce a standard profile. Trunnions could be fitted with nails and any heraldry, sculptural details to the cascabels and dolphins added, in potter's clay or wax. He recommended avoiding undercuts in the 'cornices' or mouldings as they would 'break the mould and spoil its beauty'.[5]

To the making of the moulds, he devoted much time. The heat that they would experience in the process of casting and the great weight of the bronze could lead to them cracking, which would at best produce a poor appearance

4 A board with the profile of the gun cut out.

5 Biringuccio, *Pirotechnia*, p.257.

and at worst an unusable piece: 'Thus you must proceed with thought and good judgement to prepare in every detail the quality of the materials, adjusting the forces of the weights to the forces of the moulds and fires, and also the measurements according to the things that you wish to make.'[6] Robert Norton is more informative and long-winded than Biringuccio when considering what qualities are sought for in mould making. He also considers the consequences of using poor material:

> all earths being eyther fat or leane, or soft, or grosse, or viscuous, whereof the leane soone turneth unto dust, without holding together, which also are a long time drying, and are but of little continuance for strength. The Fat and viscous Earths shrinke and chappe, and doe soone breake by their naturall brittinesse, and doe often grow crooked by the vnevennesse of their mixture or temper, whereby they become crooked in the Mould, and so warpe the patterne it selfe: so that it is rare that such a Peece should come neatly or well out of the mould: Where upon we may conclude that good Earths are neyther Fat nor Leane but betweene both, and of a fine and subtill graine or mould, which soone dryeth and remaineth firme, without breaking, being able to resist the vehemency of the fire.[7]

The clay-covered pattern for the HO Group's cannon. What is visible at the ends of the pattern is the thick rope which was bound around the wooden mandrel, to which the clay would adhere. The quality of the finish is very high, achieved by using clay 'slip' towards the end of the process. (Courtesy of Peter Hansen)

The pattern was covered with fine wood ashes and wax or tallow applied with a brush. This acted as a separating layer between the pattern and the mould. On this would be added layers of clay mixed by preference with water and wool-cloth clippings forming a single homogenous mass. The first layer could be applied with a brush but subsequent layers by hand. Other gunfounders used the dry dung of horses or other animals, finely cut straw or flax remnants but Biringuccio strongly recommended his wool-cloth clippings. The thickness of the mould varied from one inch to three inches for a 6-pdr and 24-pdr respectively.[8] On this would be placed a layer of iron wires which were again covered in clay. Onto this was attached an armature of six to eight iron rods followed by another layer of clay over which were laid iron bands along the length of the spindle. Biringuccio stressed that: 'it

6 Biringuccio, *Pirotechnia*, p.216.
7 Norton, *The Gunner*, p.70.
8 A. N. Kennard, *Gunfounding and Gunfounders* (London: Arms & Armour Press, 1st Edition, 1986), p.12.

is very important to have a strong armature for the safety of your work'.[9] The weight of iron in an armature was not inconsiderable, although of course it could be reused or recycled. Collado records that the armature for a culverin, cast in Milan at the end of the sixteenth century, weighed 230 kg.

Gates and vents had to be made for safe casting. Bronze is a heavy and relatively viscous liquid when molten. and had to find its way throughout the mould. It was essential to ensure that 'gates' were added to allow the ingress of metal and vents to allow the egress of air and gas. As Biringuccio commented: 'When it [molten bronze] is to enter the mould cavity, it encounters the air therein which would necessarily find itself entrapped and would either refuse entrance to the bronze or break the mould in order to escape.'[10] There must therefore be a means for the air to escape by the use of vents: 'the more vents you make on your moulds and the wider the entrances, the surer you will be of a good result in your casting, if you have melted well.'[11] In addition several hollow runners should be made to lead some of the metal to all parts of the mould.

The mould, with the spindle around which the model or pattern of the cannon was made, should be allowed to dry thoroughly and then heated so that the wax or tallow in the separation layer would melt. The trunnions would be removed, and the spindle would be removed by main force. This would be achieved by: 'Putting the whole mould in motion strike the protruding front part [of the spindle] against the wall, and thus the spindle which is inside will come out'.[12]

The breech part of the gun was made in the same way as the barrel, either from wood or clay or a combination and made with a 'male part of the mould', which fits exactly into the foot of the large mould 'as does a box into its cover'.[13] A separation layer would be made between the positive model and the negative mould and layers of clay carefully added. The final mould would also be bound with iron wire, reinforced with a cage of iron rings and plates, tightening the armature to ensure that it didn't crack under the weight of bronze.

The core was a positive mould to produce the empty bore of the gun. It should be perfectly straight and of the correct diameter. If it warped in the heat of the casting, shifted in the mould or fractured during the process it might make the gun at best inaccurate and at worst dangerous and unusable. The core was formed around an iron spindle, one end of which was shaped like the heel of a shoe with a hole drilled through. This was to enable the core to be withdrawn more easily after casting, providing points of purchase. The iron spindle would be wrapped with rope or hemp tow and then covered with a paste of powdered charcoal. To this would be applied layers of clay mixed with sifted horse dung and cloth clippings. A mould board could be used to ensure precision as the spindle was turned on a pair of trestles. The exterior should be finished with a fine layer of sieved soft clay. In some cases,

9 Biringuccio, *Pirotechnia*, p.236.
10 Biringuccio, *Pirotechnia*, p.249.
11 Biringuccio, *Pirotechnia*, p.249.
12 Biringuccio, *Pirotechnia*, p.236.
13 Biringuccio, *Pirotechnia*, p.244.

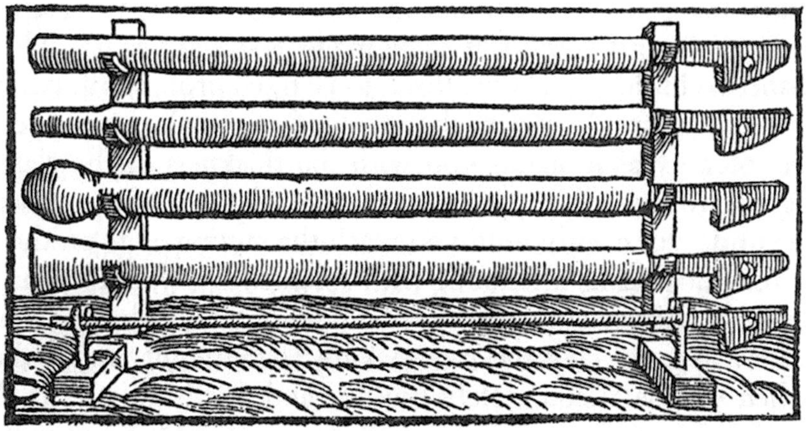

An illustration from Biringuccio's *Pirotechnia*, showing the iron core bar (bottom) and finished cores for different shaped chambers. It is easy to see how the hook and eye of the core bar could be used to help withdraw the core. The breech section like 'a grain of barley' and a 'trumpet' would have presented particular difficulties when it came to removing the core. Both recent reconstructions faced considerable problems when it came to removing their cores, even with modern equipment.

the core was bound with iron wire, as in the case of some of the *Mary Rose* guns where evidence of its use can be found in their bores. The core had to be strong enough to resist the heat of the molten metal but also, 'tender or crumbly' to permit its ready removal.

There were numerous variations that could be made to the breech end of the core. There were bulbous, trumpet-shaped and narrowing bores proposed by other gunfounders. Biringuccio is scathing about his colleagues who were also, it should be remembered, his competitors.

> Under this veil these men pretend to have a great secret and puff up their reputations by telling lies which deer could not leap over, promising that from their guns not only balls but lightning flashes will issue. But in the end they make only what others have made, and if you ask them what theory moves them, they make only a surly answer.[14]

He is very wary of the dangers of leaving an air gap between the powder and ball that might occur in tapered breeches. John Smith writing almost a century later was also wary of a tapered bore: 'which is dangerous (if the Bullet goe not home) to burst her'.[15] Biringuccio supported the plain constant diameter bore as it was easier to load safely, although he sees the appeal of the barley or trumpet shaped breech which he thinks somehow might 'increase the fire'.

The core had to be held in place using an iron chaplet, referred to as a 'collar' or 'castle' or variation on those two types. The 'castle' consisted of two or more arched irons on which was welded a ring or castle which held the core. The cylindrical collar would fit around the core and was held in place by two, four or six legs fixed in the barrel mould. The castle mould was fixed into the breech mould and held the core in the castle part of the structure. After casting, any legs that protruded could be sawn level with the bronze, or chiselled out and a bronze plug fitted. Evidence of their presence can be

14 Biringuccio, *Pirotechnia*, p.241.
15 Smith, *A Sea Grammar*, p.65.

Examples of the types of collars and a castle as well as the clay cap, the *rotella*, which would be positioned at the top of the mould. It was too easy for the core to 'float' or otherwise be displaced, as happened with the first casting of the author's gun, very frustrating! The blacksmith and his skills were an essential part of the foundry's workforce. (Biringuccio's *Pirotechnia*)

seen in some examples, with rust 'weeping' through the surface. The breech mould would be thicker than the barrel mould as it had to cope with a heavy weight of molten bronze pressing down.

The moulds had to be completely dry, if not it 'would also cause your casting to come out incomplete. If indeed it should be complete, it would be full of holes and spongy on the inside and perhaps also on the outside, and it would occasion much loss of material because of great boiling, gushing, and overflowing'.[16] At the very worst the moisture in the mould could be instantaneously turned into steam which would force the molten bronze out of the mould with disastrous results for those standing nearby.

Drying the mould should be a slow and gentle process rather than the application of a ferocious flame, as that might damage its integrity. The barrel mould could be dried with a wood and not a charcoal fire, either placed horizontally with the fire laid inside or vertically, with it placed over a purpose-built furnace, until flames emerge from the top and the mould itself glowed red. The core had to be baked in a special charcoal firepit to avoid the iron rod warping. The breech was baked held upright from the inside, with the fire maintained until the interior was fiery and red, or with the hollow held downward over a charcoal firepit. After drying any cracks should be re-plastered with white of egg, crushed brick and quicklime. The interior of the mould should be cleaned and dusted with blowpipes and bellows and then a layer of young ram's-horn ashes finely ground should be applied, using a sponge.

The barrel mould would now be placed into the casting pit, using block and tackle. If a castle was attached to the breech the breech would be joined at this point. If a collar was used the breech would be joined after the core was in place. The core was lowered very carefully into place passing through a clay disc that would centre it at the top of the mould so that: 'it meets the iron ring that you put at the foot of the mould for this purpose'.[17] When all the gates and vents are closed, and core inserted the breech should be joined to the barrel mould. The three main parts of the mould would now

16 Biringuccio, *Pirotechnia*, p.250.

17 Biringuccio, *Pirotechnia*, p.245.

be secured together: 'The armature hooks that are at the ends of the moulds that are to be joined together and are bent in opposite directions from each other must be very well bound with several turns of [Brescian] ironwire.'[18] The core was held with iron wire attached to the armature to stop the core from lifting out of the mould or shifting in it: 'as once happened to me and to many another master I could name.'[19] Then the trunnion spaces should be closed off securely with tile or brick and attached to the armature.

The casting pit would then be filled with earth, layer by layer. The instruction was to: 'press the earth down well all around it with stamps and to see that you make it hard with blows.'[20] So that it should be well-pressed and solid soil, somewhat moist but not wet. The mould was now ready for casting. Where possible and when necessary, several guns could be cast on the same occasion. The largest gun would be cast first, with shutters cutting off the flow of metal when the mould was full, and the next mould filled, until all the moulds were cast.

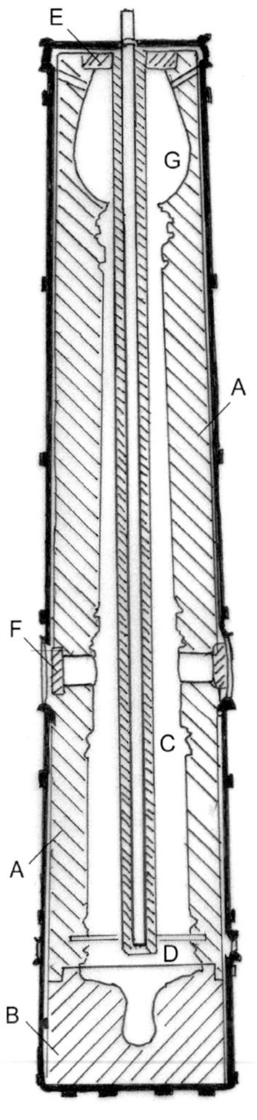

Right: a barrel mould
A – Main body of the mould
B – Breech mould
C – The core with the iron rod seen in place
D – Iron chaplet holding the core in place
E – The clay disc or rotella at the top of the mould
F – Trunnion cavities closed by bricks.
G – Feeding head
The whole structure held in place by the wire armature. (Author's drawing)

The Furnace and Casting

The temperature necessary to melt copper is very high at around 1,100 degrees Celsius and often reaching a 'crispy' 1,200 degrees before pouring. In the manufacture of iron, the ore was heated with its fuel (charcoal). This was not possible when melting copper and tin, as it would seriously contaminate the metal and the large volumes of metal that were needed for cannons could not be melted in a ladle. This made the reverberatory furnace essential. This is a furnace where the metal is kept separate from the fuel and the hot gases and radiant heat passes over its surface.

Biringuccio provides the first detailed description of a reverberatory furnace, although it was already a well understood technology by then. There was considerable variety in design: 'There are almost as many shapes of such furnaces as there are masters.'[21] The height and shape of the roof of

18 Biringuccio, *Pirotechnia*, pp.253–254.
19 Biringuccio, *Pirotechnia*, p.254.
20 Biringuccio, *Pirotechnia*, p.254.
21 Biringuccio, *Pirotechnia*, p.281.

The furnace at the Middelaldercentret approaching completion. The furnace was a very sophisticated structure which would require careful management to ensure that the molten metal reached the right temperature, or the bronze would become sluggish and this would result in an incomplete cast. This was a major problem facing the group working there. On the other hand, too high a temperature in the modern foundry when the metal was poured created honeycombing problems for the *Mary Rose* team. One cannot but be amazed at the skill of early gunfounders when getting it right. (Courtesy of Peter Hansen)

the hearth varied greatly from circular to oval and even rectangular. The numbers of flues varied as did their proportions. Loading with fuel could be accomplished from the side or even the top, although whichever was used, wood was always employed.

Biringuccio's design combined the elements from other examples as he saw fit. He first laid out the shape of the hearth which was similar in outline to the body of a lute. He ensured that it incorporated a slight slope so that the molten bronze could flow freely. The brick or stonework had to be well fitted and sturdy to deal with the temperature and weight of metal. A window was placed where the metal could be observed. There was also, set into the hearth, a conical iron plug from which the bronze would be tapped.

The fire pit was large, almost the full length of the furnace. It was also sloped and provided with a convex roof that helped deflect the flames over the hearth. The vault of the firebox had to be lower than the vault over the hearth, with the vault over the taphole especially low, in order to ensure the reverberation of the flames. Biringuccio used two little chimneys or flues for the escape of gas and flame. Before charging with bronze, the hearth was thoroughly baked and any cracks mended. The interior was coated with 'wash ashes' to protect the brick and stonework and to close up any joints which would otherwise be lifted up by molten metal. If this was not done, the charged metal would not melt: 'and thus you would suffer loss and shame'.[22]

22 Biringuccio, *Pirotechnia*, p.285.

Plate A

Above: Two late medieval field guns illustrated in the Swiss Schilling chronicle
Below: The Norwich Great Hospital falconet

See Colour Plate Commentaries for further information

Plate B

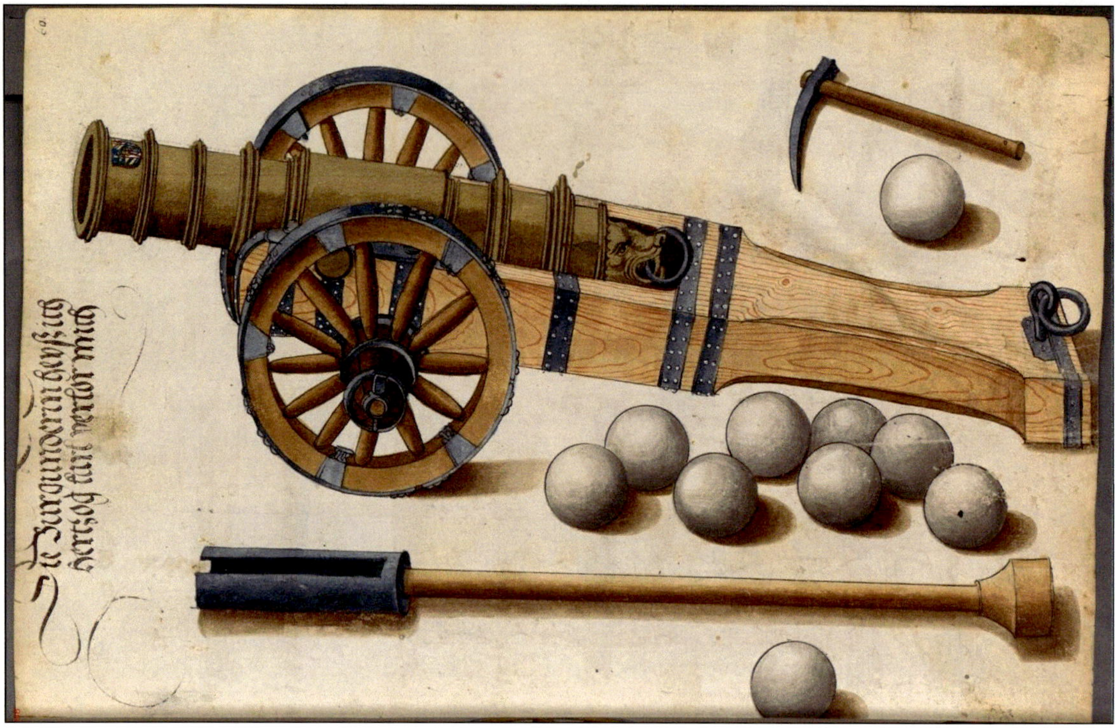

Above: A transitional bronze gun from the *Zeugbuch* of Emperor Maximilian I
Below: An early depiction of a gun with a limber

See Colour Plate Commentaries for further information

Plate C

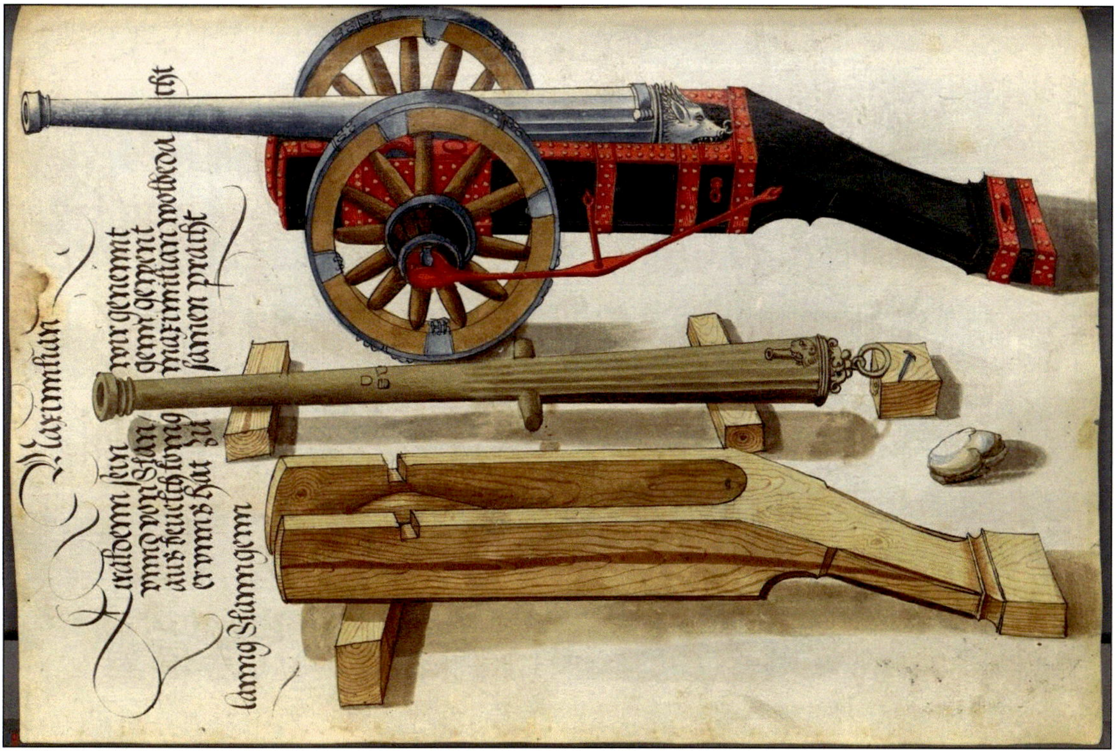

Above: A carriage in construction and completed, from the *Zeugbuch*

Below: A magnificent transitional piece and limber from the *Zeugbuch*

See Colour Plate Commentaries for further information

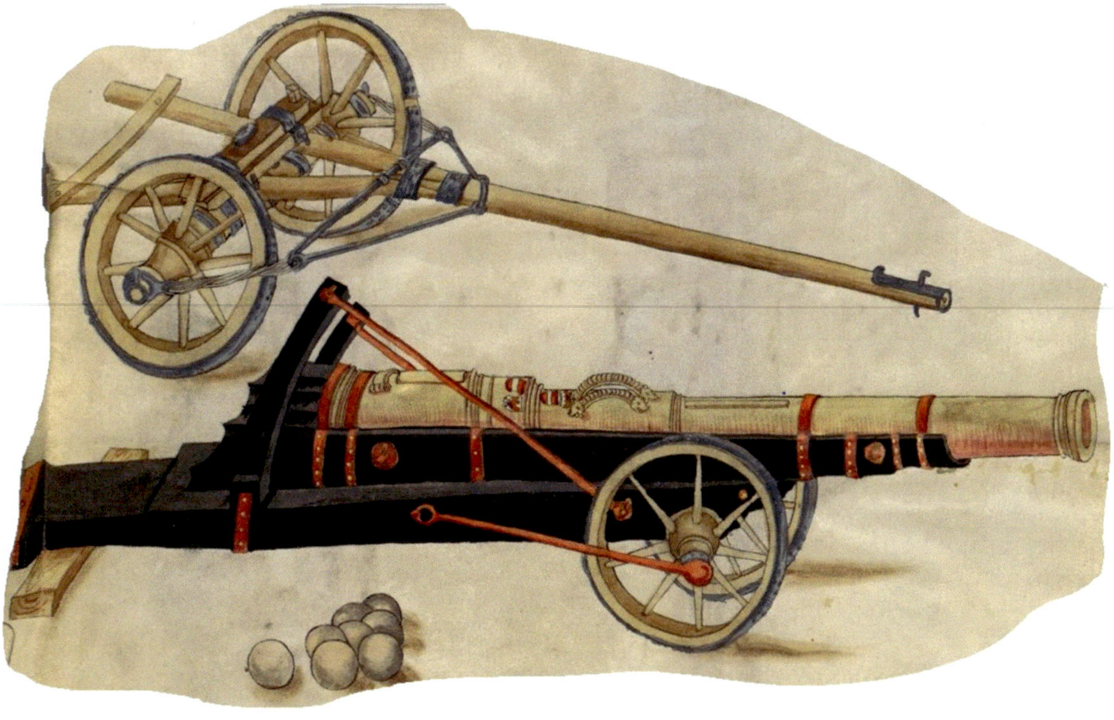

Plate D

Top: The bronze bombard *Die Katerie*
Below: A pair of bombards found in the *Zeugbuch* of Maximilian I

See Colour Plate Commentaries for further information

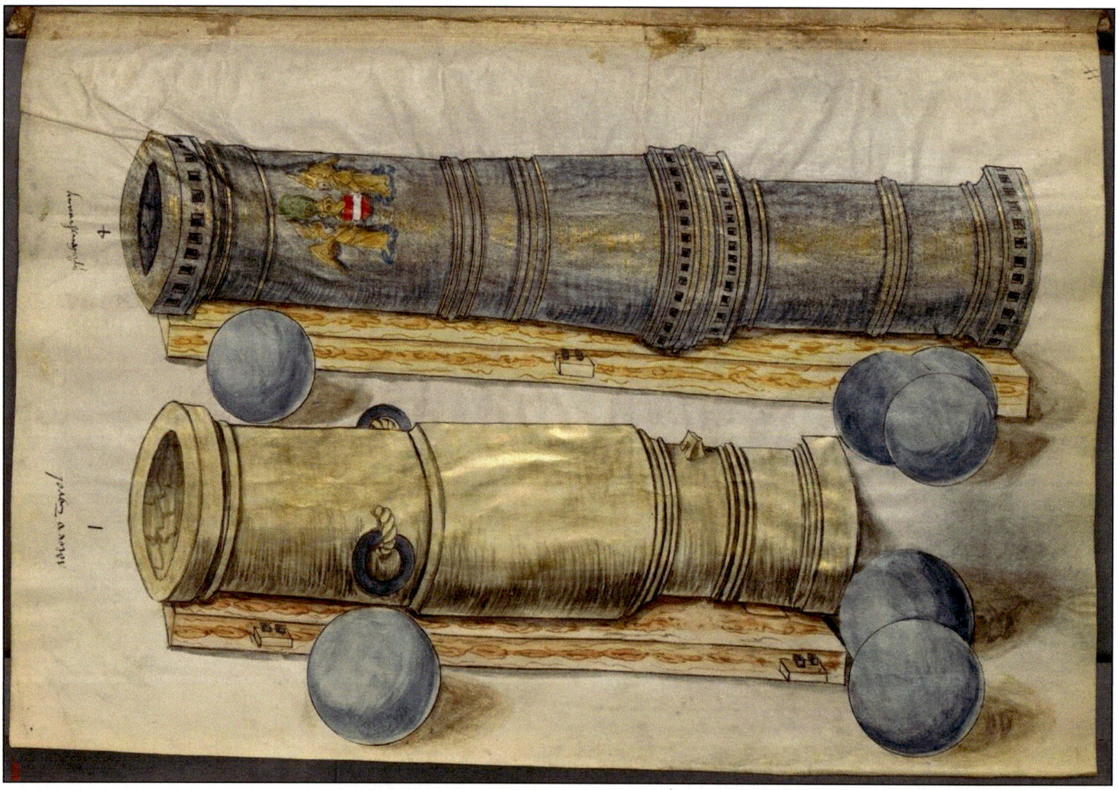

Plate E

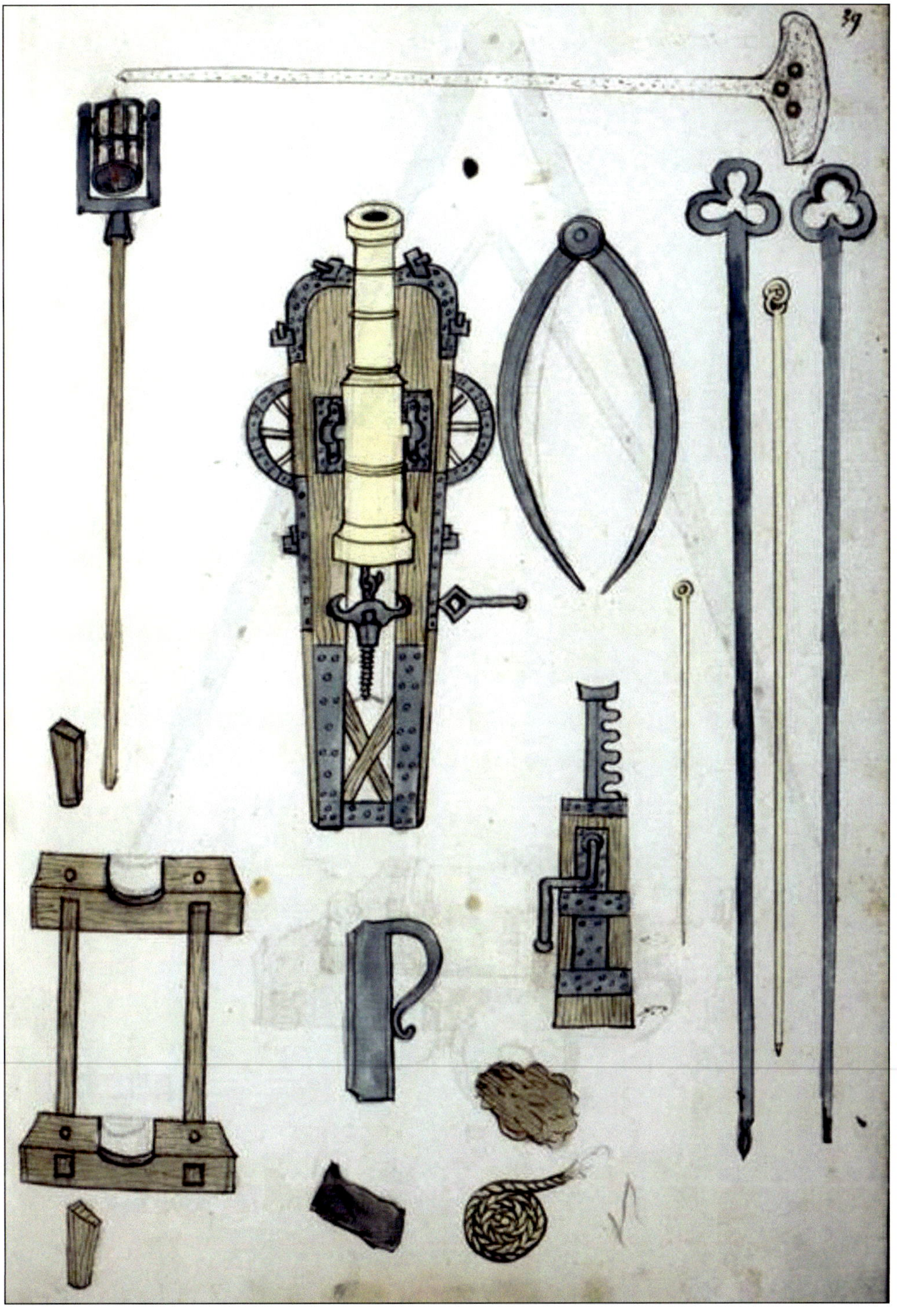

The paraphernalia of gunnery

See Colour Plate Commentaries for further information

Plate F

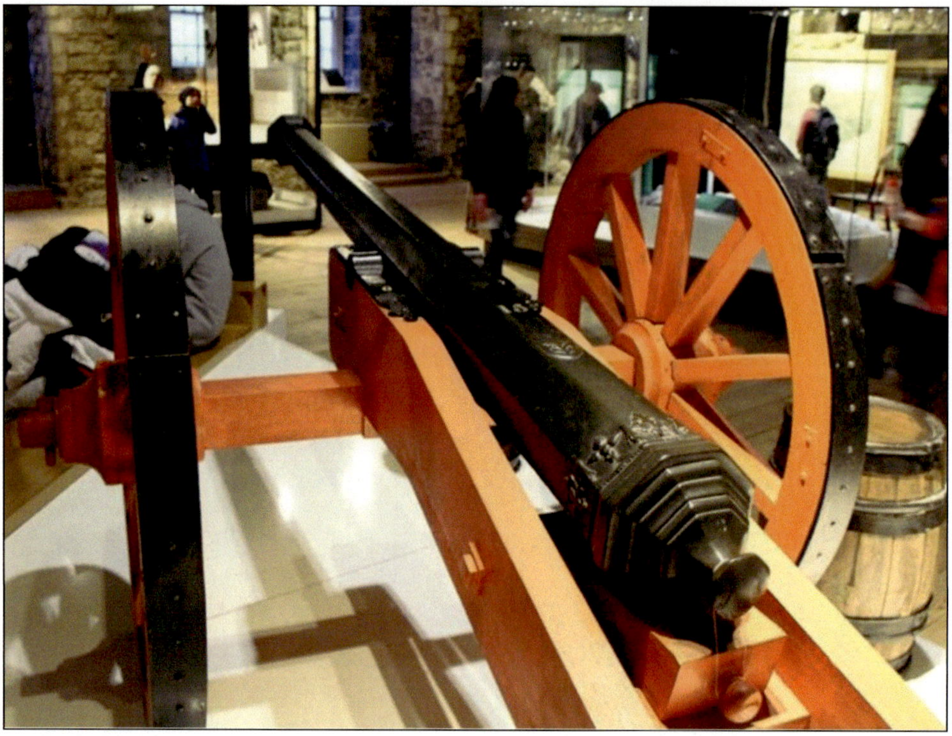

Above: An early-sixteenth century falcon on a reproduction carriage
Below: An original Armada siege piece on a reproduction carriage

See Colour Plate Commentaries for further information

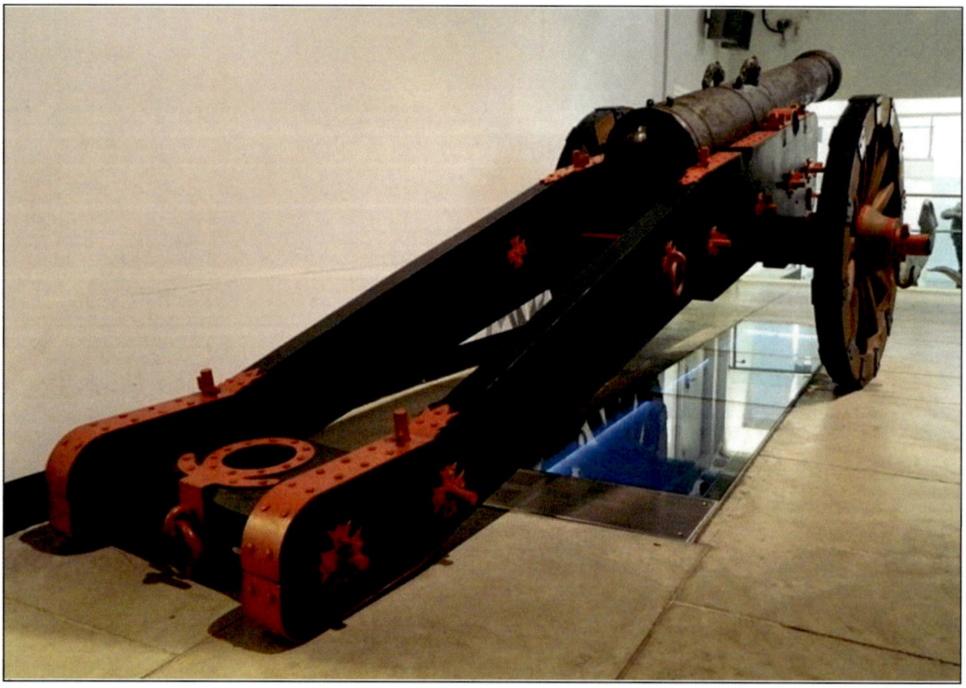

Plate G

A reproduction iron saker and carriage from the late sixteenth century

See Colour Plate Commentaries for further information

Plate H

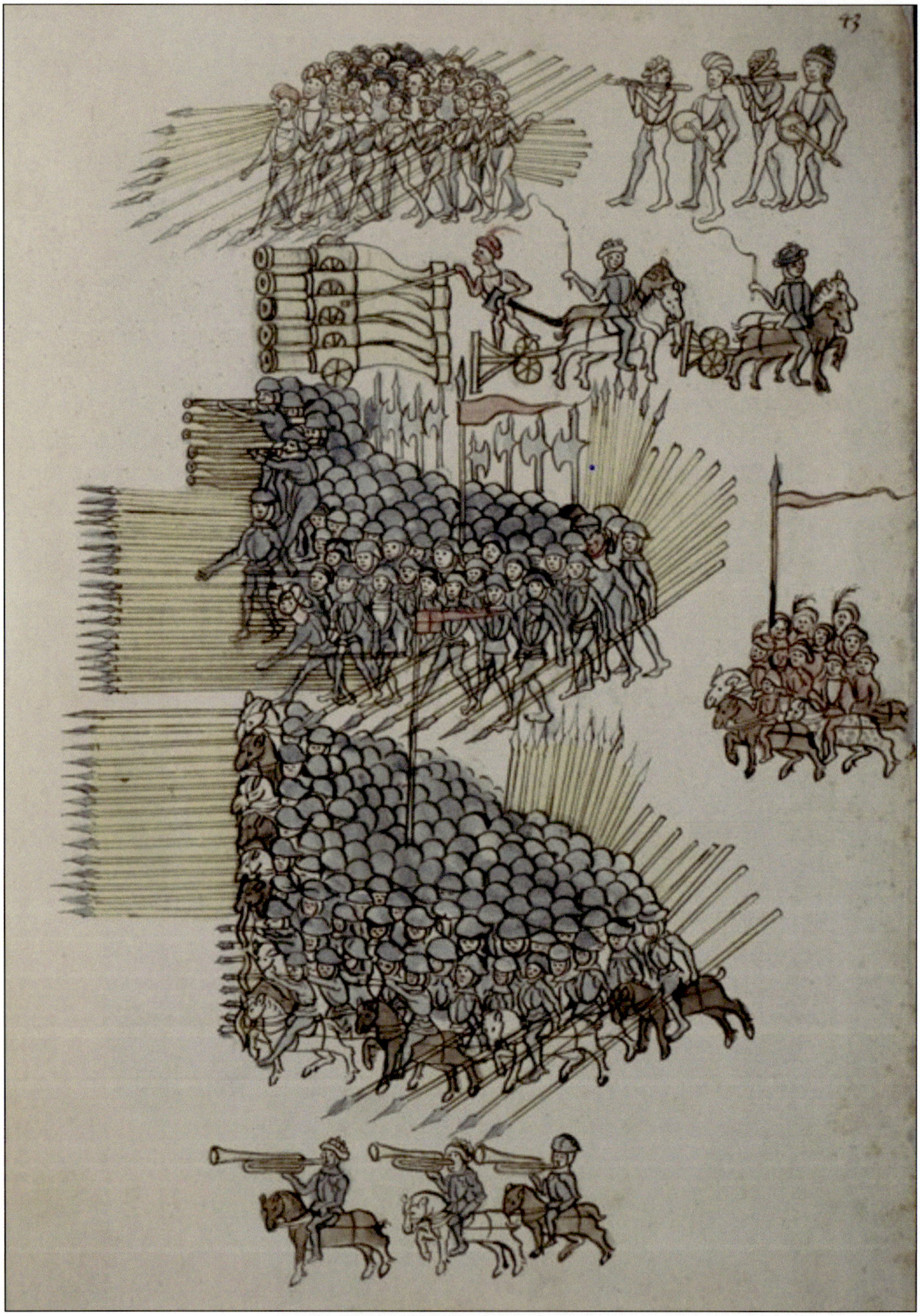

A military revolution in motion

See Colour Plate Commentaries for further information

Plate I

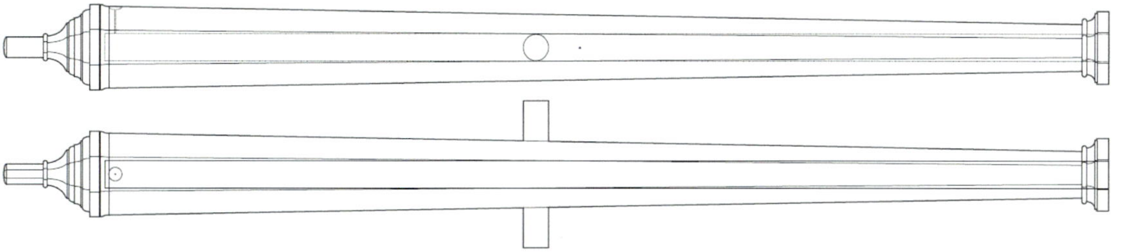

A reproduction falconet and carriage

See Colour Plate Commentaries for further information

Plate J

A bronze saker cast by Alberghetti on a reproduction carriage

See Colour Plate Commentaries for further information

Below and overleaf: Firing the *Vasa* reproduction gun

See Colour Plate Commentaries for further information

Plate K

Queen Elizabeth's Pocket Pistol

See Colour Plate Commentaries for further information

Plate L

Plate M

Accoutrements associated with the falconet. Clockwise from top left: ladle and rammer, shot box, gunsight

See Colour Plate Commentaries for further information

New Inventions of Modern Times [*Nova Reperta*], The Invention of Gunpowder, plate 3 c. 1600 Jan Collaert I Netherlandish. This illustration is of a cannon foundry with inset, above the furnace hearth, a rather melodramatic and inaccurate interpretation of the manufacture of gunpowder. The furnace is being charged with broken pieces of guns. The molten metal is being poured into a vertical mould that was buried into the ground. A breech portion of the mould can be seen on the ground near the feet of the well-dressed man. Two workmen are clearing up the castings with cold chisels. The author used chisels and files to clean up his falconet, producing identical marks to those on original guns. On the left can be seen a man in a wheel, turning a drill to ream the bore of a cannon that can be seen behind the mortar. (Courtesy of the Metropolitan Museum of Art, Open Access)

Biringuccio refers to the use of multiple furnaces, when the capacity of one was not enough, a technique that was already well known. Leonardo da Vinci intended to use three furnaces in casting the 'colossus of a horse' for the Duke of Milan, and a Flemish founder used two furnaces when casting a gun.

The fifteenth-century-style furnace built at the Danish Middle Ages Centre was designed by Dr Bastian Asmus and consisted of a furnace chamber, with the hot gas being drawn through four flues in the top of the hearth.[23] Charging ports were constructed on either side of the furnace and a tapping point for the metal made in the end opposite the hearth. Wood was used to get the hearth up to temperature (1,150 degrees Celsius) and only then was the hearth

23 The author would like to thank Dr Asmus for his generous assistance in reviewing this section of the book. His expertise has proved invaluable in correcting some assumptions and misapprehensions. Any remaining faults and weaknesses in the text are entirely the author's own.

charged while the temperature was maintained. Apparently charging on a cold hearth would lead to the metal coagulating. It was also important to maintain the temperature of the hearth during the pour. This was an issue that affected all cannon casting, as there was a tendency for the metal to cool during the process, with potentially disastrous consequences. In a recent large bell-casting factory in Austria,[24] where a 25-tonne bell was being cast for Bucharest cathedral, multiple furnaces were also used. In addition, to avoid the metal cooling, the clay channels were preheated with charcoal.

Throughout this period the preferred metal for the manufacture of artillery was bronze, although cast iron was also used in far smaller quantities. Bronze is an alloy of copper and tin. Brass is an alloy of copper and zinc and is sometimes known as latten. Bronze has twice the hardness of copper. Copper was mined in Hungary, the Tyrol, Bohemia, Morocco and Sweden, while England was the principal provider of tin with Bohemia and Spain also providing some.[25] Guns could be recast if they were worn out or deemed obsolete; they could be cast into statues and vice versa. Michelangelo's statue of Pope Julius II was melted in 1512 to cast the gun named *Giulia*. The ancient Roman pantheon provided the gilded bronze roof tiles that cast cannon to defend Pope Urban VIII. Most distressing of all, the huge statue of a horse designed by Leonardo for Ludovico Sforza, was sold to his ally Hercule d'Este to cast cannon when he faced a French invasion in 1499.

In England and Germany, the guns were often referred to as 'brasse' even though they were of bronze. It is common these days to refer to the metal as 'copper alloy' and then to provide a more detailed analysis. For the sake of simplicity, if not of precision, this work will refer to copper-and-tin guns as being of bronze. As an alloy of copper, bronze is twice as hard and more fusible (has a lower melting point) and flows better when casting than copper. It is not as hard as wrought iron but resists corrosion far better.[26]

Copper on its own would not cast well: 'if you wished to cast a fine thing from it, it would not succeed because it is not fluid, on account of its viscosity; and if it should succeed, it would be all full of holes, like a sponge',[27] so that: 'it must be accompanied with tin or brass. Some, for less expense, put in some lead'.[28] In bell casting up to a quarter of the metal could be tin while for gun barrels it was usual to have a much lower percentage of 4–10 percent. Although a higher proportion of tin would allow the metal to flow more easily, it would result in a greater chance of cracking. Bells were comparatively thin and were only struck by a clapper, not subject to the high pressures to be found in a gun's chamber. Bronze guns can still ring sweetly like a bell when struck and this was a simple test of their integrity.

24 <https://www. youtube. com/watch?v=YaVfZdw3btI>

25 Relative prices of 100 libbre of metal in 1480s Florence: bronze 24 lire; iron seven lire; tin 40 lire; copper 29 lire and five soldi. One Lire = 20 soldi

26 The Brinell hardness test produces the following results: Wrought iron –100, copper –35, bronze (gunmetal) –70, grey cast iron –200+, lead –5.

27 Biringuccio, *Pirotechnia*, p.299.

28 Biringuccio, *Pirotechnia*, p.300.

Robert Norton recorded the following proportions of metals recommended by contemporary authorities for ordnance.

Table 11. Recommended Proportions of Metals

	Copper	Tin	Latten*	Lead
Biringuccio, artillery	100 lb	12 lb		
Biringuccio, for bells	100 lb	20–23 lb		
Jerome Rosselli	116 lb	10 lb	8 lb	
Alexander Bianco	100 lb	20 lb	5 lb	
Diego Uffano	100 lb	8 lb	5 lb	10 lb
Sieur de Prissac	115 lb	5 lb		
Sieur de Prissac	100 lb	10 lb		
Cyprian Lucar §	80 lb	8 lb	10 lb	

* An alloy of copper, tin and zinc.
§ Lucar, *Appendix*, p. 34.

Lucar explained the combination of copper, tin and latten as follows: 'The tynne (as the Gunfounders doe say) causeth the said mixture to be hard, the latten helpeth much to incorporate the mettals togeather and maketh the mixture to bee of a good colour, and the copper doth much strengthen the same mixture.'[29]

The molten metal should be heated to about 1,200 degrees Celsius but there was still the possibility that not all the metal was liquid everywhere in the furnace. To help avoid this it was recommended that the metal be preheated in front of the furnace mouth and that the copper should be melted before the tin. It was also recommended that the heat should not be so great as to cause turbulence within the metal, which would affect its freezing properties. To ensure that all the metal was liquid: 'in order to make it more certain, put a dry chestnut stick through one of the openings and, touching the bottom with the point, pass over the diameter of the furnace passing it in various places and searching at the taphole plug, at the entrance of the flame, and all the sides to see whether there are any pieces of unmelted bronze.'[30] After this, a heated iron rod should be used to stir the metal to the very bottom, to make sure there was no sediment in the melt, with the rod coming out clean. A rake of wood or iron would be used to draw off 'superfluity', or surface dross. He also advised that a ladle of the metal should be taken out to test its viscosity, and if need be to add more tin to make the cast flow. If this was necessary the foundryman should add the new metal and heat and stir thoroughly: 'to incorporate the alloy throughout'. Cristöbal de Espinosa reported some 'tricks of the trade', such as adding ram's horn to supposedly increase the density of the metal, wine to change the colour and soap to make the metal run more freely!

29 Lucar, *Appendix*, p.34.
30 Biringuccio, *Pirotechnia*, p.259.

The channels of bricks or the clay conduit down which the molten metal would flow, should be covered with hot ashes and baked very well with charcoal, to ensure that the metal flow was not slowed by congealing. Once thoroughly heated, the careful foundryman should finish the conduit by greasing with a little fat. After everything had been checked, and the time and temperature judged right: 'by striking with an iron rod in the hole that was arranged as an outlet of your furnace, cause the bronze to run out like an oil'.[31]

It was vital that there was a good head of bronze above the level of the muzzle. Bronze shrinks when cooling, as liquid bronze is less dense than the solid. The cast object will shrink by 1.5 percent in every dimension, which has to be taken into account when making the mould. A good head of bronze would increase the weight compressing the metal in the breech. If you did not have a good head of metal, 'although the mouths will fill up, lower down in the thinner parts the casting will be spongy and full of hole'.[32] Biringuccio recommended adding tin at the critical end of the cast, so that the feeding head 'becomes fat with tin', as 'thus it will be close, dense, and safe, and pleasing to the sight in its beauty'.[33] If all went well the gunfounder might be justly 'rewarded with profit and honour'.

Table 12. Bronze Shrinkage

Shrinkage of the liquid metal	1.0–1.6 percent volume change for every 100°C
Shrinking during solidification	5.0–7.3 percent volume change
Shrinking of the solidified casting at 20°C	Approximately 5 percent volume change*
* Martin, *The Equipment and Fighting Potential of the Spanish Armada*, p. 260.	

Casting Flaws

> The running of the mettall not fine or too cold, and the mould not well nealed, or the mettalls not well incorporated; either of these causeth flawes, crackes, spungynes, or hony-combs in the mettall of the Peece, whereby great danger often ensueth.[34]

Norton provided a heavily partisan critique of Continental castings which were universally substandard. He outlined the culprits, the causes and the consequences in some detail:

> If we shall examine the most used foundings in Europe: namely that of Lisbon, Malaga, Barcelona, Naples, Sicily, Cremona, Milan, Genoa, Venice, Malines and

31 Biringuccio, *Pirotechnia*, p.255.
32 Biringuccio, *Pirotechnia*, p.259.
33 Biringuccio, *Pirotechnia*, p.260.
34 Norton, *The Gunner*, p.68.

> Utrecht: in which by reason of their continual practice they might easily become excellent and expert, yet whether it be by ignorance, negligence, or else by too much haste foundings, it is apparent that they commit great and absurd faults therein. Some of their pieces (and not a few) are bored awry, their soule [bore] not lying in the width of the body of the metal; some are crooked in their chase, other of unequal bores; some too light in the breech turn their mouths downwards, and so endanger their own vamures and defences … others are too heavy also in their breech … some and a great many are come forth of the furnace spoongy, or full of honeycombs and flaws, by reason that the metal runneth not fine, or that their moulds are not thoroughly dried, or well nealed … when such gunners load them, as are either ignorant or negligent in examining their defects, they will either break, split, or blowingly spring their metals and (besides that mischief they do) they will be utterly unserviceable ever after … yet this much I dare say to the due commendations of our English gunfounders, that the ordnance which they of late years hath cast, as well for neatness, as also for reasonable bestowing and disposing of the metal, they have excelled all the former or foreigners.[35]

One of the failed castings of the late medieval cannon cast in Denmark. The liquid metal has not completely filled the mould. A further problem faced when casting this gun was the movement of the core in the mould. Of course, 'failures' were naturally melted down and cast again. Although bronze was comparatively expensive, its ability to be easily recycled was a great advantage. (Courtesy of Peter Hansen)

Problems were evident when melting the alloy, when pouring as we have seen but also when the metal was cooling. The melting point of copper and tin are markedly different 1,084 degrees and 232 degrees respectively. This is not a serious problem when the metal is being melted but it is a significant issue when it is freezing (solidifying). A 10 percent tin alloy freezes over a wide range from 1,010 degrees and 840 degrees Celsius. This has some advantages over iron, but it can create problems.[36]

When the alloy begins to cool, a copper rich alloy forms and the tin then tends to form a tin-rich compound. 'Coring', or the segregation of tin into dendrites (crystalline structures) can produce porosity as the tin cools and shrinks. This would tend to produce a greater degree of porosity as the thickness of the bronze increased. A 25 mm-thick wall would be fairly sound but over 50 mm porosity would become a serious issue. Porosity has been noted in some large reconstructions of cannons which when fired appeared to 'sweat' through the metal. As the molten metal cools, its ability to absorb gas is reduced and steam and hydrogen are released. The long freezing range of bronze produces a 'mushy' stage and there is a tendency for gas bubbles and cavities to form in the centre of the cast as this is where the metal remains liquid last, where it interacts with the oxygen, hydrogen and core. The tin

35 Norton, *The Gunner*, pp.67–68.

36 The author is indebted here to Sam Murphy for his immensely helpful explanation of the casting process. 'The technology of casting cannon', *Journal of the Ordnance Society*, Vol. 13 (2001).

The muzzle of the Great Hospital falconet may have been damaged as the result of poor loading but the smoothness of the interior and the size of the flaw would most likely suggest an incomplete casting. This of itself would probably not endanger the gun, although the condition of the vent would suggest that it was not heavily used. (Photography courtesy of Thomas Davies)

rich metal would tend to gravitate towards the centre of the mould where the metal remained liquid for longest.[37] These spots of tin would coalesce and during use could erode, potentially leaving an open 'honeycomb' like structure. When eventually barrels were cast solid this portion would be removed when boring out. When a core was used there was less of a problem as the metal solidified from two sides, from the outer mould and from the core. The rate at which the metal cools depended upon the mould material, its makeup, thickness and permeability as well as the nature of the material filling the casting pit.

As the metal cools it will also shrink. In a cannon where the metal is thickest and volume greatest the cooling will be slower. As the metal at the top freezes, it blocks any more liquid bronze feeding into the spaces caused by contraction lower down and it would also stop any more gas escaping. Ideally the cannon, like a bell would be cast muzzle down to reduce these problems, as in the early days it was. This is why tin was commonly added at the end of the pour to keep the metal liquid for longer.

In the furnace the molten metal will absorb both hydrogen and oxygen, derived from the air as well as the wood used in firing and even from a damp furnace, if not properly insulated from the ground. In modern foundries degassing tubes and deoxidising rings are used to remove these. In the

37 Relative densities: bronze (with 8–14 percent tin) 7.4–8.9 g/cm^3; copper 8.96 g/cm^3; tin 7.26 g/cm^3.

Innsbruck foundry, casting the Bucharest cathedral bell, in addition to modern methods they also used a substantial elder tree trunk to stir the metal. The longer and heavier a casting is, the more problems would be created. The metal could also react with any water and organic material in the mould which is why a thorough heating of the mould before use would be essential.

Until about 1500, guns and bells were cast with the open end facing down. The reason for a dramatic change, casting breech down, was probably the contemporary observation that so much dross, spots of tin and gas floated or bubbled to the top of the casting. This would be disastrous if contained within the breech. The advantage of casting a large bronze gun is that a thick mould and heavily packed earth or sand would act as an insulation permitting a freezing process slow enough to allow the escape of gas and permit more feeding of metal to offset shrinkage. Bellfounders did not change their methods. This was partly out of tradition but also because the core of the bell is hollow, unlike that of a cannon and there is much less gas pressure in the mould, usually producing a better result.

The great advantage that muzzle-up casting gave was to increase the density of the metal at the breech, meanwhile all the dirt, dross and scum would (hopefully) float to the top. The considerable weight of bronze meant that in a long barrel the metal would be under significant pressure, stopping gas bubbles from nucleating and thereby reducing porosity. A three-metre-long cannon with a 300 mm head of bronze above it would create a pressure at the breech of 280 kN/m^2 or 40 psi (rather less than three atmospheres). The consequence of this would be a stronger and more importantly a safer piece.

The surface of the gun can show slight variations in colour, pinholes and 'blisters' in the surface, which can be filled and removed by filing and sanding. This was the case with the author's falconet but is also easily visible in contemporary castings as well. Cast decoration can also be finished with careful use of a cold chisel, files and sanding. The result is to produce a beautiful glowing, shining barrel very different from the museum examples that have lost their lustre over the centuries.

This is the muzzle of a bronze saker cast by John and Robert Owen in 1543, held in the Fort Nelson collection of the Royal Armouries (XIX.894). It is a late example of a faceted gun with a dodecahedral barrel with no reinforce. The surface of the gun was heavily abraded by a long period spent underwater, being recovered in 1992; even so, the muzzle shows the problems associated with casting, bubbles of gas and air that have been frozen in. The interior of the bore appears far less affected with this problem. The touch hole has been bushed which would suggest that the gun has seen considerable use and that the problem with the casting had little effect on its useful life. (Author's photograph)

This is a half-scale but still working copy of a Wealden blast furnace constructed and operated at the Rural Life Living Museum in Farnham, Surrey. Clearly visible are the water-powered bellows, and the substantial brick and stone structure needed to absorb the heat and manage the weight of the liquid metal produced during the melt. The furnace would be charged by hand from the top with the molten slag and molten metal tapped off at the bottom. The lighter slag floated above the heavier molten iron. (Photograph courtesy of the museum)

Casting with Iron

Casting with iron was a very different business from casting with bronze, with only the design and manufacture of the mould being the same. The iron was manufactured in the blast furnace and cast directly into the mould. The only way in which the furnace could reach the temperatures needed to produce molten iron was with a powerful blast of air produced by water-powered bellows. The availability of ore, charcoal and waterpower is what allowed the Wealden iron industry to flourish. Cast iron was not easy to use; only the best quality metal in the melt was suitable for casting. The first melt was used for pig iron or iron shot, the second for forge equipment and firebacks and only the third was suitable for guns.

Cast iron is an alloy of carbon and iron with more than two percent carbon content. Carbon-free iron would melt at 1,500 degrees Celsius, however as the temperature in the furnace rises to about 1,200 degrees the iron absorbs carbon to about 3.5 percent. The melting point of this alloy of the two materials is much lower than their individual melting points, what is called the eutectic point, and the alloy melts and flows. The ability to cast into a mould is of course cast iron's principal quality, however it is also very brittle. Grey cast iron was most commonly produced in this process, the colour caused by the graphitic microstructure. It was also important for the iron ore to have a low phosphorous content which although it would assist the flow of the iron also produced hard but weak and brittle iron.

Cast iron unlike bronze has a narrow freezing range going from liquid to solid between 1,220 and 1,140 degrees Celsius, a range of only 80 degrees rather than the 170 degrees of bronze. The temperature for pouring cast iron was 1,315 to 1,415 degrees, again significantly higher than bronze. Shrinkage of cast iron was far less than that of bronze due to the formation of low-density graphite, which offsets the effect of shrinkage, but the process of melting and casting was more difficult because of the higher temperature.

Although cast iron is significantly lighter than bronze, it gives no indication that it is under stress and likely to break.[38] Cast iron has a crystalline structure that made it potentially brittle, a disastrous quality in any gun. In addition, the problem of rust inside the bore was never effectively solved, a serious issue at sea. Thus, for safety iron guns were made on a more massive scale than their bronze equivalents. To reduce stress further, the weight of powder allocated to cast iron ordnance was often less than for a bronze equivalent. The great advantage of iron was its comparative cheapness. The cost of bronze guns in England was seven times that of iron in 1581 and 12 times by the end of the century. It was the availability of cheap English cast iron guns that had a devastating effect upon the Genoese gunfounders in the mid to late sixteenth century. The English gunfounders were to find that the market for their ordnance also went into a steep decline in the early seventeenth century after the end of the war with Spain.

Ammunition

It is noticeable that the earliest cannons used *garros* or gun arrows rather than round shot. It was thought that they were the best shape to travel through the air on the basis of the evidence of existing projectiles shot from longbows, crossbows and springalds. These were replaced by round shot of stone by the end of the fourteenth century.[39] There was clearly a practical limit to the size of a *garros*, whereas there was little to limit the size of the bombard and its shot. The trebuchet in its various forms used stone balls and it was clear that they flew with great precision. Gunpowder artillery would inherit what was already an effective and readily manufactured projectile. Wrought iron and bronze were expensive materials and their weight would put increased strain on early breeches and barrels.

It was universally acknowledged, incorrectly as it turns out, that spherical projectiles were the ideal shape for artillery. Tartaglia the principal authority of the age in matters relating to ballistics, explained that:

> we define the spherical figure or shape as the most appropriate among all possible shapes in order to avoid the mentioned opposition of the air in the frame of each kind of motion. This [spherical] shape is most appropriate for the motion on all of its sides and it is equally heavy on all of its sides as well.[40]

The development of the blast furnace meant that relatively cheap cast shot could be made on a very large scale. The casting of iron shot was certainly an easier task than casting barrels and was in use on a large scale far earlier, with wrought iron guns such as the Sling using iron shot in the late fifteenth century. Biringuccio claimed that iron shot was unknown in Italy until the

38 Relative densities: bronze (8–14 percent tin) 7. 4–8.9 g/cm^3; cast iron 7.15 g/cm^3.

39 Jonathan Davies, 'Gun Arrows: "An amusing novelty or intelligent innovation"?', *Journal of the Ordnance Society,* Vol. 27 (2020).

40 Tartaglia, *Nova Scientia* 1558, first book, 1r–1v. p.83.

Bonawe charcoal fired ironworks, Scotland, built in 1735. It is the most complete remaining early type of blast furnace in the United Kingdom. It exploited the availability of cheap charcoal in the Highlands, although iron ore had to be imported by sea. Operating continuously for perhaps nine months a year, it produced about 2.5 tons of cast iron a day and a maximum of 700 tons a year. The waterwheel and bellows were sited in the partially demolished building on the left. and the iron and slag were tapped in the entrance facing the camera. The furnace was charged with charcoal, iron ore and limestone by hand. (Author's photograph)

arrival of the French in 1494. According to him the first moulds were made from bronze and then later from iron. He goes on to recommend making individual or group moulds in plaster of Paris from a pattern in wood, clay or metal. The moulds still in existence were be made as pairs with: 'little openings for the gates and vents and likewise four holes for pegging the parts together'.[41] The two halves would be held together with a pair of tongs.

The quality of the iron was not high, as was shown by some of the iron shot recovered from Armada wrecks. Biringuccio recommends that it should be made from: 'some of the crude, corrupted kind [of iron] that has been sent through the furnace in order to purify it of earthiness, or some of these old rusty bits of scrap iron'.[42] Using waterwheel-powered bellows, the scrap would be melted and using a little iron channel the molten metal directed to the moulds held by the tongs, which would be filled one after the other. Shot moulds of very different sizes have been found aboard excavated shipwrecks as well as examples in the museums at Neuveville and Churburg Castle. These have moulds made of bronze varying from one to three and a half inches in diameter with a mould thickness varying from one sixteenth of an inch to half an inch.

The evidence from the *Mary Rose* shot was that it was cast into a cool brass mould resulting in chiller crystals.[43] On the *Mary Rose* a two-part sandstone mould was found. It was impossible to cast iron aboard ship, but casting lead was certainly possible. Shooting lead shot from a gun designed to use iron was not a good idea and would severely test a barrel used to a much lighter projectile (a practice occasionally used to proof a gun). Cast lead shot was used by smaller calibre lighter pieces such as falconets, robinets, slings and bases, although a demi-culverin in Portland Castle in 1544 was provided with shot of both iron and lead.

41 Biringuccio, *Pirotechnia*, p.319.

42 Biringuccio, *Pirotechnia*, p.320.

43 Hildred, *Weapons of Warre*, p.125.

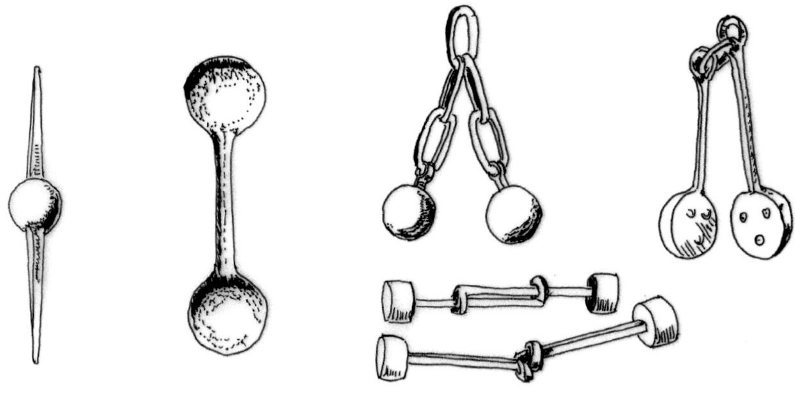

While varying in nomenclature the purpose of these projectiles was clearly the same, the destruction of rigging, sails, masts and yards of an enemy vessel. To avoid damage to the bore, crossbarshot could be wrapped in soft material, or as in the case of the *Vasa* trials provided with a wooden tampion affixed to either end, also a contemporary practice. (Author's drawing)

The provision of well-fitting shot seems to have been a major problem. We have noted already that the *Vasa* carried shot with a windage that varied from 4–11 mm, rather than the standard 7 mm. In the *Batavia* wreck there was a similar variation, as can be seen in the table below.

Table 13.

Gun	Bore diameter	Preferred diameter of shot*	Range of diameter of majority of shot
5 punt	90 mm	85.5 mm	80–88 mm
8 punt	103 mm	98 mm	95–102 mm
12 punt	116 mm	110 mm	110–116 mm
24 punt	148 mm	141 mm	138–146 mm

* Using the usual rule that the ball's diameter should be 1/20 less than that of the bore.

It is easy to see why any proficient and sensible gunner would use a shot gauge to ensure the safety and accuracy of the piece.

Other shot carried aboard the *Batavia* and the *Vasa* were expanding bar shot designed to damage rigging, while the *Batavia* carried cleve shot. This consisted of two cast iron hemispheres attached by a wrought iron bar, a munition designed to damage standing rigging, as it would have little effect on a warship hull as was shown by the failure of scissor shot against the *Vasa* reconstruction. Chain shot which was made from two iron shot connected by a chain was also used against rigging, although one assumes with a reduced powder load if the gun was not to be endangered. John Smith in his *Sea Grammar*, provided a full description of the variety of projectiles found aboard ships at the end of this period:

> Round Shot is a round Bullet for any Peece: Crosbarshot is also a round shot, but it hath a long spike of Iron cast with it as if it did goe thorow the middest of it, the ends whereof are commonly armed for feare of bursting the Peece, which is to binde a little Okum in a little Canvasse at the end of each Pike. Trundle shot is onely a bolt of iron sixteene or eighteene inches in length; at both ends sharpe pointed, and about a handfull from each end a round broad bowle of lead

> according to the bore of the Peece cast vpon it. Langrell shot runnes loose with a shackell, to be shortened when you put it into the Peece, and when it flies out it doth spred it selfe, it hath at the end of either barre a halfe Bullet either of lead or iron. Chaine shot is two bullets with a chaine betwixt them, and some are contrived round as in a ball, yet will spred in flying their full length in bredth; all these are vsed when you are neere a ship to shoot downe Masts, Yards, Shrouds, teare the sailes, spoile the men, or any thing that is aboue the decks.[44]

Lead shot was used for some small calibre pieces and sometimes for proofing purposes. In the 1490s lead shot was being cast in Florence for large calibre pieces. In cases of emergency, when Florence was besieging Pisa in 1499, cast bronze shot were manufactured. Bonnacorso Ghiberti cast 364 bronze cannonballs with an average weight of 52 libbra (17.65 kg).[45] Two of these were engraved by goldsmiths with a suitably offensive message. The siege was a dismal failure due to lack of nerve rather than a failure of the artillery. The few bronze cannonballs that remained in the state arsenal were sensibly used to cast new guns.

Finishing and Proofing

It would take between five to eight days for the mould of a large casting to be cool enough for the gun to be removed safely. The separation of the gun from its mould was itself no mean feat as the armature and wirework, which was essential for its integrity, was difficult and time consuming to remove. The core had to be pulled out, again using main force where the heel and hole on the end of the core bar served its purpose. Once the core was removed the extra head of metal, essential for a good casting, had to be sawn off, another long labour. The exterior of the gun had then to be cleaned using files, abrasives and chisels to produce a crisp and polished finish. Leonardo da Vinci recommended a wetted iron scouring wire. A fine finish was not always applied. The Norwich Great Hospital gun was left rough and unfinished on the underside, with numerous visible file marks. Guns were typically hammered out or planished, to make the surface smooth and hide cavities as well as harden the metal, as both Biringuccio and Collado recommended. This was also an opportunity to fill any small pinholes and blisters that had emerged as was the case with the Great Hospital gun and the author's piece.

The touch hole could now be drilled either vertically to the end of the breech or at an angle. The vent for the *Vasa* guns were set back at an angle of almost 45 degrees, perhaps to save the low ceiling of the warship form scorching. This proved difficult to prime with loose powder during the recent *Vasa* trials as well as producing a significant back blast. The Löfflers, Owens and Herold family of Dresden drilled their touch holes at 30 to 40 degrees starting at the upper edge of the cascabel. The hole could be of a

44 Smith, *A Sea Grammar*, p.67.

45 1 libbra = 12 ounces = 333 or 339.5 g.

This image shows two horizontal boring machines, both hand-powered. The gun is moved onto the boring tool on a sled drawn forward by using a windlass system. It is clear that such a technique could only be used to ream out a barrel rather than drill it. 'Machines for boring guns' by Vanoccio Biringuccio, from *De La Pirotechnia* (1540).

constant diameter or trumpet shaped, in either case it would be subject to considerable wear over time.

The bore itself was often imperfect, some of the *Mary Rose* guns have evidence in the bore of the iron wire that was wrapped around the core to ensure its integrity. The bore although true may still be roughly finished. A failure of the core in situ could produce bronze shards in the bore, which would have to be removed using a boring tool which could be man or water-powered. The bit was centred in the bore and in the absence of a bearing it could slip off course, especially if it met a soft honeycombed area or hit a piece of hard slag. Leonardo da Vinci drew a picture of a horizontal boring machine *c.* 1495–1497, but a vertical boring machine is illustrated in a drawing from the late fifteenth century, which uses the weight of the barrel to assist the operation as it is lowered onto the bore and the barrel revolved around the boring bit.[46] Another illustration in the series shows vertical boring but with the gun fixed while the bit was powered by two horses.[47] Vertical boring machines were used with guns cast solid in the early eighteenth century to drill out the bore but it appears that these were a development of these earlier machines rather than an innovation in their own right.

Although each gun was very much the product of individual craftsmanship and the mould was broken, the pattern could be reused. By the early sixteenth century there is evidence of early series production with guns being numbered. A gun from 1505 is inscribed 'N° 1DE 13' (Number 1 of 13) and pieces cast by Gregor Löffler, Hans Christoff and Cosimo and Giovanni Maria Cenni, all have indications that they were part of a production run. Some, such as Henry VIII's *Apostles* and James IV's *Seven Sisters*, were quite

46 BSM 356 p.132.
47 BSM 356 p.132.

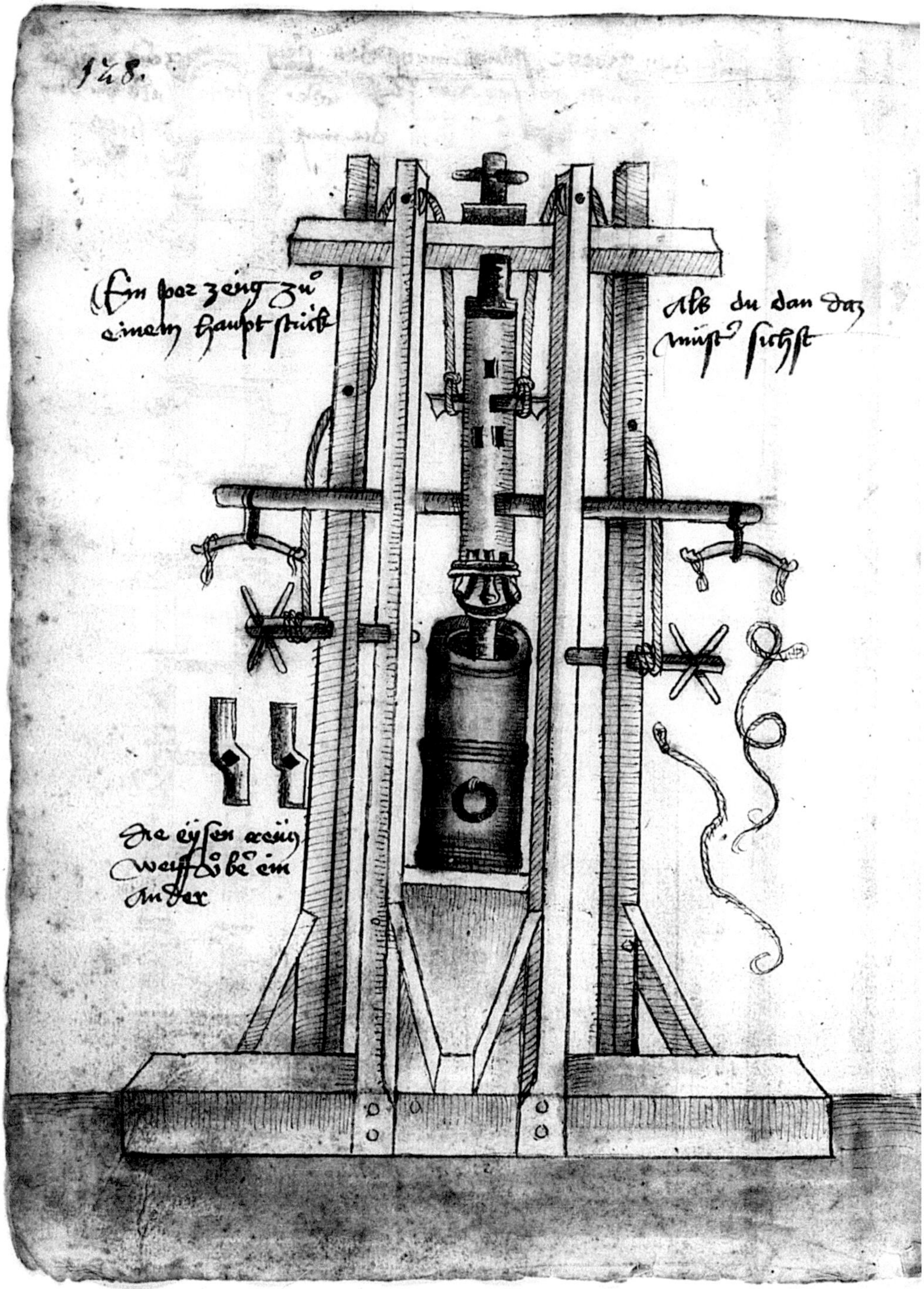

This image of barrel boring dates from a half century before Biringuccio. In many respects this seems a more heavy-duty and sophisticated piece of kit than that illustrated by him. The barrel was being reamed vertically and the bit was turned by a pair of horses. (*Feuerwerk- und Büchsenmeisterbuch* – Shelfmark, Fol. Cgm 356 p. 131. End of the fifteenth century. Courtesy of the Bavarian State Library)

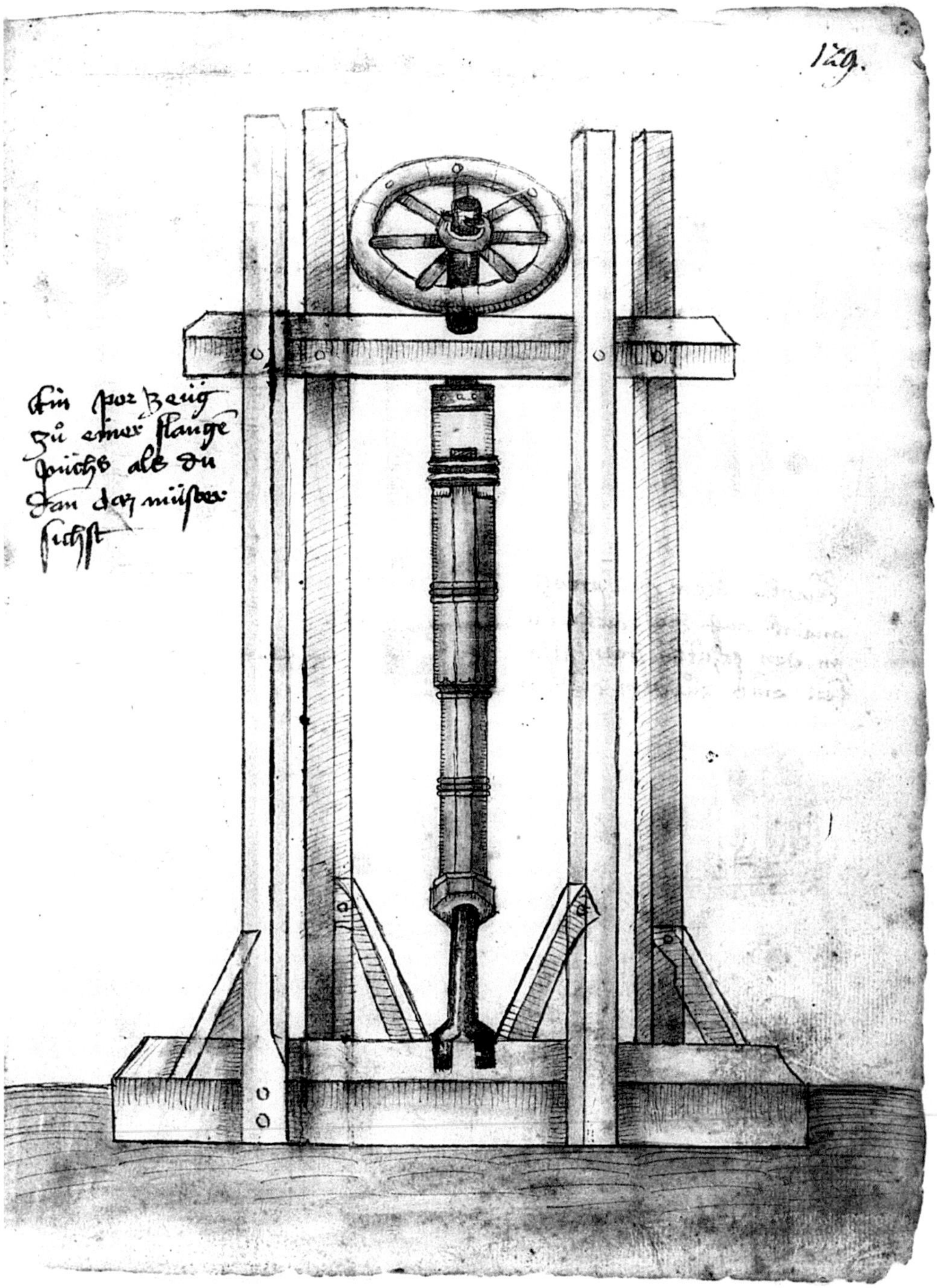

The weight of the cannon is being used here to drive the reaming tool up the barrel. There is no attachment here for a horse, so perhaps simple manpower was employed? The bore was being reamed which meant that if the bore was cast off-centre in the first place, reaming would make no difference to its straightness. The barrel is faceted, the breech reinforced and additional bands provided, which would serve very little purpose. The cascabel is flat and a tiller is fitted, there are clearly no trunnions. It is what may be described as a 'transitional' piece, where the barrel is of the 'new' form with a constant bore, but it would still be mounted on an older style carriage. (*Feuerwerk- und Büchsenmeisterbuch* – Shelfmark, Fol. Cgm 356 p. 132. End of the fifteenth century. Courtesy of the Bavarian State Library)

This is an excellent drawing of guns being proofed, made by the Dutch artist Willem Buytewech in the early decades of the seventeenth, entitled *Firing Cannons*. These barrels were not mounted on carriages but laid in a row on a dyke pointing seawards. There is a failed breech section in the foreground and one cannon has been unloaded onto the road, possibly waiting its turn. Carriages were constructed to match the barrel and there would be little point in making and risking a carriage before proofing. The drawing is undoubtedly taken from life, those besides the guns are standing in the cold and rain and are clearly just getting on with a noisy job in miserable conditions. Norton recommended proving barrels at an elevation of 20 to 30 degrees, which suggests that they were unlikely to be on carriages. (Courtesy of RKD)

clearly meant to be matching, although in the case of the *Apostles* they were two groups of six.

Gunfounders were clearly highly individualistic but like all masters of a craft commanded a workshop containing journeymen, apprentices and labourers, which meant that they could work on the preparatory stages of several guns at once, with the master supervising the critical moments in a gun's life. Workshops could be remarkably prolific. Master Bartolomé in Malaga in 1504–1505 with six workmen in nine months produced 67 pieces of which eight had to be returned after failing proof. Several decades after Bartolomé the famous Gregor Löffler was able to cast 66 guns in a single month. Remigy de Halut as part of another imperial contract in 1551–1552, cast 37 of the largest *canon-de batir* as well as nine demi-cannons, 24 culverins, 48 demi-culverins, six bastard demi-culverins and 41 falconets, 165 pieces in total. The introduction of the moulding box in the second quarter of the sixteenth century clearly increased the rate of production.

Proofing

> '... a peece which breakes doth most commonlie breake at the breeche or neere unto the mouth and seldom tymes in the middle ...'[48]

Guns are dangerous for their crews as well as to their targets. Guns could explode because they were overloaded with powder, loaded with the wrong powder, if the ball was incorrectly placed or became jammed. Weaknesses in the barrel caused during casting, a bore out of true or misshapen, could all cause catastrophic failure. It was not only the humble gunner who could be the victim. King James II of Scotland in 1460 was killed at the siege of Roxburgh, when a wedge flew from the breech of the Lion and shattered his thigh, which was according to a contemporary record 'dung in twa':

The proofing or testing of guns was therefore important. This was normally achieved by firing a charge in excess of its 'service' load. This procedure was established as one of the duties of the English Master of the Ordnance in 1456. Ordnance was proved in the 'myle end' in London with stocks of gunpowder set aside for the task. Manufacture and proving did not always result in a safe piece. In 1542, at the behest of Berwick's Master Gunner all the base pieces were proved once more and only one passed. Arthur Darcy complained that: 'the forlowkes breakys and pynes even bye the chambres, so that noon off them wyll serve'.[49] There is a description of proofing guns in Prague in 1517, a project that did not end happily:

> In this year the burghers of Prague were shooting with a larger carthaun [culverin probably] and with nine smaller field cannons from the Zlíchovská Meadow to the Kavčí Mountains. When they fired the larger carthaun for the first time, the iron cannonball broke into three pieces; and when they fired the smaller cannons, one field cannon burst and a splinter hit a tailor Jiřík of Fronta behind Klíč at his neck and tore a piece of skin with flesh. And he barely survived it.[50]

The Venetians who were responsible for some of the finest artillery proofed their guns using lead balls. Thus, a gun shooting in service an iron ball of 9 lb, would shoot a lead ball of the same diameter weighing 12 lb for proof. It was more usual practice elsewhere to vary the quantity of powder rather than the weight of shot. In 1518 a culverin cast in Fuentarrabía for Henry VIII, shooting a 14 lb shot, underwent the following proofing process, first: 'For 12 lb. powder, put in at 2 times to cleanse him, 12 lb. for 1st shot, 13 lb. for the 2nd, and 14 lb. for the 3rd.'[51] Lucar suggested the following proofing regimen for all guns which might be described as of culverin type:'for the first shoote of proofe with

48 Nicolò Tartglia, *Three Bookes of Colloquies*, colloquie 22, p.41.

49 Raymond, *Henry VIII's Military Revolution*, p.169.

50 Piotr Strzyż, 'Characteristics of medieval Artillery in the light of written sources from Bohemia and Poland' (Fasciculi Archaeologiae Historicae Fasc. XV), Ze starých latopisů českých, ed. J. Porák, J. Kašpar, Praha 1980, p.406.

51 Letters and Papers, Foreign and Domestic, Henry VIII, Volume 2, 1515–1518 on the 23 April. 4108. ORDNANCE.

the whole waight in gunpowder of his pellet, for the second shoote of proofe with 5/4 in gunpowder of the waight in his pellet, & for the third and last shoote of proofe with 3/2 in gunpowder of the waight in his pellet.'

As usual Norton provides some detailed recommendations:

> Q. 5. How much Powder is it fit to allow each Peece for proofe, and for Action?
>
> A. For the Cannons proofe 4/5 and for service ½ of the weight of her iron shot, for the Culuering and lesser Ordnance, the whole weight of their proper shots for proofe, and for Action for the Culuering ⅔, and for the Saker and Falcon 4/5, and for lesser Peeces the whole weight in Action untill they grow hot, for then must abatement be made with discretion. But in proofe the small Peeces should have once and ⅓ of Powder that the shots weigheth.[52]

It is interesting that the culverin, a gun which may be considered by virtue of its design a strong piece, is therefore proofed using a higher proportion of powder than the cannons. He also points to the danger of overheating the piece which could also put the integrity of the barrel at risk. He recommends proofing with three rounds increasing the weight of powder each time. For a gun shooting a ball weighing 10 pounds and under, he recommended a powder charge first, equal to the weight of shot, then one and a quarter and finally one and a half times its weight. For heavier guns up to 50 lb he recommends, a proofing charge two thirds, five sixths and finally the same weight as the shot. Proofing could benefit the gun. Eighteenth-century experiments suggested that proofing strengthened the bore by increasing the density of the metal.[53] After proofing the gun could be weighed and the number incised or stamped on the gun. The gun would now be paid for by weight by its new owner, now utterly confident in its reliability and safety.

Longevity

Bronze guns are not indestructible but if they are not mishandled, they come pretty close. Wear and tear will occur in the barrel due to the corrosive effect of black powder and the passage of the ball careering around inside. The bottom of the barrel would become scored by what the Dutch called a *kogelleger* or 'ball-groove'. This was a particular problem with bronze guns rather than iron because of the softer metal from which they were made. Guns could be bored out to a larger diameter to restore the regularity of the tube, but this would require them being used with a lighter charge, or bronze guns could simply be melted down when they were considered worn out, no doubt the fate of the vast majority.

The most obvious evidence of heavy wear and tear was to be found at the touch hole or vent. The vent, as suggested in the name, vents hot gases

52 Norton, *The Gunners Dialogue*, p.3.

53 Martin, *The Equipment and Fighting Potential of the Spanish Armada*, p.265.

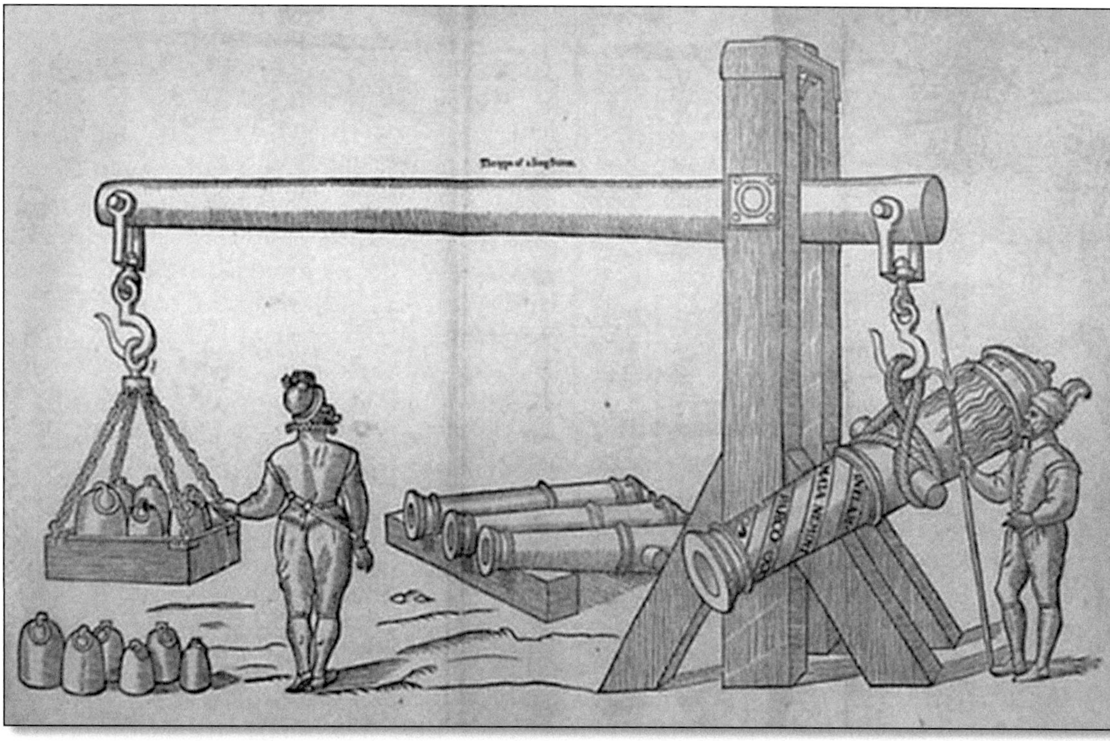

Weighing such a heavy piece was not a matter to be taken lightly. The steel yard was able to manage the task because of the mechanical advantage given by the long lever. Weight mattered as the foundry was paid, as it still is, by weight. The figure was usually recorded precisely, on the barrel or trunnions. Nicolò Tartaglia (1500–1557), *Three Bookes of Colloquies* (1588), pp. 100–101, in appendix; caption, 'The type of a long Stater'.

from the breech (easily visible in photographs and video recordings) which will wear the hole from its usual 3–5 mm. This widening of the vent is disconcerting for the crew and of course reduces the pressure behind the shot. When the cannon's vent had become too large it was common practice to drill it out, thread the new hole, screw in a replacement bronze or possibly iron pierced screw bolt, known as a *dado* a process called *bouching* or *bushing*. The touch holes of the nine pieces of sixteenth-century ordnance in the Museu de Angra do Heroísmo have touch holes that vary from 0.2 of an inch (5 mm) to in one case 1.2 inches (30.5 mm).

Henry Hexham recounted in some detail how massive the problem could become and how it was resolved:

> To conclude, with Touch-holes, I have seene in Ostend upon the West Bulwark, that some touch holes of Canon, were blowne so great with often and continuall shooting, that I have putt my fist into them. Now such a touch hole being blowne some three or foure incles, maye easely be remedied. For if yow bore the hole round, and drive in a screw of yron into it, as thick as your finger, and in the midst of the screw above said, bore a small Touch hole in yt, yow shall finde this to last longer unblowne, then any other way, which hath bin invented, for now when ordnance are cast a new at this day, for the preventing of this two much blowing, they use now to make such a touch hole with a screw, as is here mentioned which will not so soone be blowne great, and as your brasse or Copper Touch holes will.[54]

54 Hexham, *The Third Part of the Principles of the Art Military*, pp.3–4.

The touch hole of the Genoese falcon XIX.15 in Fort Nelson has clearly been bushed, with clear evidence of new threading. (Author's photograph)

The French expression 'Faire les quatre cents coups', which is the rough equivalent of the English expressions 'to raise hell', or 'sow one's wild oats', is ostensibly related to the vent life of a cannon, able to withstand 400 shots before replacement. During an Ottoman siege of Candia in 1669 the siege guns became unusable and apparently had to be recast after only 500 or 600 rounds.[55] Adrian Caruana considered that a lifetime for an iron gun in later centuries could be up to 10,000 rounds with the vent requiring replacement every 2,500.

Bronze guns had a remarkably long life. Many survive either as trophies or archaeological discoveries, unfortunately very few have their active history recorded. The fortress of Sarzanello in an inventory of 1679 recorded the presence of an octagonal falcon almost certainly cast for the fortress by Gregorio I Gioardi in 1499. A piece cast by Hans Poppenruyter with a dragon muzzle, at the turn of the sixteenth century, was captured from an Austrian vessel during second siege of Vienna in 1683, an active life of almost two centuries. An English bronze saker cast by Richard Phillips in London in 1601 and now on display at Fort Nelson was captured by British forces in the First Opium war in 1840/1 after goodness knows what travels. On it was inscribed the following Chinese testimonial from 1800 almost two centuries after its casting: '... found to be a gun of good workmanship and of intimidating power. At each discharge it sent forth a thundering sound ...'

Caruana records the history of an English culverin cast in 1590 that served in three warships and 20 actions up to its retirement in 1716, a mere 126 years of active life. A bastard falconet from the island of Sark, provided by the Tower for the island in 1565, is still to be found in the Seigneurie on Sark. This gun had remained in use for almost three centuries until a decade or so before 1875. With such a long and active life, it is hardly surprising that there have been four vent holes drilled into the breech, with the previous ones plugged with bronze. An Italian Esmeril cast before the mid sixteenth century was last fired 300 years later, and in anger, during the Italian Wars of Independence in 1861. The official record must go to an Ottoman gun cast in 1524 when the 'Solomon of the Age', Suleiman the Magnificent, ordered this example of one of his 'dragon guns' to be made. Mutilated at some point in its history, losing several feet off the muzzle, it was captured by the British at the battle of Alexandria in 1801, 277 years after its casting.[56]

55 Rhoads Murphey, *Ottoman Warfare 1500–1700* (New Brunswick: Rutgers University Press, 1999), p.215 fn. 5.

56 Royal Armouries exhibit XIX. 243. Length 15 ft 11 in (4.851 m); bore 7. 5 in (19 cm).

6

Gun Carriages

Land Carriages and Limbers

Initially guns were mounted on wooden beds which would enable them to be aimed, provide a stable platform and whose weight would help absorb recoil. This would be perfectly adequate whilst guns were engaged in siege warfare and when they played little part in field engagements. To permit elevation to be adjusted, the gun's bed could be pivoted between a pair of A-frames. An elevation arc could be passed through the rear of the bed, with pegs used to fix its position. These are well illustrated in an illuminated Zurich manuscript dated to 1420–1440.[1]

For greater flexibility in the field, guns could be mounted on carts or provided with their own wheels, as illustrated in the same manuscript. There are illustrations of four-wheeled carts with a gun mounted to the rear of the cart, providing a secure base for the gun as well as storage space for powder and shot. This was an attractive option but did not permit much flexibility on the battlefield.

With the gun mounted in a substantial bed to absorb recoil, the front of the bed could be pivoted on a separate trail that would be attached to an axle and wheels. An elevation 'arc' could then be passed through, or on, either side of the cannon bed from the trail. The introduction of the trunnion did not make the gun carriage fitted with an elevation arc obsolete. The trunnion would now be positioned at the front of the trail where the pivot for the gun bed had been. A lug or extension on what had previously been the flat face of the breech could be inserted between the arms of the elevation arc and pegged or wedged in place. This is probably how the Norwich Great Hospital gun was mounted. The invention of the trunnion meant that the gun would be easily elevated and depressed. The gun could be mounted in a solid tree trunk as illustrated in the *Zeugbuch* of Kaiser Maximilian I. This system would only provide a very limited degrec of elevation or depression.

The carriage developed into a standard and entirely practical form. Two sizeable timbers formed sidepieces or the 'cheeks' of the carriage, replacing

1 Zürich Zentralbibliothek, e-codices_zbz-Ms-Rh-hist0033b_070v_ma War technology.

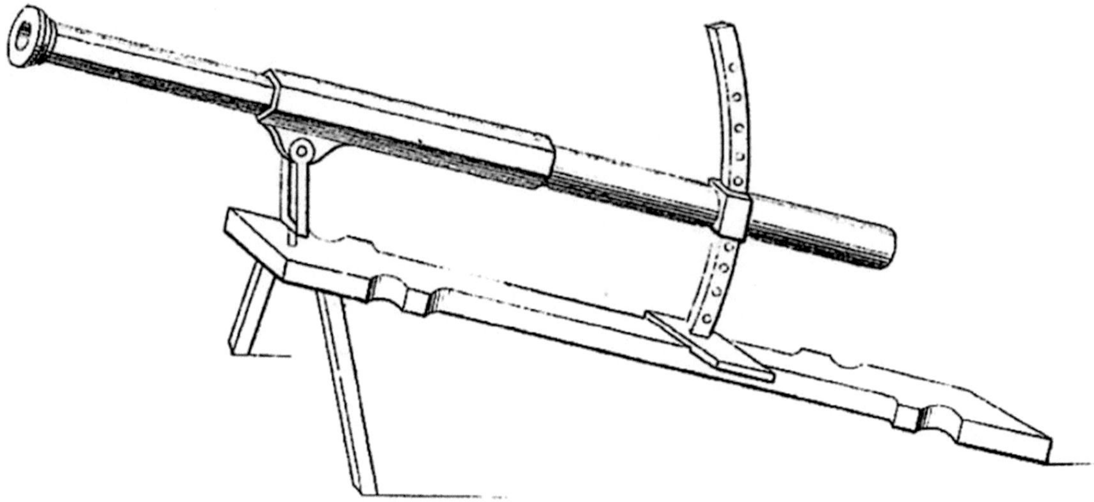

Above: a simple mounting from a fourteenth-century Swiss manuscript, showing an early use of the elevation arc in a very light piece. (Napoleon III, *Études sur le passé et l'avenir de l'artillerie*, Volume 1. (Paris: J. Dumaine, 1846)

Below: This is a small piece designed for use in a fortification rather than the field. It has a sophisticated pinion and arc rack mechanism. This was designed to enable the barrel to be depressed rather than elevated making it suitable for shooting down from high walls upon potential attackers. Note that the back of the breech is flat which enables the mechanism to be used but is redolent of a time when this area of the gun transmitted the recoil of the piece into the gunframe. This gun is however fitted with trunnions. The reinforcement rings spaced along the barrel perform no useful purpose but recall the appearance of wrought iron guns where the system of hoops and rings was critical to the design. (*Feuerwerk- und Büchsenmeisterbuch* – Shelfmark, Fol. Cgm 356 p. 31. End of the fifteenth century. Courtesy of the Bavarian State Library)

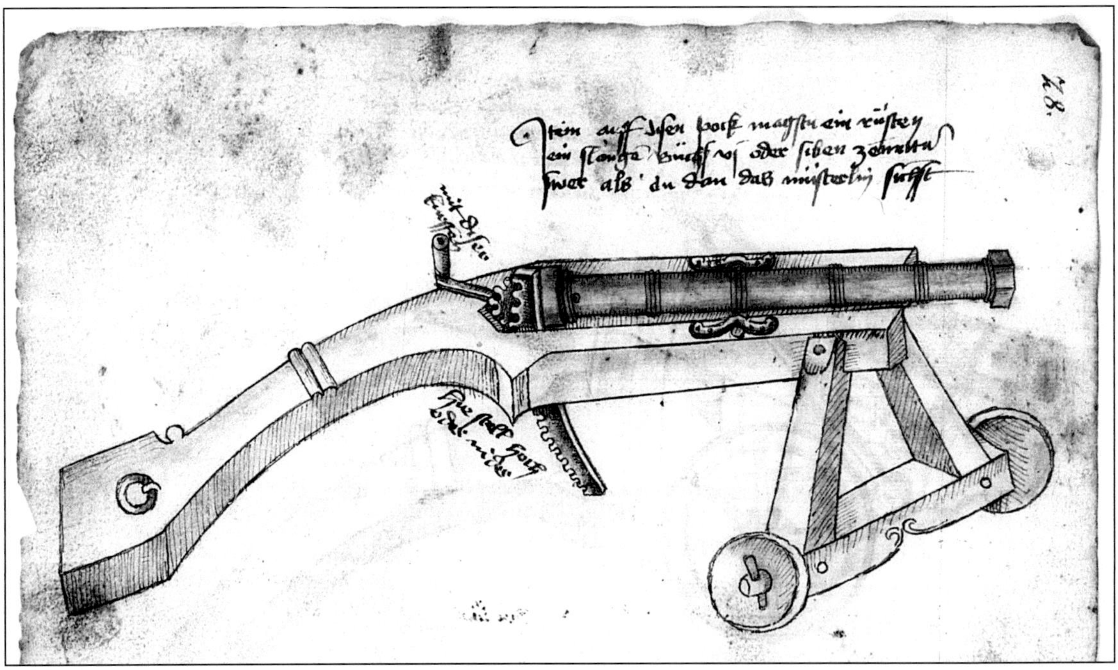

the solid trail. These were connected together by three or four transoms to form a sturdy box like structure. The axle was attached to or through the cheek pieces. The trunnions were almost fully contained in the cheeks of the carriage and were covered by a well secured iron band or a capsquare which could be hinged to allow the quick removal of the barrel. The trunnions made it much easier to adjust the elevation of the piece using quoins (wedges). Other means of elevating the gun, possibly experimental or simply notional, used a vertical screw thread, while others employed a rack and pinion system as in the case of some of Emperor Maximilian's guns.

The defeat of the Charles the Bold of Burgundy at Grandson and Murten (Morat) in 1476 by the Swiss were catastrophic defeats for the Duke's invincible army: it would not be the last. One of the outcomes was the capture on both occasions of his extensive artillery train. This would have amounted to several hundred pieces. Most were lost again, rusted away or recast but a small number remain in Swiss hands in museums in Basel, La Neuveville and Murten. The collection consists of guns and in eight cases their carriages as well, a feature which makes them of special importance. The survival of these carriages is extraordinary and gives an insight into a period of transition in the development of both guns and their mountings.

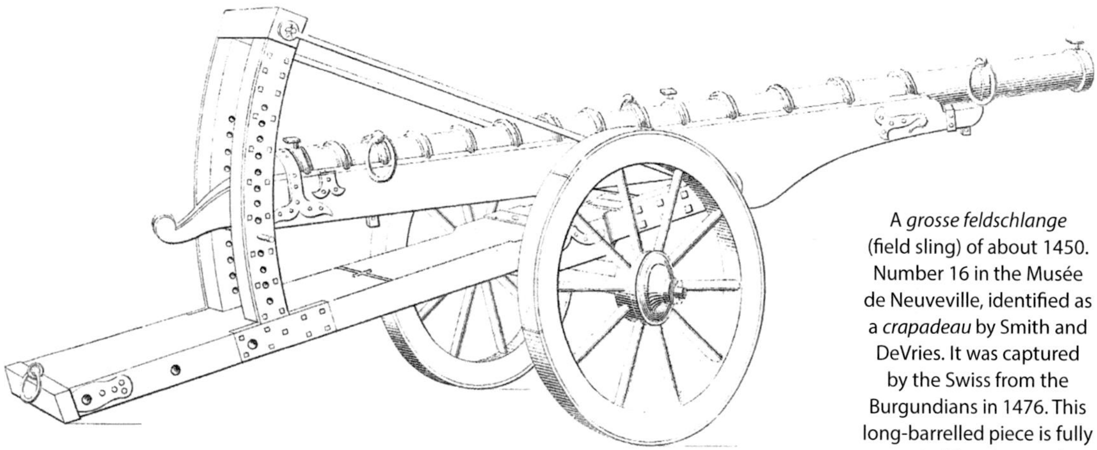

A *grosse feldschlange* (field sling) of about 1450. Number 16 in the Musée de Neuveville, identified as a *crapadeau* by Smith and DeVries. It was captured by the Swiss from the Burgundians in 1476. This long-barrelled piece is fully mounted into its wooden bed, hinged at the front to permit elevation. The recoil is transmitted through the breech end and elevation and depression is achieved using the traditional arc. The army of the Duchy of Burgundy clearly contained many traditional as well as more 'modern' pieces. (Napoleon III, *Études sur le passé et l'avenir de l'artillerie*, Volume 1, Paris: J. Dumaine, 1846)

Some of the guns are clearly meant for the carriages that they are mounted on, as they fit them comfortably. They conform to two principal types. The first, where the barrel has a flat end to the breech and no trunnions, are mounted on a wooden bed which is pivoted on a solid trail. An elevation arc extends upwards from the trail and a portion of the wooden bed behind the bed passes through the arc and can be pegged into place. The other type is a box trail, made up of two cheeks held together with crosspieces or transoms. The barrel has to have a trunnion and elevation is achieved using a quoin. How representative a group these guns are of late medieval artillery is impossible to say.

Table 14.

Ref No.	Date	Barrel diameter/ length	Material/ method	Loading	Trunnions	Cascabel shape	Mounting
9 N Crap	1460–70	72 mm/1,595 mm	Wrought H&S §	ML/BL ¥	No	Flat	Pivoting on trail
10 N	1460–70–1500	58 mm/ 2,213mm	Wrought Forged	ML	yes	Flat	Box trail
11 N	1460–70	70 mm/1,525 mm	Wrought H&S	ML	yes	Cone	Box trail
12 N	1460–70	73 mm/2,190 mm	Wrought H&S	ML ¶	yes	Pin/cone	Box trail
13 Murten	Booty	35 mm/1,388 mm	Wrought H&S	ML ¶	no	Flat	Pivoting on trail
16 N Crap	Booty	60 mm/2,925 mm	Wrought H&S	ML	no	Flat	Pivoting on trail
27 N	Booty	75 mm/ 837 mm	Bronze Cast	BL	no	Flat	Block trail, no elevation

* The reference number and technical details are from Smith and DeVries, *The Artillery of the Dukes of Burgundy* (Boydell Press, 2005)

§ Hoop & Stave construction

¥ muzzle or breech loading

¶ Converted breech-loader

Below: Burgundian artillery captured in 1476 held in Musée de Neuveville and Musée de Morat (Murten). This field gun at the museum at Neuveville (10 N) is of a style that would remain common for almost 400 years. The ironwork near the trunnion is there to stop the carriage splitting with the recoil. The barrel which appears at first inspection to be a cast piece is of wrought iron. It is unsurprising that so few early carriages survive, especially if they are left outdoors, as time quickly takes its toll. Carriages often seem to have been built for specific barrels at the beginning of a campaign, with only a small proportion of an arsenal's holding of guns being fitted with carriages at any one time. (Napoleon III, *Études sur le passé et l'avenir de l'artillerie*, Volume 1. Paris: J. Dumaine, 1846)

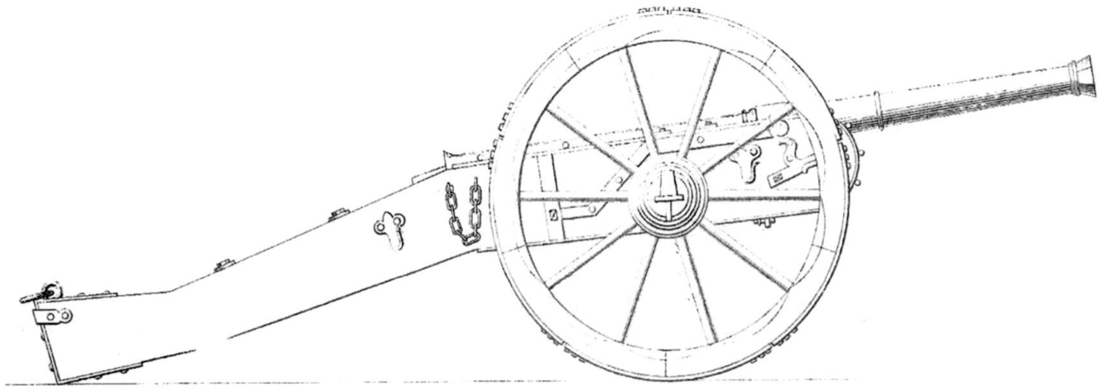

The carriage was heavy and soundly constructed to cope with uneven roads and the stresses of the gun's recoil. The complexity and cost of the carriage was considerable. In 1566 Fronsberger gave the following breakdown for the cost of a *nachtigall*, a large siege piece weighing over three tons.[2]

Table 15. The Cost of a *Nachtigall* (Florins)

60 cwt of metal for the gun	1,080 fl
The cradle inclusive of wood and metal	80 fl
The carriage	150 fl
The limber	24 fl
The limber spike	5 fl
Chains	5 fl
Two sets of wheels for limber and carriage	10 fl
Ladle, sponge etc.	2 fl
Total	1,356 fl

The cost of the equipment to provide carriage and limber was about 20 percent of the total.

Biringuccio was, as always, critical of amateurishness when it came to the construction of carriages. They should according to the irascible Italian be made of well-seasoned oak and elm and constructed on the basis of fixed proportions. Thus the: 'beds of the heavy guns must be made as long as the length of the gun and of wood as thick as one-third of the ball ... make the diameter of the wheels seven times the diameter of the thick part of the gun at the breech'.[3] The carriage should be: 'strong, well proportioned, and neat, and gracefully wrought'.[4] There was some debate as whether longer or shorter carriages were preferred. Norton cites Louis Collado as preferring the longer type, as he thought it was stronger and ran more easily when recoiling. Alexander Bianco preferred the shorter as it took up less space when recoiling and was easier to return to the firing point, he also considered that it would shoot further.

Smith provided a more detailed explanation of how a carriage should be designed. He began by instructing the gunner to first measure the length of the bore of the gun. The carriage should be one and a half times that length. The planks should be four times the diameter of the bore in height at the front of the carriage, three and a half times in the middle and two and a half times at the end. The width of the planks should be one diameter. The two cheeks should be held together with four transoms held with 'strong yron boltes'. The trunnions ought to be seated three times and a half the diameter

2 G. Embleton and D. Miller, *The Landsknechts* (Oxford; Osprey, 1976), pp.13–14.
3 Biringuccio, *Pirotechnia*, p.314.
4 Norton, *The Gunner*, p.129.

of the bore from the front of the carriage and should be sunk into the wood by two thirds the depth of the barrel.[5]

Robert Norton provides a pretty standard description of how the system of proportion was applied to the construction of a carriage:

> First, for the Culuering, the Cariage shall be once and a halfe of the length of the Peece, so that if the Peece be 32 dyametres, the Cariage shal be 48 in length, and the Wheeles 11 dyametres high. The Nave 4 high and 5 in length, the Spoakes 4 besides the one halfe let in at each end. The Fellowes two, and the Arming one. The Axis shall be in length 13 dyametres ⅓, and at the cutting of the Cariage therewith to bee 2 in breadth, and 2 in thicknesse. The Limber Planks or sides of the Cariage must be 4 and a halfe, or 5 dyametres broad, one thicke, at the Trunions 4, and at the Tayle 2 and a halfe.[6]

There were four transoms specified; the *fore*, the *quoin*, the *bed* and finally the *tail*. The transoms were sturdy pieces in their own right one and a half calibre broad and one calibre thick except for the trail transom two calibres broad and five long. The extra width of the transom at the rear meant that the cheeks were not parallel but narrowing towards the front of the carriage. This creates many, many problems for the poor carpenter, not only with the transoms but with the alignment of the trunnions.

The gun carriage made from oak or elm was usually left unpainted. There is evidence from the *Kaiserbuch* that they were sometimes painted black, probably with a preservative oil, with the ironwork painted red. A red paint was used on medieval wrought iron guns to protect them from corrosion. One possibility was to use linseed oil-based paints with a natural earth colour, red ochre or yellow and an early modern red-painted gun carriage has recently been excavated in the Ukraine. The author used three coats of red ochre-based oil and turpentine paint on his falconet, which was very sticky and took a long time to dry.

The wheels were and had to be substantial and heavy. A good guide was that their height should be half the length of the bore, although for smaller guns they needed to be taller. Either way they should enable the gun to shoot over the parapet of whatever fortification they were emplaced. The key to the strength of the wheel was its 'dished' or saucer shape. The spokes were not set vertically but at about seven degrees, which led to a more even stress being imparted to the wheel and improved its stability and produced a characteristic dished appearance. Too much dishing could weaken the wheel and produce calamitous results. This system provided great resilience as well as performing the role of a modern differential on a car. Although there was much debate about its efficacy, it developed after 1500 and was widely adopted.

The continuous iron tyre was unknown in this period and instead they used: 'yron strakes, grasped and nayled with two or three Rankes of great head Nayles'.[7] These produce a very different appearance to the modern tyre

5 Thomas Smith, *The Arte of Gunnerie*, pp.30–31.

6 Norton, *The Gunner*, p.131.

7 Norton, *The Gunner*, p.131.

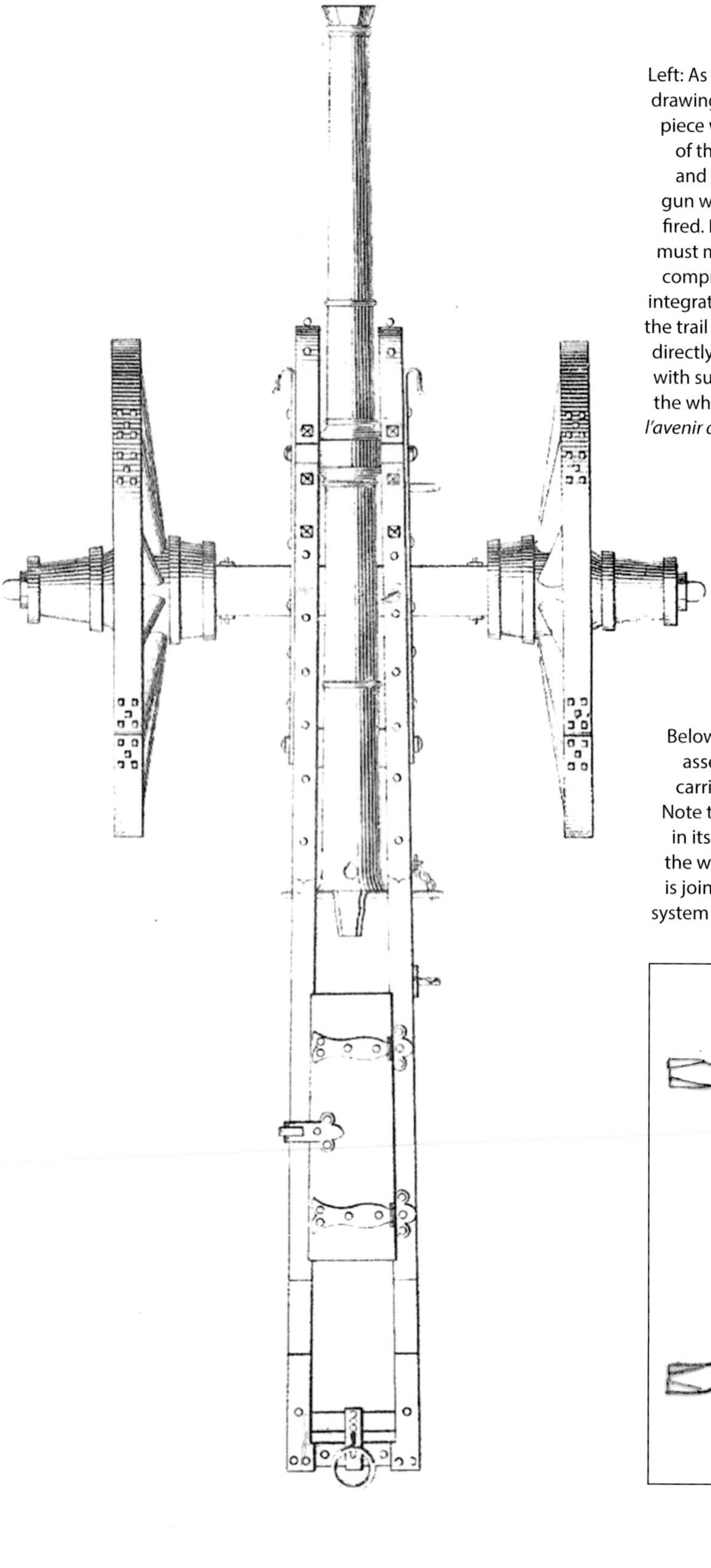

Left: As can be seen in this not entirely accurate drawing of Neuveville (10 N), the cheeks of the piece were not parallel but follow the profile of the barrel. The importance of strength and stability was obvious both when the gun was being drawn as well as when it was fired. However good the barrel, the carriage must match it or accuracy would be seriously compromised. The carriage has a useful box integrated into the trail and a ring at the end of the trail but no pintle hole. Two trails were fitted directly to the gun as no limber was necessary with such a light gun. Note also the dishing of the wheels. (Napoleon III, *Études sur le passé et l'avenir de l'artillerie*, Volume 1, Paris: J. Dumaine, 1846)

Below: A drawing of the undercarriage assembly for one of the siege guns carried aboard *La Trinidad Valencera*. Note the substantial timber pieces used in its manufacture and the 'dishing' of the wheels. Each of the felloes or fellies is joined to the hub by two spokes. The system of strakes held the felloes together in pairs. (Author's drawing)

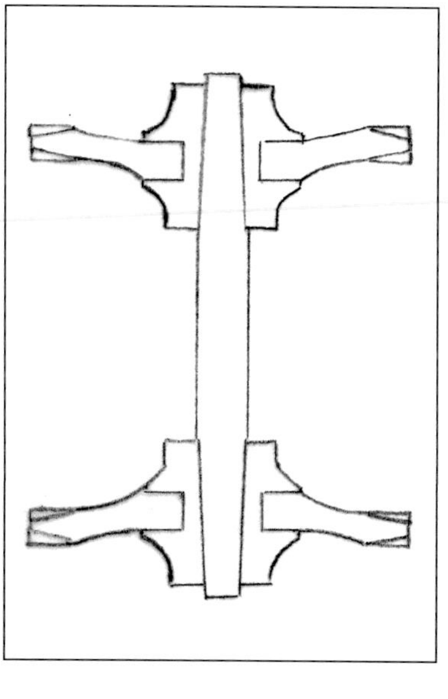

and enabled sections of the wheel to be more readily replaced if damaged. To further strengthen the wheel, cross clamps or strake bands could be fixed that covered the ends of adjacent strakes. Continuous hooped wheels only began to be used in England in the last quarter of the eighteenth century, iron strakes continued to be used on heavy wagons until the late nineteenth century.

What is particularly noticeable about the large gun carriages, in the first half of the sixteenth century, is the complex iron interlinkage from the naves of the wheel axle to the main body of the carriage. Its purpose was to transmit the shock loads at the end of the axle, where they were most severe, to the carriage. This system was cumbersome and added to the weight and expense of the assembly.

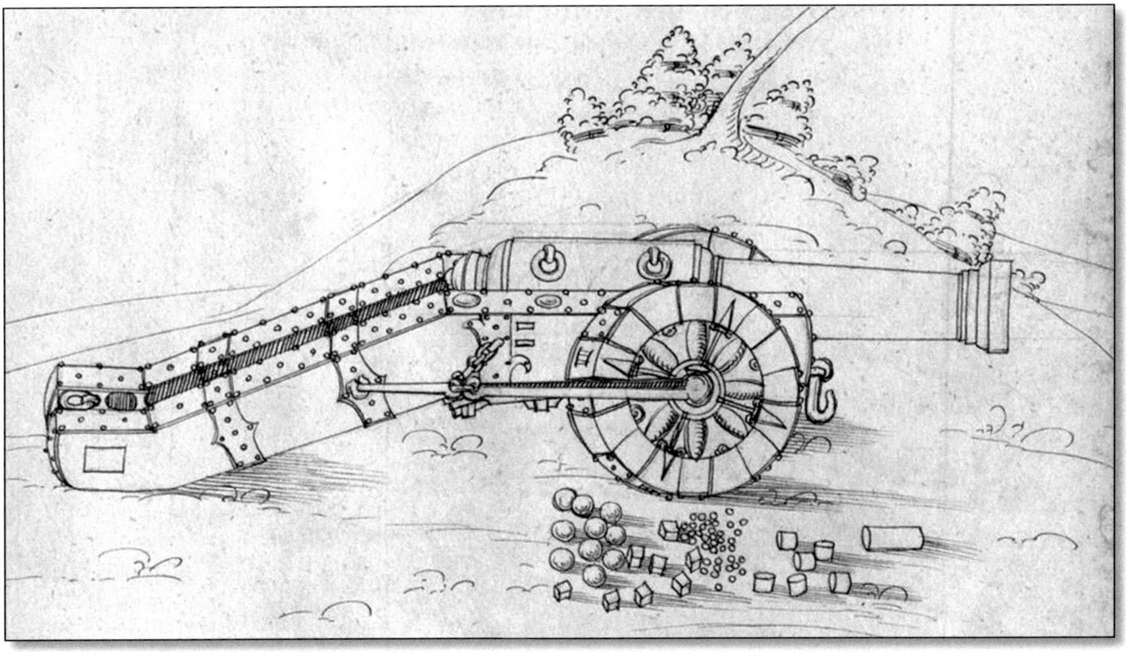

Above: This drawing gives a good idea of the quantity and complexity of the ironwork required for a heavy field carriage. The attachments from the wheel hub to the carriage are well illustrated as is the iron strapwork that covers both of the cheeks. In the foreground there is a collection of shot, round, dice and small as well as cylindrical 'clod' shot in two sizes. (Walther Litzelmann, *Vortrab zu der Arckalay – Büchsenmeisterbuch*, Shelfmark, Fol. Cgm 909 p. 303. Courtesy of the Bavarian State Library)

The excavation of *La Trinidada Valencera* revealed guns and carriage parts that were to have been part of the Armada's siege train, if it had landed on the coast of England. These guns lacked the interlinked ironwork of earlier pieces. An examination of the axles showed that the axle tree had an iron counter-axle running along its underside (see facing page, top). This axletree and the chamfered axle bed dealt with the shock loads at the junction of the axle and nave. With all the appropriate ironwork fitted:

> [the] Cariage is locked fast, as into one entire body, and is plated and bound strongly with yron, that the ioynts open not with the vehemencie of the reuerse [recoil]. And thus is the Cariage in his perfection, so that if it were armed with his Axtree and Wheeles, the Peece might be thereon mounted, and ready to doe seruice.[8]

8 Norton, *The Gunner*, p.130.

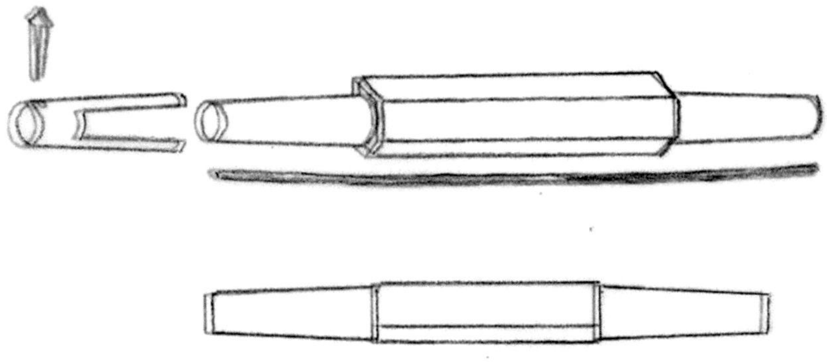

Left: The main carriage axle from the siege gun carriage excavated from *La Trinidad Valencera*, showing the iron counter-axle and the chamfered axle bed. This was a simple but very effective means of strengthening the axle and doing away with the complex supporting iron framework that had previously been used. (Author's drawing)

Below: Gunners had to be expert riggers, able to lift valuable and often very heavy barrels on and off their carriages. A block and tackle were suspended beneath shear legs, with the lugs or dolphins providing the lifting points. It was obviously vital that the dolphins were placed at or very near the point of balance. (Walther Litzelmann: *Vortrab zu der Arckalay – Büchsenmeisterbuch* – Shelfmark, Fol. Cgm 909 p. 313 .Courtesy of the Bavarian State Library)

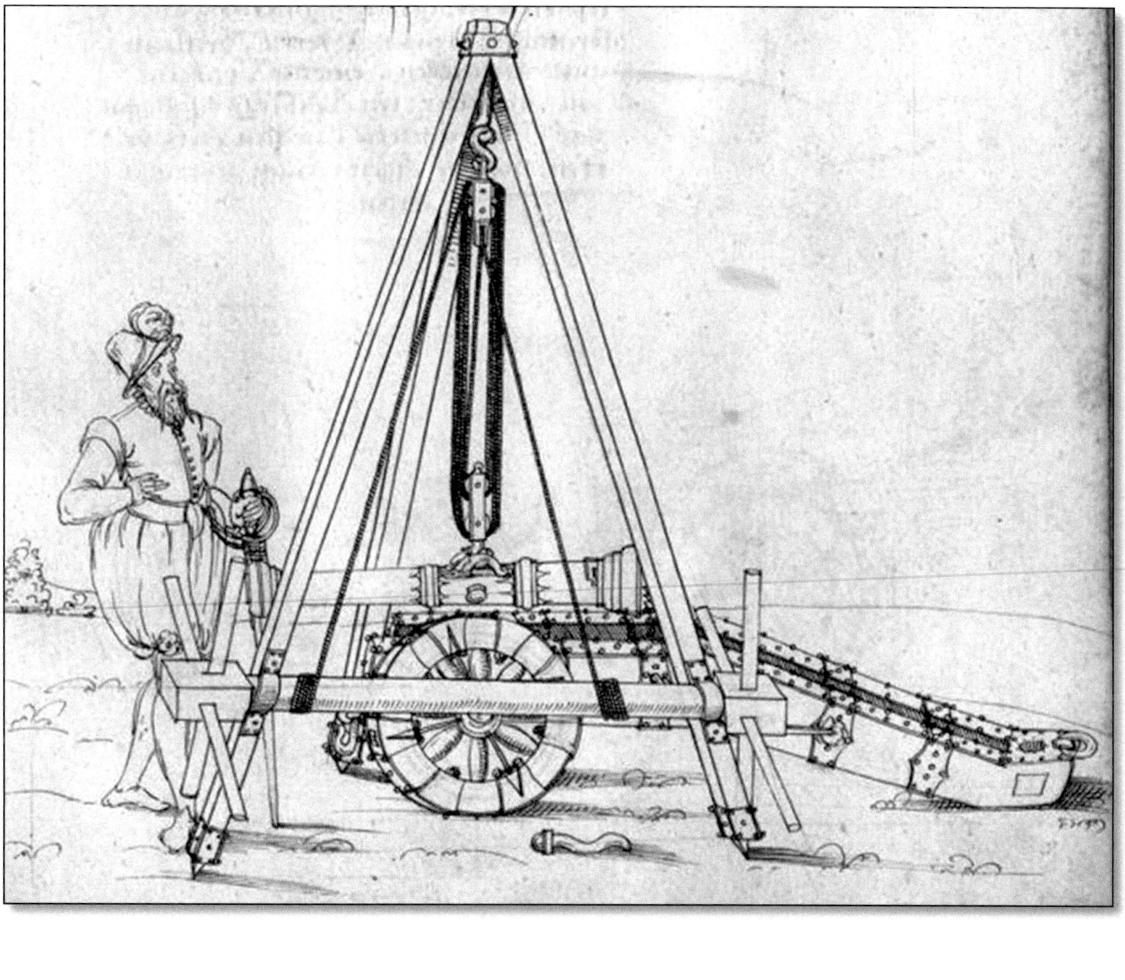

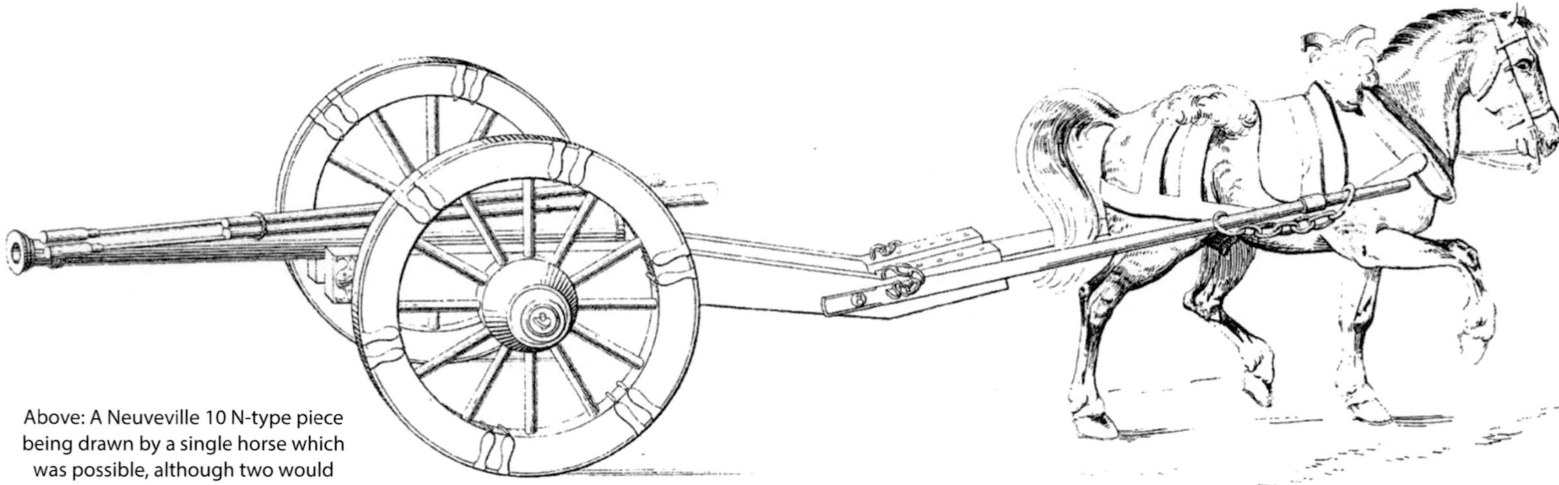

Above: A Neuveville 10 N-type piece being drawn by a single horse which was possible, although two would have been preferable. As with several other captured Burgundian pieces, it was provided with two trails that could be swung forward when not in use. It was a simple and practical means of making such a light gun quick and easy to transport. (Napoleon III, *Études sur le passé et l'avenir de l'artillerie*, Vol.1. Paris: J. Dumaine, 1846)

Above: This image was taken from the Cowdray Engraving showing the transport of heavy guns from Calais to the siege of Boulogne in 1544. It shows the use of a block carriage to carry a dismounted barrel. The alternative to this system was to site a second trunnion space further back along the carriage, with the gun then sitting within the cheeks and not able to rock up and down. This was the case with the *La Trinidad Valencera* carriage for the heavy siege gun. (Napoleon III, *Études sur le passé et l'avenir de l'artillerie*, Volume 1. Paris: J. Dumaine, 1846)

It is easy to see how the second axle on a four-wheeled wagon could become a detachable limber. It is quite clear that both these guns were designed to have the second axle removed so that the heavy wooden trail could rest on the ground. There is no evidence for a pintle in either illustration, perhaps the pintle hole was inside the heavy trail? The lower gun barrel is characteristically faceted for the period and has the usual swollen midriff, suggesting a sub-calibre powder chamber. There are also hints of a trunnion, although positioned really far too low for the period. The upper gun clearly pivots on the heavy wooden bed in which it is securely fixed, very much the older system. It does on the other hand have a pair of dolphins, which are missing from the lower piece. (Heidelberg, Mönch 1496 cpg 126_0058. Courtesy of the University of Heidelberg Library)

Above: This example of a late medieval short-barrelled field piece is equipped with a fixed trail suitable for horses, or if needed manhandling. The clever design includes two chests, either side of the barrel, no doubt for items of equipment and ready-use ammunition. Gun chests are often seen either included in the design or as removable items carried on the trail or even over the breech. (Heidelberg, Mönch 1496cpg 126 0060. Courtesy of the University of Heidelberg Library)

One of the most important but least commented upon aspects of the 'new' artillery was the introduction of the limber, an axle fitted with two wheels to which the gun was attached through a pintle or upright pivot pin. The limber transformed the transport of guns which with the aid of the limber could be turned into a strong and manoeuvrable four-wheeled carriages.

The first illustration of a gun with a limber that the author has found is in the *Feuerwerksbuch* of Martin Merz.[9] It is clearly no prototype but already highly practical and demonstrating the same design as would be employed for centuries. The limber wheels are slightly smaller in diameter, but the width of the axle is the same as that of the gun, an important factor as the author has discovered when making models of carriages and limbers.

The first illustration of artillery with its limbers in action that the author has found can be dated to 1496 and is in the *Kriegsbuch* of Philipp Mönch. It shows what we could refer to as batteries of field guns emplaced wheel to wheel. They have their limbers behind them, perhaps preparing for a quick getaway? The artist is a little sketchy when it comes to details such as the wheels on the limbers but clearly, he is illustrating something that was not uncommon. The blocks of pikemen, handgunners and mounted arquebusiers all demonstrate how different European warfare was becoming when compared to the scene that had taken place in England at Bosworth a decade before, where bill and bow still predominated.

9 *Feuerwerksbuch*, Martin Merz – Bayerische Staatsbibothek München, Shelfmark, Fol. Cgm 599, [Sl] North Bavaria/Franconia, I: 2nd half of the fifteenth century; II: 1473.

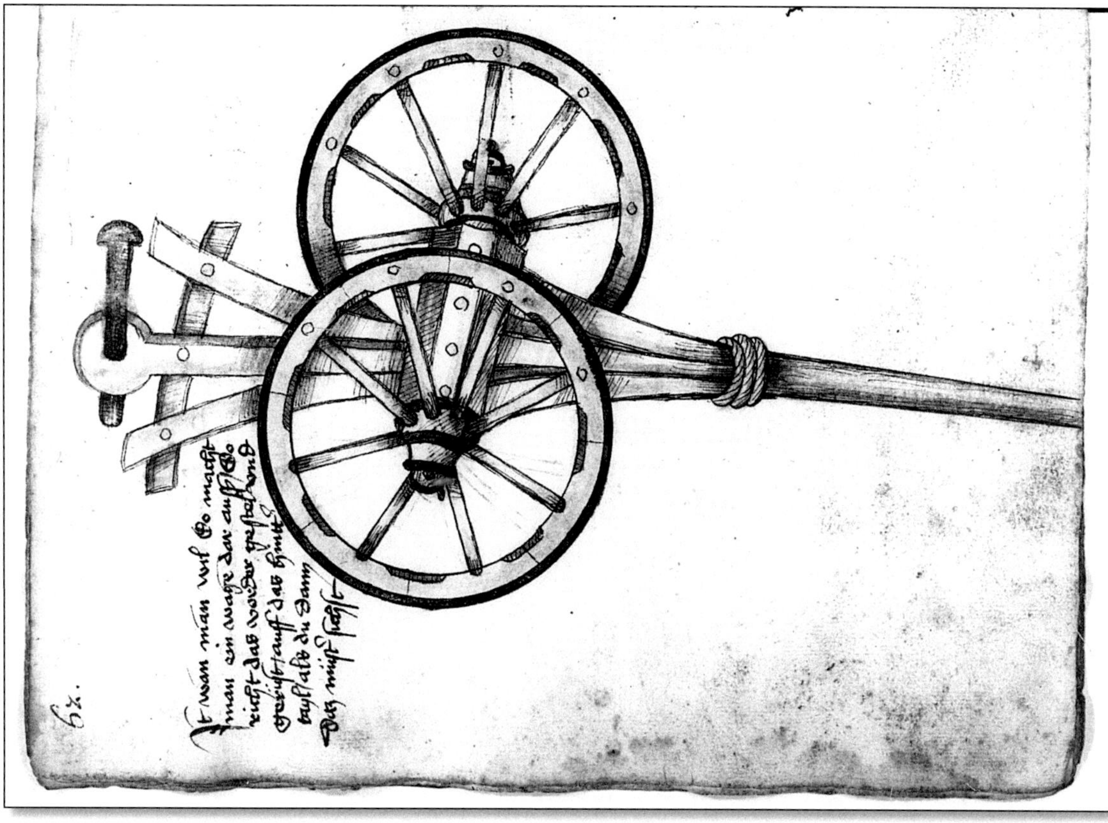

Above: The humble limber was thought worthy of inclusion in a wonderful but anonymous collection of images originating in Bavaria and dated to the end of the fifteenth century. It is of a much better design than the other examples, with the pintle well behind the wheels. The reinforced V shape was probably meant to stiffen that section as it would be subject to strong twisting forces if the gun it was towing was moving on uneven ground. (*Feuerwerk- und Büchsenmeisterbuch* – Shelfmark, Fol. Cgm 356 p. 65. Courtesy of the Bavarian State Library. Picture rotated sideways here, for clarity)

Below: A large gun with limber ready to roll, perhaps too fast. The carriage and limber are swathed in chains. Those around the barrel were perhaps there to stop it rocking on its trunnions when moving over rough ground. The chains connecting the limber and trail would have acted as a form of breakaway chain. Such precautions must have been necessary on occasion, another matter of concern for the gunner and his overworked commander. Note the barley twist and faceted barrel, together with the renaissance decoration just behind the neck of the muzzle. The gun was a contemporary work of art as well as a technological marvel. (Walther Litzelmann: *Vortrab zu der Arckalay – Büchsenmeisterbuch* – Shelfmark, Fol. . Cgm 909 p. 23. Courtesy of the Bavarian State Library)

One vehicle referred to in the context of the invasion of Italy in 1494, and used by the Florentine forces in 1498, was a *carrette alla franzese*. These seem to have been two-wheeled light caissons loaded with *pale de fierro* (iron shot) and *sache de farina* which were undoubtedly ready-made cartridges. They were deemed very useful, especially for the movement of the light artillery which can be seen preceding them. (*Fasciculus temporum*, Italy, Naples, *c.* 1498 MS M. 801 fol. 109v. The Morgan Library. Author's drawing)

Many barrels were evidently stored without carriages or limbers or 'fore carriages'. The record of ordnance held at the Tower of London in 1620 shows that only three of the 235 guns had both carriages and limbers and a little over 40 percent had carriages. Perhaps it was assumed that limbers could be constructed quite quickly.

Table 16. W.O. 55/1679, List of Ordnance Stored in the Tower in 1620

Name	Bronze	Iron	Total	Carriages	Limbers
Cannon	20	nil	20	20	3
Demi-cannon	16	nil	16	6	nil
Cannon Perrier	7	nil	7	4	nil
Culverin	21	10	31	4	nil
Demi-culverin	33	34	67	23	nil
Saker	17	37	54	13	nil
Minion	13	1	14	11	nil
Falcon	9	2	11	2	nil
Falconet	15	nil	15	13	nil
Totals	151	84	235	96	3

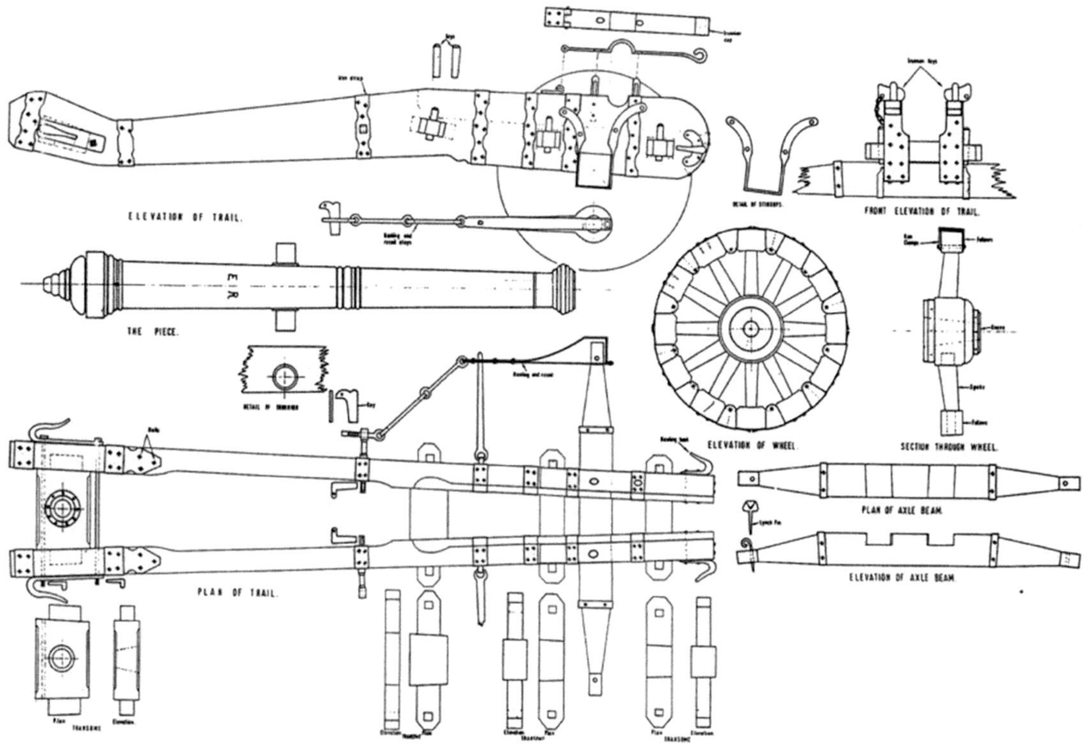

A scale drawing showing the main component parts of a wooden field carriage for an Elizabethan culverin *c.* 1565, designed by Austin Carpenter, an expert in the refurbishment and manufacture of historic gun carriages. This drawing gives an excellent impression of the size and complexity of an Elizabethan gun carriage. The four transoms are easily identifiable and the splay of the two cheeks is apparent. The ironwork protected the wooden framework and clenched it together, in order to withstand the stresses caused by the gun's recoil transmitted through the trunnions. Designed to cope with the rigours of the road and the recoil of the barrel, the carriage had to be weighty and well reinforced with ironwork. (Austin C. Carpenter, *Cannon: the Conservation, Reconstruction and Presentation of Historic Artillery*. Courtesy of Halsgrove Press)

Naval Gun Carriages

The late medieval naval gun mountings seem to have been simple wooden beds with very little means of elevation for the larger pieces. The predominance of breech-loading meant that guns did not have to be withdrawn to be reloaded and could be secured to the gunwale. The barrel did not project far from the hull and guns themselves were predominantly relatively small. The transition from clinker to carvel in warship construction, which occurred in Northern Europe around 1500, meant that gun ports could be cut into the hull without compromising the integrity of the structure. Heavier guns could now be mounted lower down in the vessel thereby not reducing the stability of the ship. At the same time heavy large calibre cast bronze guns were beginning to dominate artillery development; they could also now be mounted shipboard.

There is no available evidence to suggest when or where the innovative truck carriage appeared, let alone the humble genius who was responsible. The examples of this type of carriage were found in the *Mary Rose* and their workmanlike construction would suggest that such mountings were well

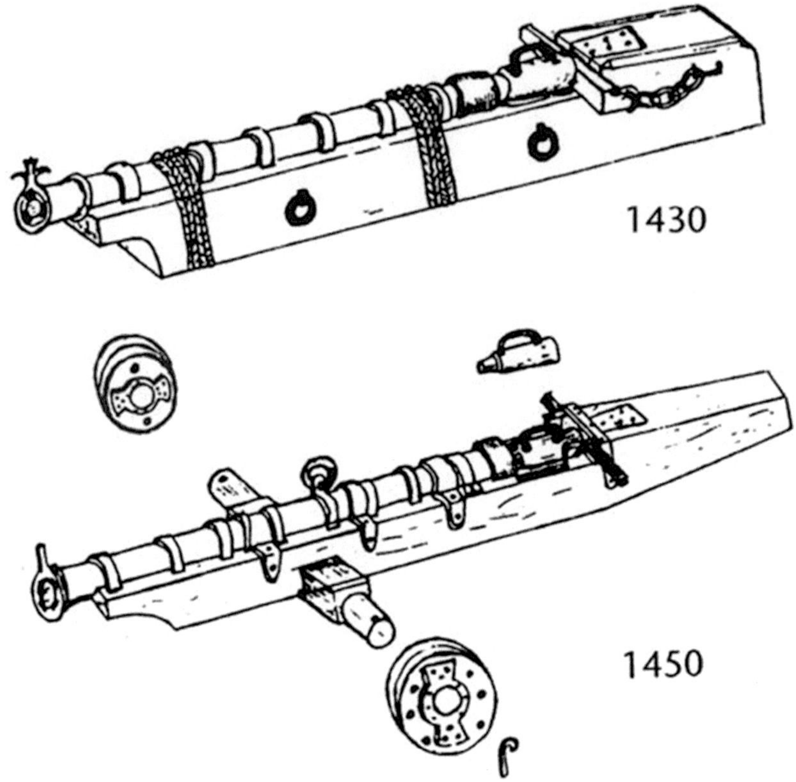

Wrought iron breech-loading pieces were mounted on large elm beds, provided with a pair of wheels and elevated using a tiller at the rear of the bed. This arrangement was entirely unsuited to bronze muzzle-loaders and a four-wheel truck carriage was invented for their use. (Austin C. Carpenter, *Cannon: the Conservation, Reconstruction and Presentation of Historic Artillery*, p. 24. Courtesy of Halsgrove Press, 1993)

known, and their handling understood well before the 1540s. Colin Martin provides an excellent description of an English truck carriage pointing to its design advantages.

> Two short wooden cheek pieces were fastened to a base-board to which were fixed axletrees fore and aft, upon each of which was set a pair of small wooden wheels or trucks. The front trucks were rather larger than the rear ones, partly to accommodate deck camber of approximately three to five degrees and partly because they took the main weight of the whole assembly. The gun sat on its trunnions close to the fore part of the carriage so that its centre of gravity lay just to the rear of the front trucks. Because of the small wheel diameter almost the whole of the gun's chase could be thrust through the gunport, while the rear of the carriage did not extend beyond the breech. A gun mounted on such a carriage would therefore protrude further through its port than one mounted on the two-wheeled type, and it had no awkward trail or wide wheels to obstruct its sides and rear. Adequate working space for the crew was therefore available beside and behind it.[10]

The Queen's ships at the time of the Armada were provided with all the equipment needed to run the guns out, pull them inboard and secure them when not in use. The gun carriages of their opponents were far less well designed for the purpose.

10 Martin, *The Equipment and fighting Potential of the Spanish Armada*, p.378

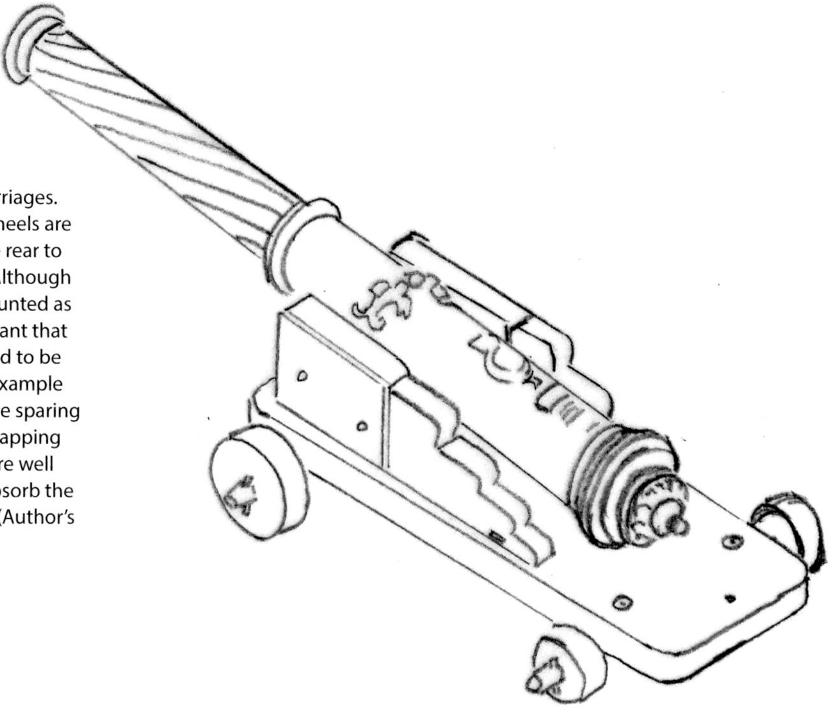

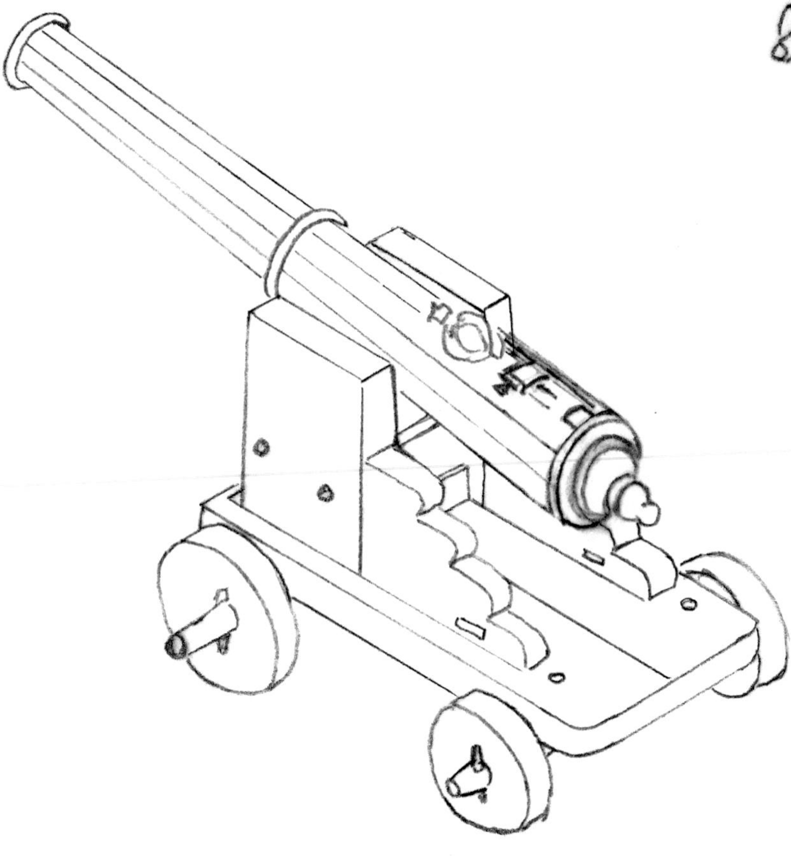

Two *Mary Rose*-style gun carriages. Note that the front pairs of wheels are of a larger diameter than the rear to suit the camber of the deck. Although ideally the gun should be mounted as low as possible, necessity meant that sometimes the mounting had to be higher as in the case of the example below. Clever carpentry and the sparing but intelligent use of iron strapping mean that the carriages were well built and strong enough to absorb the fearsome recoil of the pieces. (Author's drawings)

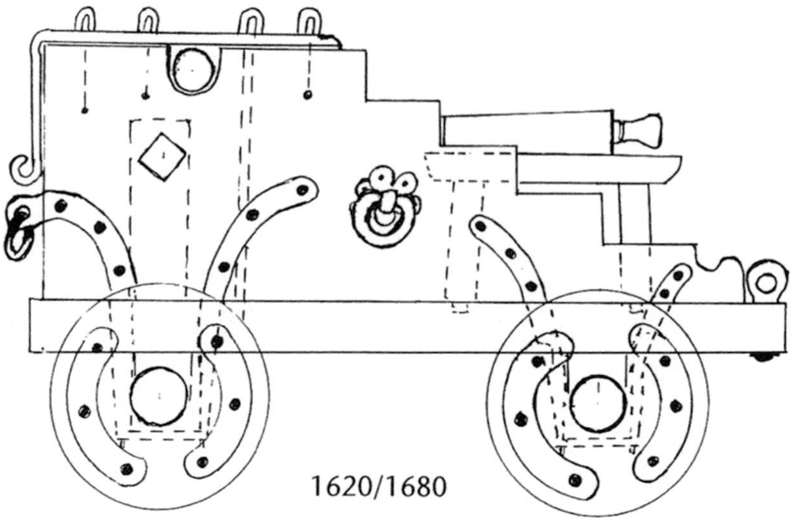

A sea or more likely garrison carriage from around 1620 (the wheels are all the same diameter). The trunnion is set fully into the cheeks, whereas in field carriages they are mounted slightly proud. The simplicity and economy of this design, in use in English ships in the first half of the sixteenth century at least, contrasts with the massive and unwieldy carriages that were used in Spanish Armada ships. (Austin C. Carpenter, *Cannon: The Conservation, Reconstruction and Presentation of Historic Artillery*, p. 25. Courtesy of Halsgrove Press, 1993)

Spanish, Portuguese and Italian carriages were very different, being little more than reduced versions of land carriages. Built lower, to permit their use on ship through gunports, the trails upon which they were mounted were shortened but still added significantly to the overall length. The large and ungainly wheels together with the axle made them very wide. A reconstruction of a two-wheeled carriage from *La Trinidad Valencera* has demonstrated what a difficult beast it must have been to manage. The gun would have been 3.69 metres long, but the overall length of gun and carriage was 5.8 metres and its width across the axle 1.65 metres. With the gun pushed as far forward as possible against the hull the carriage would still project 4.13 metres across the deck.

Operating such a gun in the congested space inside a warship would have been a nightmare. In 1988 comparative trials between a truck carriage and a Spanish sea carriage were made using the same barrel; both had to be withdrawn, loaded and run out. It was surprising that the Spanish carriage only took twice as long as the other. Why the truck carriage was not universally employed it is difficult to say. A Portuguese document of 1582, now held in the Bodleian Library, has illustrations of stepped truck carriages with both three and four wheels and the Dutch certainly employed truck carriages as well.[11] Perhaps ships more used to service in the Mediterranean and the Atlantic were tied to tactics where the provision of firepower was not perceived as being of paramount importance?

Moving the Guns

The siege guns of the artillery train would have been the heaviest objects ever regularly moved over land; a task made more difficult if not quite impossible

11 Richard Barker, 'Gun Carriages in a Notebook of Portuguese Field Artillery', *Journal of the Ordnance Society*, Vol. 14 (2002), pp.19–30.

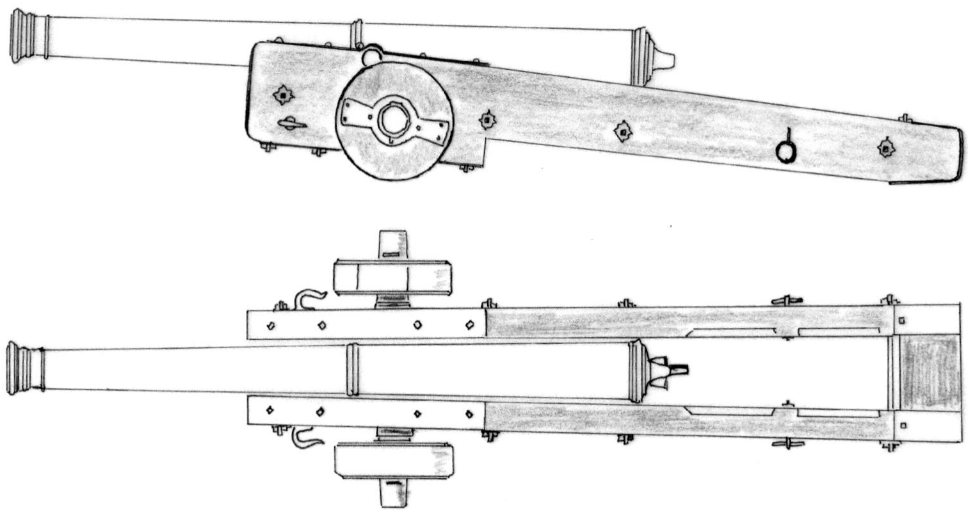

Reconstruction of a two-wheeled sea carriage of the type discovered on the wreck of *La Trinidad Valencera*. It mounted a long-barrelled piece of artillery similar to the Alberghetti gun mounted in St Mawes. The barrel was almost four metres long and the overall length of gun and carriage was almost six metres. It would fire a ball of about 6 kg or 13 lb from its 30 calibre barrel. Clearly it was not as light or manageable as the English truck carriage. This type of mounting reflected a different tactical approach to the presence and use of guns aboard ship. (Author's drawing)

by the invariably poor roads that criss-crossed the Continent. The preferred draught animal was the horse rather than the slower ox. Italians commented on the French use of horses rather than oxen when they witnessed their descent upon Italy in 1494.

Below: three tables showing the numbers of horses required to move guns, and the weight of metal allocated to each beast from 1544, 1568 and 1639.

Table 17. Details of the Ordnance Supplies for the Invasion of France, 1544

Type*	Horses	Weight of metal §	Ratio of horses to metal
Cannon	35	8,000	1:228
Demi-Cannon	25	6,000	1:240
Culverin	25	4,800	1:192
Demi-Culverin	16	2,700	1:168
Saker	10	1,800	1:180
Falcon	6	750	1:125

* Details of the Ordnance supplies for the invasion of France 1544, R O 13

§ William Bourne, *The Arte of Shooting in Great Ordnaunce* (London, 1587), pp. 65–71

Table 18. Drawn from French Sources from 1568

Type*	Horses	Weight of metal	Ratio of horses to metal
Cannon	23	8,000	1:348
Grand Culverin	17	6,500	1:382
Demi-Culverin	13	4,400	1:338
Culverin Moyenne	9	2,200	1:244

* James B. Wood, *The King's Army* (Cambridge: Cambridge University Press, 1996), p. 157, drawn from French sources from 1568.

Table 19. Du Praissac, *The Art of Warre*, 1639.

Type*	Horses	Weight of metal	Ratio of horses to metal
Cannon	20	5,600	1:280
Grand Culverin	17	4,000	1:235
Demi-Culverin	13	2,500	1:192
Minion	9	1,350	1:150
* Sieur. Du Praissac, *The Art of Warre, or Militarie Discourses of Leavying, Marching, Encamping* (London, 1639), pp. 107–121			

The figures above covering the period 1544 to 1639 show that while the weight of artillery pieces had decreased the horse was still required to draw a fearsome load. The numbers do not include those extra horses needed in case of sickness or lameness and they would all require to be fed and watered every day and well cared for. There is a distinction between siege and field weapons where there was a clear attempt to provide more horses for the field guns for the weight of metal, presumably in order to better enable them to keep up with the field army. The tables above discount the weight of the carriage itself and the author assumes that the carriage added the same amount proportionately, irrespective of the weight of the barrel.

Horses were preferred for draught purposes but oxen and even men were used. Smith made the following calculation:

> It is to be noted that 3 yoake of oxen is thought to draw as much as three horses, and that 3 yoake of oxen is sufficient to draw a Saker of 1400 weight … If you devide the weight of the whole Cannon being 8000 pound weight by 34, the oxen appointed to draw the same, the quotient is 235 pound 5/17: so much did every oxe draw.[12]

Lucar, when discussing the weight that different beasts of burden could manage, considers that a man may draw 100 lb, a horse 143–150 lb and an ox 176–210 lb.[13] So much would still depend on the suitability of the horses for draught use, the weather and state of the roads, the distance to be covered, the availability of fodder and especially water as well as the enthusiasm and competence of the carters and their captains.

The numbers of draught animals required in bad weather were even greater. While 16 horses may suffice to draw a gun in good weather a further eight would be needed in bad. Oxen may be substituted if necessary, but the 16 horses would need to be replaced by 17 yokes of oxen (34 beasts) and probably six yokes more in winter. If men were required then it was calculated that one man could draw about a hundredweight (112 lb) of metal or less. Manhandling guns was only possible over short distances and good ground unless the lightest field pieces were being used.[14]

12 Smith, 'All Manner of Peeces', pp.78–79.

13 Lucar, *Appendix*, p.38.

14 Nye, *Gunnery*, p.35.

Moving the guns was the least of the problems faced by any Master of Ordnance. The massive quantity of ammunition that had to be carried as well as all the ancillary stores were the main difficulties he faced. In 1544 at the siege of Boulogne two cannons required 70 horses to draw them, but a further 132 horses to draw the 12 carriages containing 19,000 pounds of powder and a further 21 wagons for the 360 shot. Almost a century later to bring a cannon into action with sufficient ammunition for three days bombardment at 120 rounds per day would require 18 wagons and 92 horses. Each wagon drawn by four horses carried 1,500 lb of powder or shot.

In the autumn of 1568, during the Wars of Religion the Royal French army mobilised eight cannons, six great culverins and six bastard culverins in its siege train. It was provided with a substantial provision of 91 *miliers* of powder and 5,000 shot. The requirement for drawing the guns was some 348 horses, the 192 wagons carrying powder and shot required a further 782 animals. The tools, forges, extra wheels, limbers etc. required a further 41 wagons and the pontoon bridge required 30 more wagons, while the baggage of the Grand Master demanded 20 more. The total requirement was for 283 wagons and 1,550 horses, as well as 2,600 men. Averaged out, each gun therefore required 14 wagons, 75 horses and 130 men to serve its needs.[15] At the siege of La Rochelle in 1573 an even greater number of pieces, some 42, were brought to bear on that unhappy city.

It was calculated that a single wagon drawn by four horses carrying between 1,000 and 1,200 pounds, would take up 50 feet of road with the wagon accounting for 12 feet in total.[16] The guns, carriages and men supporting the enterprise in 1568 would have filled almost three miles of road, a good half day's journey from one end to the other. If the wagon did not carry powder and shot it could carry, for example, 14 hurdles each 10 foot long and six foot broad, or 12 dozen 18-feet pikes or 300 assorted shovels, pick axes, or spades.

The civil war in France exhausted the vast resources available to the French monarchy, as the conflict spread further from the principal sources of ordnance, powder and personnel. Water transport was more of an option in France than it would have been in Britain but even that was not enough to sustain a large train of artillery over long distances for a prolonged siege. The failure at the end of campaigns to restore guns to their arsenals for care and maintenance and the lack of money and skilled personnel to refurbish them, led to the collapse of the French Royal train of artillery by the end of the Wars of Religion. The bronze barrel was only part of a much greater project that demanded administrative and technical competence and tremendous wealth:

15 James B. Wood, *The King's Army* (Cambridge: Cambridge University Press, 1996), p.160.

16 Du Praissac, *The Art of Warre*, pp.108–109.

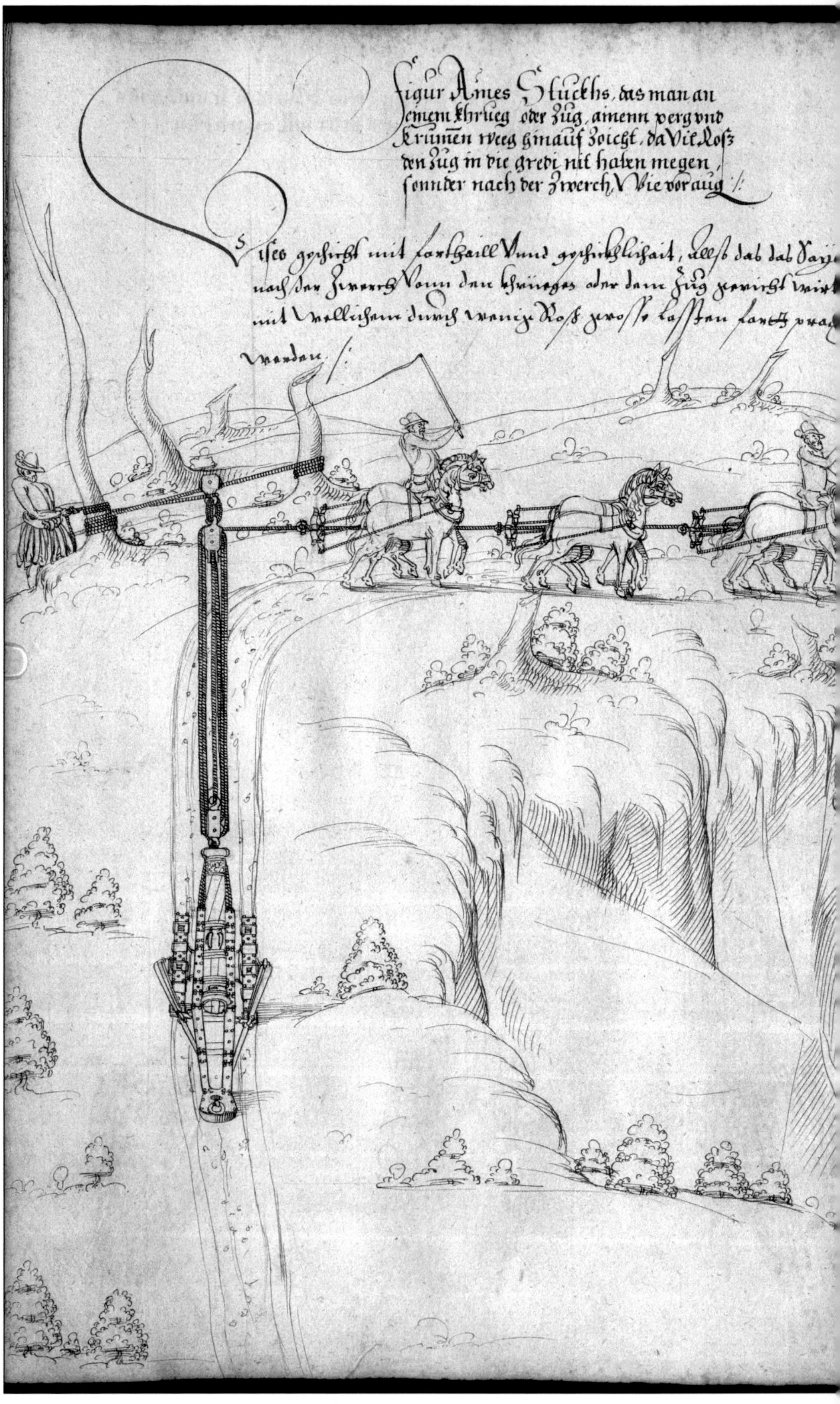
Figur Aines Stuckhs, das man an
einem khrieg oder Zug, ainenn berg vnd
krümen weeg hinauf zoicht, da Vil Rosz
den Zug in die gredi nit haben megen,
sonnder nach der Zwerch, Wie voraug ./.

Facing page: Dragging a heavy gun up a hill with the aid of some convenient trees, lots of horses and some clever rigging. What is missing are the heroic but poorly paid fellows who would be standing at the ready to place chocks behind the wheels, either when the horses decided not to play, or the ropes broke. It is one of the few drawings that show something like the number of horses that were needed to draw these heavy guns. (Walther Litzelmann: *Vortrab zu der Arckalay – Büchsenmeisterbuch* – Shelfmark, Fol. Cgm 909 p. 336. Courtesy of the Bavarian State Library)

Left: What goes up must come down! Preferably in as controlled a fashion as possible. Managing and maintaining the sheer quantity of rope, as well as the blocks and pulleys, land anchors, levers and chocks that were also needed must have been a major effort and expense. It is easy to ignore the logistical needs of artillery when the 'sound and fury' of the gun in action dominate the popular imagination. (Walther Litzelmann: *Vortrab zu der Arckalay – Büchsenmeisterbuch* – Shelfmark, Fol. Cgm 909 p. 337. Courtesy of the Bavarian State Library)

> **My good man, I am beside myself that if only I had six cannons and enough ammunition to fire two thousand rounds, this town was ours' (Duke de Guise at the siege of Orleans)**[17]

> **The artillery be a great, cumbersome, and heavy contrivance, and if anything is lacking, the whole thing is useless. Indeed, any service and utility one claims to derive from an army resides in its artillery, without which all its expense and force are useless.**[18] **(Grand Master of Artillery Sieur de Biron to Henry III of France 1577).**

As important as the equipment was, it required an expensive and sophisticated infrastructure to maintain and prepare it in peacetime and then deploy it in war. France could boast of the administrative competence of the Bureau Brothers whose work at least in part explained the expulsion of English force from France after 1453. They were succeeded as masters of the artillery by Jean Cholet (1477) and then Jacques Galiot de Genouillac (1479). Under Charles VIII five brigades of artillery would be formed and it was from this that the impressive artillery train of 1494 that so amazed Europe emerged. In England the Master of Ordnance had overall responsibility for the artillery on campaign and was supported by a lieutenant who was responsible for ensuring the serviceability of equipment and the supervision of gunners. The master gunner had the responsibility of supervising and appointing gunners, ensuring their competence and supervising their training. The titles of the master gunner included such variations as Master Gunner of the Ordnance, Master Gunner of the Tower and even Master Gunner of England.

In an army a master of ordnance would be appointed to manage all aspects relating to artillery and engineering. The Ordnance was a self-governing institution somewhat separate, even aloof, from the remainder of the army. It had its own rank structure, administration and disciplinary procedures a situation that continued in Britain in some respects for centuries. An artillery train consisted of five main and distinct parts.

Firstly, the officers and administrators responsible for the management of different aspects of the train together with their staff and servants. The second were actual gunners, some were part of the permanent establishment, others taken on as supernumeraries and then of course there were mercenaries. The next group were the craftsmen, wheelwrights, carpenters, smiths, powderers, foundrymen and the like. The carters responsible for managing the multitude of wagons in themselves formed a small army and finally the large number of pioneers and their officers. The pioneers were usually the worst paid but one of the most essential elements as they would have made and mended the roads and bridges that took the siege train to its objective. At the siege they would build the mounts and defensive fortifications that were essential to maintain a successful bombardment. Finally, they would assist in manning

17 Wood, *The King's Army*, p.153.
18 Wood, *The King's Army*, p.155.

the guns, carrying powder and shot and moving the guns back into position after each shot.

The phenomenal effort required to maintain, transport and support even the smallest artillery train was a task that could exhaust the coffers of even the wealthiest. It was generally considered that the artillery would absorb a third of the total cost of any army. When referring back to the French Invasion of 1494 and afterwards, although the French did succeed in bringing their guns to Italy, very few Italian states could manage the resources needed to match them. The administrative infrastructure, the skilled craftsmen and officers and number of draught animals needed to support them were simply lacking.[19] The French artillery which had provided Europe with the model of administrative and operational efficiency was itself exhausted and diminished by civil war. Its magazines were ransacked and empty, her guns abandoned and left in a state of decay. For artillery to be of use it needed administrative and technical competence in a skilled workforce and above all money.

19 The author is indebted to Dr Fabrizio Ansani for generously sharing his article on this subject: '"This French artillery is very good and very effective". Hypotheses on the Diffusion of a New Military Technology in Renaissance Italy', *The Journal of Military History*, 83 (April 2019), pp.347–378.

7

Gunpowder

> A great and incomparable speculation is whether the discovery of compounding the powder used for guns came to its first inventor from the demons or by chance.[1]

Gunpowder is a devilish concoction that has caused untold misery. Sulphur that emanated from the roasting bowels of the earth was combined with the efflorescence of faecal decay and mixed with charred wood. This demonic black mixture was clearly brought from hell to drive mankind to self-slaughter. It was also a remarkable discovery, almost certainly coming from China. It was difficult to categorise according to the doctrine of the four elementary qualities, that were used at the time to explain the natural world. To taste it was *hot* and *dry* but as a propellant it was considered *hot* and *moist airy*, its colour and melting point and weight suggested that it was primarily a thing of a *watery nature*; as it crushed and breaks, it could also be *earthy in nature*: 'Thus it seems that every one of the elemental qualities is predominant in it'.[2] It was if not a blasphemous miracle at least a hellish mystery.

Even so, after prolonged criticism of the power and malice of gunpowder, the natural philosophers of the Renaissance had discovered what lay behind the phenomenal power of this mixture of elements. It was explained that:

> One part of fire occupies the space of ten parts of air, one of air that of ten of water, and one of water ten of earth. Therefore, since gunpowder is a corporeal and earthly thing composed of four elemental powers … it makes a great multiplication of air and fire, producing with the moisture and subtle earthiness a thick burning vapour … it produces the effects we see which seem to be miracles to one who does not understand the cause.[3]

Thus, a small volume of solid was transformed into a very large volume of very hot gas. Bert Hall cites the impressive figure that one gram of gunpowder

1 Biringuccio, *Pirotechnia*, p.409.
2 Biringuccio, *Pirotechnia*, p.404.
3 Biringuccio, *Pirotechnia*, p.411.

A cheerful saltpetreman surrounded by miniature examples of his utensils. The scales indicate the importance of ensuring the correct proportions of the three materials when manufacturing gunpowder. His relaxed pose is not one normally associated with those engaged in this business! (Walther Litzelmann: *Vortrab zu der Arckalay – Büchsenmeisterbuch –* Shelfmark, Fol. Cgm 909 p. 14. Courtesy of the Bavarian State Library)

will produce between 274 and 360 cubic centimetres of gas at a temperature between 2,100 and 2,700°C.[4]

Gunpowder was made of three ingredients. Saltpetre that provided the oxygen for burning, charcoal that provided the fuel and sulphur that assisted the process of ignition. Sulphur was found occurring naturally as a result of volcanic activity and charcoal was already made in very large quantities. Charcoal production was designed to produce specific forms of fuel. Straight dense hardwoods produced fuel for smelting metals, while twisted and gnarled wood and tree stumps produced fuel for the forge. Alder, poplar and willow produced less dense charcoals which were more suited to gunpowder manufacture.

> The best quality charcoal was obtained from 'straight, knot-free, branches about the thickness and length of a man's arm. Harvested from soft-wood trees before the sap rises, cleaned of any adhering dirt and stripped of its bark the wood was aged for a year or more.[5]

4 Bert S. Hall, *Weapons and Warfare in Renaissance Europe* (Baltimore: Johns Hopkins University Press, 1997), p.67.

5 J. Tomkinson, 'Gunpowder and the Introduction of Industrial Charcoal', *Journal of the Ordnance Society*, Vol. 28 (2021), p.98.

This produced a water content from 40–50 percent. Biringuccio preferred willow charcoal for artillery gunpowder and hazelnut twig charcoal for the small arms. He also mentioned the use of grapevine twigs, laurel, pine cone shells, osier twigs, elder and reeds for its manufacture: 'In short, all charcoals that are made of soft wood with much pith and woods that are thin, young and without hard knots are good for this process'.[6] Sulphur was essential to gunpowder as it lowered the ignition temperature of gunpowder (280–400°C), while charcoal provided the fuel and saltpetre the oxygen. Saltpetre was the *master* or key to the mixture and either occurred naturally as a process of decomposition of organic material and harvested from midden, byre and stable, or was manufactured in what was called a plantation.

The Manufacture of Saltpetre[7]

Biringuccio was refreshingly honest when he considered the origin of saltpetre: 'It is my belief that it is engendered in these soils from an airy moisture that is drunk and absorbed by the earthy dryness',[8] but 'considering its effects, I do not know how much to decide and say exactly what the nature of saltpetre is'.[9] Saltpetre does occur naturally in efflorescence or 'brushes' of crystals. Ulrich Bretscher managed to source 95 percent pure saltpetre from 'brushes' found in an old cellar which had backed on to a stable.[10] Geoff Smith has powerfully argued for a very pure form of saltpetre from early on in its manufacture.[11] In pursuit of this domestic or rather privy industry, saltpetremen were given carte blanche throughout Europe to collect saltpetre, mining earth from dovecotes, stables, latrines and cellars, reducing the material in copper boilers. Biringuccio recommended pig dung as an excellent source of material but 'manurial soil' of any sort would do.

Biringuccio describes in detail the method of leaching out the saltpetre. First the saltpetreman had to make a pile of the earth containing saltpetre and beside it a pile, slightly less than half that size, made up two parts of quicklime and three parts of oak ashes. The pile should be mixed thoroughly and loaded into casks or shovelled into the casks in alternating layers, space should be left above the material for water (see B on illustration, facing page). Water should now be poured into the casks through the open top and drained through holes drilled in the bottom which had cloth or straw covering them to strain the liquid (D). This liquid contained all the 'substance and virtue of saltpetre' which would then be collected in another tub where it should be tasted for saltiness. If the water was 'sharp and strongly salt', all well and good.

6 Biringuccio, *Pirotechnia*, pp.413–414.

7 For a very helpful detailed account of the manufacture and use of gunpowder, see articles by Geoff Smith.

8 Biringuccio, *Pirotechnia*, p.404.

9 Biringuccio, *Pirotechnia*, p.404.

10 Note to Fig. 2, <https://web. archive. org/web/20110520015329/http://musketeer. ch/blackpowder/saltpeter.html>, accessed 20 December 2021.

11 Geoff Smith, 'Saltpetre in Medieval Gunpowder; Calcium or Potassium', *Journal of the Arms and Armour Society*, Vol. XIX, No. 3., p.121.

A—CALDRON. B—LARGE VAT INTO WHICH SAND IS THROWN. C—PLUG. D—TUB. E—VAT CONTAINING THE RODS.

The manufacture of saltpetre, with all the various stages in the process described by Biringuccio and well illustrated by Agricola. The equipment was in itself neither complex nor especially expensive, what was essential was the skill of the craftsmen. The text below identifies the equipment being used. From Georgius Agricola, *De Re Metallica* (1556).

If need be, water would be poured in the tub back through the casks until it was salty enough. Then second and third washings should be carried out, the results of which were to be kept separate and used as a primer in subsequent productions. The liquid should now be placed in large copper kettles, the sort used for dyeing and filled two thirds full. The volume was reduced by a third by slow boiling (A). The liquid should then be covered and left to settle, and the sediment removed.

The material then collected was dissolved in water and boiled. The salts particularly sodium chloride remained in solution when the liquid cooled; the potassium saltpetre precipitated out. After this was repeated most of the chlorides would have been separated. The nitrates that remained included not only those of potassium but also calcium, sodium and magnesium. These had markedly hygroscopic properties, absorbing water from the atmosphere, properties which were highly undesirable. The clear liquid should now be boiled again to which would be added a strong lye of three quarters soda, or ashes of oak or olive wood, with a quarter of lime. In addition, for every 100 pounds of water should be added four pounds of alum. The lye should be added as the saltpetre-rich water boiled and seethed. This process removed most of the remaining calcium and magnesium nitrates.

An excellent illustration of a late sixteenth century saltpetre plantation. Although open to the air, the piles of 'black' earth are carefully tended. The vegetation on the surface would protect the saltpetre being generated from leaching away, while maintaining the dampness necessary for its growth. In the building the process of separating out the saltpetre would take place. Note the water barrel that would collect fresh rainwater for the purpose. (From Lazarus Ercker's *Beschreibung all fürnehmsten mineralischen Ertzt*, Frankfort, 1598. Fig. 22 in Guttmann's *Monumenta*.

The colour of the water should turn bluish and over time the water would 'congeal', at which point it should be removed and left to cool in wooden or pottery containers where sticks had been placed for the saltpetre to 'congeal' (crystallise) on (E). After a few days the water should be poured or drained off but saved for recycling into the process. Biringuccio attributes the 'clarity and beauty' of the product to the 'magistry of lye' (lime, ashes and alum) added during the boiling process. When the saltpetre had been chiselled from the sides of the vessel it would be put on boards to drain and dry. To improve the quality if wished, a longer process involving sand filtering and reboiling with a judicious addition of aqua vitae could be employed: 'In this way you will have your saltpetre very white and beautiful and much better than in the first cooking'.[12]

The first major improvement in the history of gunpowder manufacture in the West was the development of saltpetre plantations. This is where the 'natural' process of decomposition was managed on an industrial scale. The

12 Biringuccio, *Pirotechnia*, p.408.

increasing availability of saltpetre and its halving in cost in France between 1380 and 1420 can only be attributed to a great increase in domestic manufacture. Biringuccio referred to using the saltpetre dissolved in water as a starter for a pile of the 'soils already used in the making of saltpetre', that is the refuse from the first part of the saltpetre extraction process.[13] The pile should be covered to avoid rainwater washing through it, and after five or six years the earth so produced could be used again for extraction of the saltpetre. There is a detailed contemporary account of saltpetre manufacture from 1561, although it seems probable that that earlier manufacture followed a similar pattern, the following information is contained in a manuscript by Gerard Honrick.

The ingredients were: black earth, that is composted waste including faecal material, urine, dung, quicklime and oyster shells. The urine provided the essential ammonia; the black earth, dung and oyster shells provided a growing medium for the bacteria and the quicklime offset any acidity in the mixture. The manufacturers were remarkably precise as to the source of their raw materials and for good reason. The urine should be 'namely of those persons which drink either wine or strong beers', the ammonium levels of urine increase three-fold when the body metabolises alcohol while sodium excretion decline. The urine of a wine-drinking prelate was thought to be especially efficacious! The dung was 'specially of those horses which be fed with oats', as oats contain more potassium and less sodium than meadow or clover hay. The resultant heap should be placed in a dry, brick-lined building and turned fortnightly with the walls and floors swept of their saltpetre, which 'will hang like snow upon them'. These conditions occurred naturally on the walls of stables and privies, although less prolifically.

The introduction of saltpetre plantations led to a rapid decline in the cost of powder. Frankfurt, which was quite clearly a centre for production saw a reduction from 41 florins per hundredweight in 1381, to 16 in 1410, and nine to 10 florins in 1439–1440. In England the price reached eight to 11 pence per pound by the mid fifteenth century, a price that remained constant until the end of Elizabeth's reign despite debasement and inflation. In the Tower in 1559, serpentine powder was purchased at eight pence a pound and corned powder at 10. The consequence of the major reduction in the price led to the use of much larger weapons with heavier charges.

Powder Manufacture

The simplest method of manufacture was to mix the dry ingredients in the appropriate proportions. This process of incorporation could be done by hand with a pestle in a mortar, using materials unlikely to produce a spark. This would only suit the manufacture of small quantities as it is very labour intensive and time consuming. Grinding with stone mills by hand as with a flour mill, was far too dangerous. The best method was to use a water or man-powered stamp mill, using stamps and mortars made from oak, with the mortars fitted with a copper bottom.

13 Biringuccio, *Pirotechnia*, p.409.

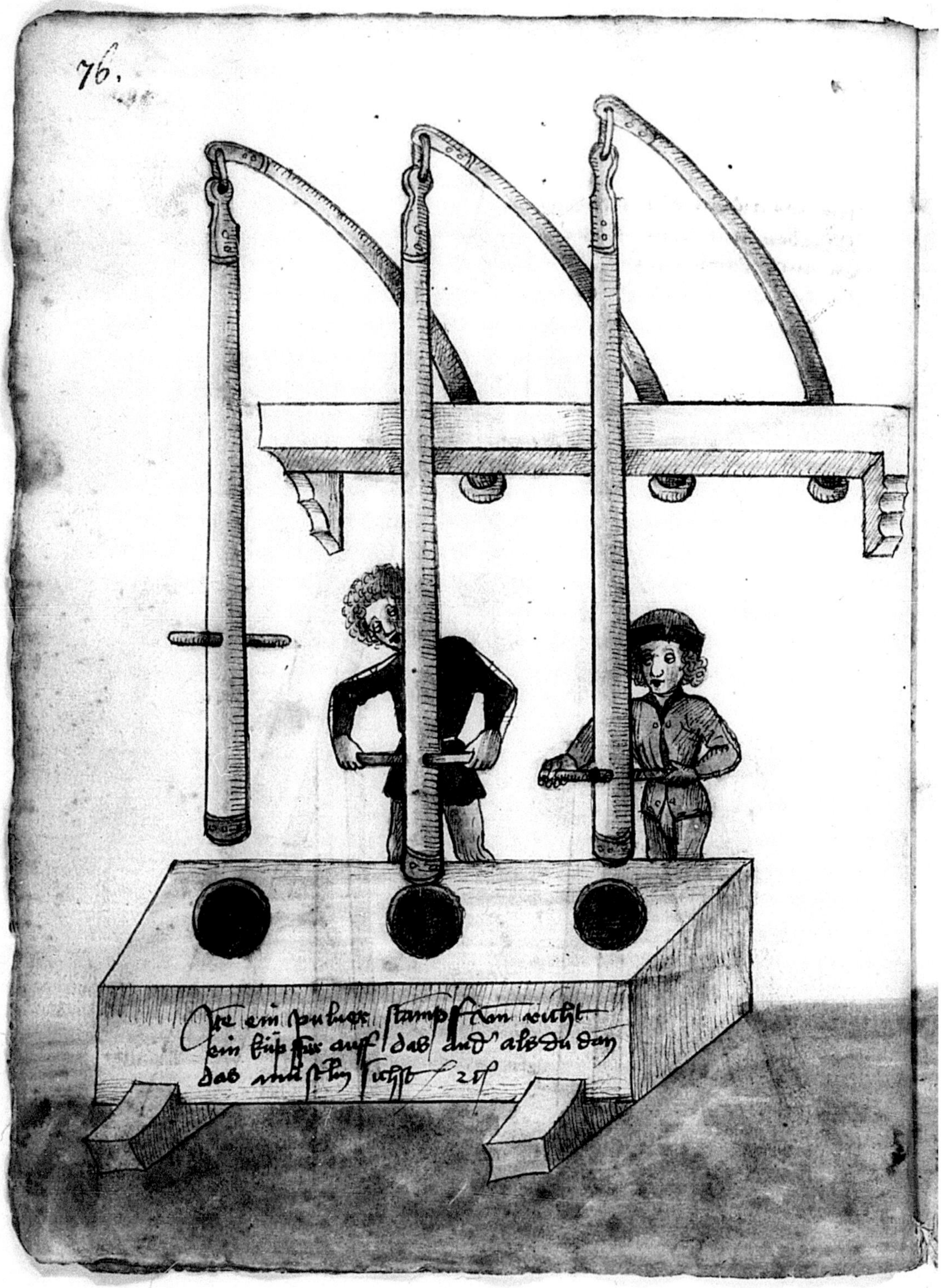

Labourers incorporating the three ingredients of gunpowder. The clever arrangement of springy laths would greatly speed their efforts. The materials in the equipment used were designed to avoid possible sources of friction. Moistening the dry ingredients would further help the manufacturing process as well as increasing safety. (*Feuerwerk- und Büchsenmeisterbuch* – Shelfmark, Fol. Cgm 356 p. 79. End of the fifteenth century. Courtesy of the Bavarian State Library)

Biringuccio suggested making a wet mixture starting with the saltpetre solution to which would be added the appropriate proportions of saltpetre, sulphur and charcoal. Stir with a stick 'as is done with cheese' and then allow to dry and the sift it, then: 'moistening it again with ordinary water or with some vinegar, force it through a riddle or sieve to granulate it'.[14] The moistened gunpowder could then be made into cakes referred to in German as *knollen* or dumplings. These could then be stored until ready for use, when they could be broken into crumbs using either a pestle and mortar or a crushing machine. Corned powder was produced in a number of grades dependent upon the size of holes in the parchment through which it was sieved. Biringuccio describes the process of powder manufacture and corning in some detail. His use of quicksilver (mercury) and distilled orange water added value and mystery to the final product but made no difference to its performance but were slavishly copied by others. Norton provided an excellent account of the process:

> First, put your Composition into a brasen [brass or bronze] Morter, or of wood with a brasse bottome. And take also a brasse pestle, and beat it so well together for the space of sixe, seaven, or eight houres, that in cutting the same with a knife, there can none of the Materialls be discerned a part, but all well incorporated into one body, alwayes moystning the same with a little cleare water, strong Vinegar, or Aqua-vitae, so much that the coales dust not, and so little, that the composition in working become not paste. But if you would have your powder very subtill, moysten the same with the distilled water of the outer rynes [rinds] of Oranges, and let your sulpher be cleare Sulpher vine which in the melting had ¼ of Quicksilver put therein, and incorporated, finely beaten, and searsed with coales, quenched with cleare water, wherein quickely he hath beene slaked, and either let it stand to cleare, or else be filtered: and this moystning will make the cornes thereof become hard, and not yeeld to moysture sodainly.[15]

The next step in the history of gunpowder was to produce grains of a consistent size, a process which is difficult to date but was certainly widespread by 1550. This was done as Norton described through pressing the wetted gunpowder through a sheet of hide or parchment with uniformly sized holes using a pair of rollers:

> After the composition is thus well beaten and wrought together (the more the better) take a Syve [sieve] with a Velome [vellum], or Parchment, or Leather bottome, made full of holes, of the bignesse you desire your cornes, and put the Receipt or composition therein, with 2 or 3 short Rowlers, a little moystening it, that the pousier or dust thereof flye not away. And sifting them vpon a long Rowler, ouer a halfe Tub, so will the short Rowlers driving the composition through the holes, make the cornes come out round yet of severall syzes, which to bring to have all the cornes of one bignesse, you must passe through severall

14 Biringuccio, *Pirotechnia*, p.415.

15 Norton, *The Gunner*, pp.144–145.

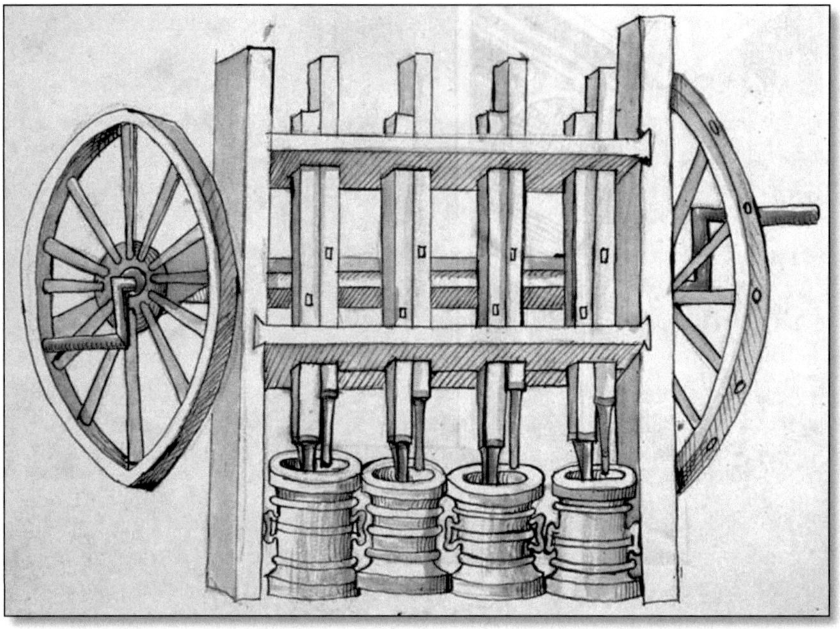

Below: The individual on the left is incorporating gunpowder in a large brass mortar. The workman on the right is corning gunpowder using the short rollers recommended by Norton to press the moist powder through the vellum sieve into a half-tub. A collection of weights and a balance are behind him, together with another sieve. It is clearly well before the idea of wearing personal protective equipment when engaged in hazardous work! (Drawn from the *Pyrotechnie* of Hanzelet *c.* 1630; fig. 47 in Guttmann's *Monumenta*)

Above: Although still dependent upon manpower to drive it, this stamping mill for incorporating the raw ingredients of gunpowder would have been a handy addition to the manufacturing process. The stamps are clearly of some metal, probably brass or bronze and the chambers containing the mixture are also of cast construction, again 'brasen'. The driving wheels turn an axle which would catch the cams on the stamps, ensuring that they worked alternately, stamping and pushing the contents to the other side of the chamber. This would have been much more efficient than a single stamp. Even so it would be hard and monotonous work but readily converted to water power. (Heidelberg Mönch 1496 cpg 126_0019. Courtesy of the University of Heidelberg Library)

> siues, as they vse to syze Hayleshott [small shot for anti-personnel use]: So shall your Cornes of every syze bee equall.[16]

Although the process of corning gunpowder is well understood, its purpose is not. It has been argued that its role was to avoid the separation of the dry constituents of gunpowder when shaken during transport. This is not the case, as the specific gravity of the three constituents is almost the same. The real problem that had to be overcome was the hygroscopic property of saltpetre. In *knollen* and then corned form, the powder presented a much smaller surface area for water to be absorbed for its volume, thus ensuring a longer active life by keeping the gunner's powder dry. Powders of this type were originally referred to as being 'durable' which points to the original purpose of the process. John Smith pointed to the weakness of serpentine compared to corned powder, for: 'Serpentine is like dust and weake, and will not keepe at Sea but be moist'.[17]

The unforeseen consequence of corning was to produce a much more powerful propellant. Consistent size of grain also produced a smooth surface which encouraged a much more rapid combustion, causing much higher and more rapidly achieved peak pressures. In tests in Austria of musket loads of powder, 335–454 atmospheres of pressure were reached in two milliseconds. It was considered that corned powder was up to twice or even three times as powerful as ordinary mealed gunpowder, or *serpentine* as it came to be known.[18] Contemporary opinion certainly recognised the difference between serpentine and corned powder. Smith rather unhelpfully suggests two different assessments as to the relative power of the two propellants.

> Any peece of Ordinance first discharged with the whole weight of the bullet in Serpentine powder, & after discharged with ½ the weight of her bullet, in such corne powder as shall cause the peece to range the same ground.[19]

> You must note for a generall rule, that 2 partes of corne powder will doe as much as 3 partes of Serpentine powder.[20]

Corned powder could be used in artillery but only in reduced quantities or with a reduced saltpetre content. The 1560 recipe for cannon powder of Peter Whitehorne, who served in the army of Charles V used only 50 percent saltpetre. This was a recipe proportion also recommended by Biringuccio for heavy artillery in his *Pirotechnia* of 1540. Tartaglia in his description of the corning of powder gave three recipes; for cannon powder, handguns and priming powder, with 67, 71 and 75 percent of saltpetre respectively, with the amount of charcoal and sulphur making up the remainder, being equal. Henry VIII's artillery at the siege of Boulogne 1544–1545 rather conservatively only employed serpentine to ensure the safety of the guns and

16 Norton, *The Gunner*, pp.144–145.
17 Smith, *A Sea Grammar*, p.68.
18 Bourne considered that 2 lb of corned powder was the equivalent of 3 lb of serpentine.
19 Smith, 'All Manner of Peeces', pp.101–102.
20 Smith, 'All Manner of Peeces', p.38.

their crews, although by the 1580s corned powder seems to have been widely employed by England on land and sea. Another advantage of corned powder for artillery was that it could be readily poured into paper or linen cartridges.

Gunpowder could be kept in store for long periods although it was prone to damp and decay. If the gunpowder was spoiled it could be revived or remade, either by the addition of dried saltpetre or making a solution and sprinkling it over the poor gunpowder and then allowing it to dry and remixing. If on the other hand it was too far gone it would be layered in a tub over twigs and straw which were soaked in fresh water. After resting for 10 or 12 hours, the water could then be drained off as it would contain the dissolved saltpetre, which could then be recycled in the manufacturing process.

There was great enthusiasm for producing coloured gunpowder for *feu de joie* or signalling purposes. To this end it could be mixed with iron and copper filings, dry field fungus and wild flowers. To improve the pyrotechnic quality of gunpowder, sal ammoniac (ammonium chloride) and zinc sulphate could produce clouds of white smoke. Camphor and the sulphides of arsenic would provide a brilliant white light, again suitable for fireworks or to provide illumination. The addition of camphor to gunpowder and ammonium chloride and corning with brandy lowered the pre-ignition temperature and created a larger exothermic event.[21]

The Management of Powder

The quality of powder was one of the prime concerns of every gunner, and their manuals devoted much time to its manufacture, improvement and testing. This could be done by taste, colour and by examining the quality of the residue after a small quantity was ignited. Gunpowder should taste salty (saltpetre is after all a salt). Agricola, the author of key works on science and engineering of the period, declared that 'saltpetre is made from a dry, slightly fatty earth, which if it be retained for a while in the mouth, has an acrid and salty taste'.[22]

Norton provided a short and memorable guide to the qualities of good powder by sight, feeling and fire:

> The first, is by the sight; the second, by the feeling: and the third and surest is by firing it. By sight if it be not blacke and darke, but bright, and enclining to a tawny blewish colour. By feeling grasping it in the fist, if it runne through the fingers quickly, and avoydeth the handling, and cling not together. By fire, if fired it rise quick, and spread in a moment, smoake little, but riseth in a cleare flash, vnto which adde further, if it leave no faeces but carrie all away, the powder is good, or else the contrary is to be judged.[23]

21 Tessy S. Ritchie, Kathleen E. Riegner, Robert J. Seals, Clifford J. Rogers, and Dawn E. Riegner, *Evolution of Medieval Gunpowder: Thermodynamic and Combustion Analysis* (ACS Omega), p.7. <https://pubs. acs. org/doi/pdf/10. 1021/acsomega.1c03380>, accessed 28 September 2021.

22 Georgius Agricola, *De Re Metallica* (New York: Dover Publications Inc., 1950), p.561.

23 Norton, *The Gunner*, p.145.

Biringuccio recommended that if good quality powder was placed on a sheet of paper, it would not leave any stain and if crumbled between fingers, there should be no evidence of sulphur. Finally, if you took two three-fingers-worth pinches of powder and placed them two *ditta* (73 mm) apart, one when set on fire would ignite the other but not burn the paper beneath. It would also not leave any residue of saltpetre.

William Bourne explained the use of an early éprouvette or proofing device, which was 'an engine of bore, and burning it [to test] whiche force of Poudre doth blowe, or lift the lid of the bore highest and you shall knowe by howe muche, by the teeth of notches that do stay the lidde of the engine of bore'.[24] He later explained in tortuous detail how to make and use this helpful bit of kit:

> First make in mettall or yron a round boxe of an ynch and a halfe in breadth more or lesse at your discretion and of two ynches deepe more or lesse, at your discretion, and then let that be placed so, that it may stand vpright, and have a little tuch-hole at the lower part thereof, and then let the vppermost part at the mouth thereof, have in mettall or yron a lid or cover that may goe with a ioynt vpon the one side thereof, and the cover or lid to be of a reasonable weight, and the other side of the cover or lid right against the ioynt to have a square hole fitted of pupose, and then vppon that side that the ioynt of the cover or lid is of, there must bee raised a thing that must have of yron or other mettall a part of a circle, and the other ende to goe through the square hole in the lid cover, and the other ende to goe with a pinne or ioynt right over the ioynt of the cover or lid, and the sayd crooked thing or part of a circle, to have teeth or notches like unto a Sawe, and the teeth to stand vpwards, and then it is finished, and then whensoever that you lift to proove the strength of powder, and you having of divers sortes of powder then wey some small quantitie of the powder and then put that into the Boxe, and then let downe the cover or lidde, and then give fire vnto it at the touchhole, and then the powder will blowe vp the cover or lid, and then the teeth or notches being well filed or trimmed of purpose, will holde vp the lid at the highest, and yet not staying in the blowing of it vpwards, for that the teeth standeth to, to stay it that it shall not come downwards, and then trying or proving divers sorts of powder, you shall knowe which is the best or strongest powder, or weaker powder, by the blowing vp of the lid of cover, you putting in the powder by weight &c. And this is the forme of the engine or instrument.[25]

Other suggestions to prove powder included loading a pistol then shooting the ball into a bank of clay and measuring the depth of penetration. A small mortar shooting an iron or lead ball could be used to test the power of different powders by measuring the ranges achieved with identical charges. The purpose of all this effort was not to increase the range of the piece but to ensure the safety of the gun and the gunner. Over-charging a gun could be disastrous as the evidence from the remaining Tudor guns in Jersey demonstrates.

---

24 Bourne, *The Arte of Shooting*, p.28.

25 H. G. Muller, *A Brief History of Powder Testers*, <http://www. reenactor. ru/ARH/PDF/Muller_H. _G. pdf>, pp.208–209.

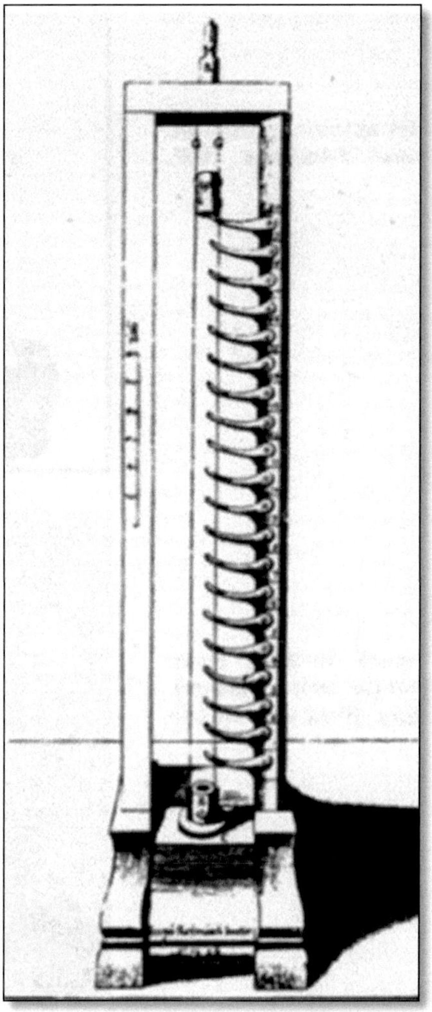

Bourne's éprouvette (above). It was number 54 in his *Rare Inventions and Strange Devises Very Necessary for all Generalles and Captaines*, published in 1587. Johann Furttenbach devised a more reliable éprouvette in 1627 (left). This consisted of a barrel that fired a weight, guided by a wire which passed through its centre. During his tests he discovered that cannon powder shot the weight only four inches, sporting gunpowder nine and very fine priming powder an impressive 12 inches. (Muller, *A Brief History*, p. 208)

Gunpowder was stored in and dispensed from wooden barrels. We are fortunate in that some of the barrels from the Armada wreck *La Trinidad Valencera* were recovered. The kegs were 61 cm tall and 35 cm wide containing 39 litres by volume. The weight of fine-grained powder at 1.14 g per cc, would mean that 46.05 kg (101.5 lb) of powder would be carried in this barrel. This equates almost exactly to the Castilian quintal of 100 lb. The English small measure of powder was a little barrel or firkin containing 100 lb of powder.[26] The cask was expected to weigh 12 lb. Twenty casks constituted a 'last' of 2,000 lb of powder.[27] To the total weight should be added the weight of the cask at 12 lb, making in total 2,240 lb or one ton. There is also a reference to a 'graund barrell', weighing 300 lb.

26 A firkin contained nine imperial gallons, approximately 41 litres.

27 Smith, 'All Manner of Peeces', p.41.

It was vitally important to keep the powder dry as it was considered that dry powder had double the strength of the 'dankish sort'. Powder could become damp because the powder casks were not closed tightly, because of storage in damp conditions or during carriage in bad weather. Damp powder had to be dried, carefully! William Clowes the surgeon records a horrible accident where two men were drying powder too near the fire using their hands as scoops. The Gunpowder plotters also blew themselves up drying gunpowder.

This image illustrates the real importance of modern health & safety protocols. The gunner is using a bowl to take some powder from the three small opened barrels, to put in a well-designed wooden 'powder boat', ideal for pouring it into the (stiff canvas?) powder bag, the neck of which forms a useful funnel and can be securely tied. Such details are of great importance to historians when so much gunnery ephemera has been lost. (Walther Litzelmann: *Vortrab zu der Arckalay – Büchsenmeisterbuch* – Shelfmark, Fol. Cgm 909 p. 318 Courtesy of the Bavarian State Library)

Slow Match and Powder Measures

All ordnance and most small-arms were ignited using a slow match. It was recommended that it should be made of hemp and as thick as a man's little finger. It seems to have been from 20 mm in diameter at the beginning of this period to 10–15 mm by the middle of the sixteenth century. It was made from old rope with the fibres opened by beating with a mallet. The fibres should then be soaked in a cauldron of strong lye, ashes and saltpetre and heated until two thirds of the liquid evaporated. The fibres would be twisted, dried on lines and then drawn through a hole, in the same way that wire drawers reduced the diameter of wire, so that the newly manufactured cord would be even.[28] Surviving examples are few but not unknown. There are some bundles of match cord in the Fortress Museum in Coburg. They are 35–40 m long in total, in looped bundles about 80 cm long. A musketeer may have been expected to be issued with a bundle of only 2–3 metres in length overall. Vast quantities were needed as it burned at a rate of 60–80 cm an hour. In the Armémuseum Stockholm there is a well-preserved bundle of seventeenth century match cord which weighs 3.4 kg and measure approximately 10 m.[29]

The *Batavia* has provided some unique gun accessories for naval use. The powder measures are unsurprisingly large, varying from nearly four litres in volume to slightly over seven. They are made out of copper with simple handles riveted on the side. The cartridge containers although differing in size are uniform in design. As with the powder measures, they are made of copper sheet with a longitudinal fold-over seam and a circular base, the closely fitting cap is of similar design. Two tubular holes on the container, made from a strip of copper riveted to the body, are matched to two on the lid. A string would pass through the holes, ensuring the lid would not be lost and that the cap would drop into place when carried. Such cartridge containers were essential on a ship when the charges had to be brought through the vessel along quite a long route. The linen cartridge recovered from the *Batavia* had a rounded bottom with the seam carefully sewn, while the top of the cartridge was tied shut.

Pyrotechnics

One aspect of the gunners' art rarely mentioned by historians was the manufacture and use of pyrotechnic devices. This contrasts with the copious references and illustrations of pyrotechnic devices in contemporary manuscripts, and they constitute a subject worthy of an entirely separate book. Biringuccio establishes the purpose of such devices in his preface to his tenth book, where he refers to the materials and procedure, 'in offensive and Defensive Warfare and for Festivities on Holidays'.[30] He encourages his readers with these words, 'You will be pleased to have this information if for no other reason than that you can teach it to those who take delight in celebrating holidays with festivities and with fires'.[31] He provides an impressive compendium of such weapons including: mines, firetubes or trunks, fire pots, squibs, fireballs, rockets, shells, grapeshot, incendiary arrows and 'bronze balls that … carry fire and break in many pieces when they hit.'[32]

28 Norton, *The Gunner*, p.147.

29 <http://www.vikingsword.com/vb/showthread.php?p=140402#post140402>, accessed 23 May 2021.

30 Biringuccio, *Pirotechnia*, p.403.

31 Biringuccio, *Pirotechnia*, p.403.

32 Biringuccio, *Pirotechnia*, p.403.

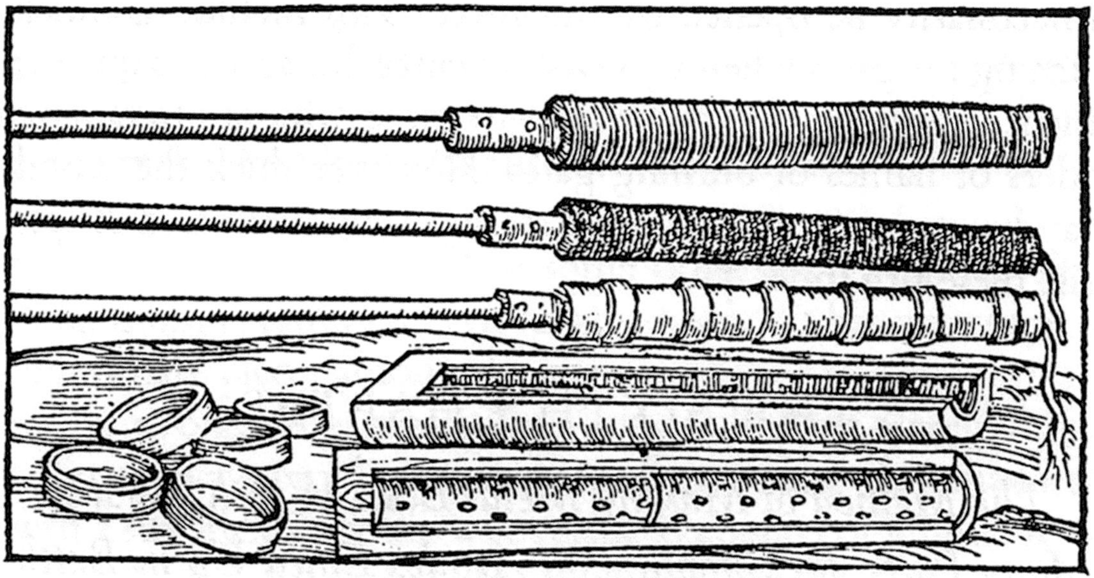

Fire tubes or trunks from Biringuccio's *Pirotechnia*. Ignited from the muzzle these examples seem to have reinforced barrels, the upper one clearly bound with wire. Presumably this was in order to ensure that the casing did not break, showering pyrotechnic madness everywhere. They were in effect militant 'Roman candles'. (Biringuccio, *Pirotechnia*)

Fire tubes were short-range flame throwers, useful for 'holding a gate' or 'forcing passage on a guarded bridge'.[33] In 1567 Richard Wright provided a detailed description of their manufacture:

> Item first for a tronk take serpentine powder 4 pounds, and of saltpetre one pound, and of sulphur one pound and of linseed oil one pint and of Rosin 4 pounds and a little smith's dust [iron filings?] and beat all these together small and then store them all rolled together and then take one pound of turpentine and melt it on the fire and then stir it with your hands again and so let it stand till it be dry and so lay them at length against that your trunk be surely bound for breaking. [For the casing] First you must see that your tronk be 20 inches long, 10 inches in compass, one inch and half high in the mouth. Then to lay first corn powder, then fill him 3 inches high and next take rosin 3 pounds and half and half a pound of corn powder and a little sulphur as small as a bean and then wrap canvas about it and so let it settle and thus you must order your trunks.[34]

Fire lances were small incendiaries that could be attached to lances or pikes and performed a similar role. Rockets could be made from rolled paper and one end folded to make a simple venturi tube. These could be fired at infantry or cavalry: 'since it is a fire that runs along the ground, and also endangers kegs of powder or barrels of munitions wherever they are found'.[35] Biringuccio illustrates incendiary weapons being shot from large calibre artillery pieces in order to scatter enemy formations. There were also recipes to enable different coloured smoke to be produced, though whether for entertainment or communication is not clear.

33 Biringuccio, *Pirotechnia*, p.425.

34 Ruth Rhynas Brown, 'Troncks, Rockets and Fiery Balls: Military fireworks of the early modern period', *Journal of the Ordnance Society*, Vol. 17 2005, p. 26; Society of Antiquaries MS 94, 1 7r; 21.

35 Biringuccio, *Pirotechnia*, p.432.

Although clearly meant to engender panic amongst cavalry, the rather random scattering of the fire tubes would suggest that they would cause at the very least confusion to their own side. Where gunpowder barrels and stores of powder were to be found across the battlefield, to employ a weapon that scattered sparks everywhere would have been looked upon as a somewhat mixed blessing. (Biringuccio, *Pirotechnia*)

The notebook of John Lad, an Elizabethan apprentice gunner, includes directions on how to make incendiary arrows, fire pots for hurling onto the decks of a ship, trunks (fire lances) and balls of wild fire.[36] Fire pots were usually made from ceramic pots that were provided with handles for both suspension and the attachment of smouldering match cord to act as a fuse. Fireballs could be hurled and rolled at an enemy and could be made from metal or canvas. Richard Wright again provides a helpful description:

> First take an earthen pot and one pound of sulphur and one pound of rosin and scome [mix] them together in a mortar and then seeth them in the same pot, then take your ball and dip them in twice or thrice. So take them out of the pot and then make your touchhole and then hang them up on the sun. In the binding and trimming of them be sure to make them strong for it is very dangerous.[37]

The young King Edward himself recorded an attempt by the French to use incendiaries in an attack on Boulogne: 'They [the French] … had made faggots with pitch, tar, tallow, rosin, powder [gunpowder], and wild fire to burn the ships in the haven of Boulogne'.[38] Incendiary devices were prepared for use by the English at the siege of Calais in 1558 but resolutely refused to burn, contributing to the disaster.

36 See Caruana, *Tudor Artillery*.

37 Brown, 'Troncks, Rockets and Fiery Balls', p.28.

38 W. K. Jordan, *The Chronicle and Political Papers of King Edward VI* (London: George Allen and Unwin,1966), p.12.

A drawing by Francisco di Giorgio Martini (1439–1501), courtesy of Lund Humphries.

One aspect of the gunner's craft rarely considered was his role as a demolition expert. His knowledge of gunpowder and fuses would prove invaluable when the artillery had not succeeded, as at Boulogne in 1544, when mines achieved what a prolonged bombardment had failed to. Di Giorgio was a true Renaissance man who was an artist, sculptor, engineer, civil and military architect, and developed the use of mines and gunpowder in the Franco-Italian war 1494–1498 in the service of Ferdinand II the King of Naples. In the image above he has placed a large gunpowder charge with a fuse and charcoal brazier ready to light it beneath the fortification. More interesting is his use of a dioptra or surveyor's transit, egyptian levels and plumb lines to survey the castle mound. These were the skills that differentiated the 'professional' gunner from the mere artisan. A good grounding in mathematics and geometry were considered essential for the up-to-date master of the military arts. Peverone in his treatise on geometry declared, 'How would the bombardier without it determine his shots, which now violently leap up, and now by natural course come down? … How would the engineer know how to measure the inaccessible places, or how to draw up the plans of countries and cities? Or designs of bastions, cavaliers, trenches …?' (Alexander Keller, 'Mathematics, Mechanics and the Origins of the Culture of Mechanical Invention', *Minerva*, vol. 23, no. 3; Springer (1985), pp. 350; *Due Brevi e Facili Trattati, il Primo d'Arithmetica; l'Altro di Giometria* (Lyon: G. de Toures, 1558), pp. 63–64).

8

Gunfounders

Bronze Gunfounders in Europe

Gunfounders were craftsmen and adhered to the traditions of their craft. The mystery that they practised was highly valued and would be shared through apprenticeship and practical experience not through book-learning. This is one of the reasons why Biringuccio's detailed account of gun founding is so rare and so precious. He makes it clear that his fellows were jealous of their trade secrets, which they thought gave them a particular edge over their commercial rivals. It is hardly surprising that as with other crafts the family would be important, with dynasties of gunfounders emerging, sometimes in the service of a single state. Others worked for a variety of state and noble employers over their lifetime. The author has tried to include in this chapter some of those whose work has been mentioned earlier.

However valuable their skills to their employers, their status was well below that of those they served. This is why ennoblement or granting of denizen status, if rare, was so important when it did take place. A letter from a Genoese gunfounder Jacobo de Lerreuci de Palma to Henry VIII was an outstanding exercise in ingratiation and highlighted the dependency of the craftsman upon his magisterial patron:

> And I declare that I would not have made the said culverin for any one else for 200 ducats of gold, yet I am content to have made it for you for the said 60 ducats, for the friendship I have for you, and for the honour and benefit I daily receive from your worship, and in order that the most serene King of England may see the work of my hand.[1]

Traditionally the centres of industrial production were to be found in the Low Countries, Germany and the vastly wealthy Italian cities, whether duchies, republics or kingdoms. Because of their wealth, developed industrial base and endemic rivalries, the manufacture of arms of all sort was

1 Letters and Papers, Foreign and Domestic, Henry VIII, Volume 2, 1515–1518 23rd April. 4108. Ordnance.

an intrinsic part of their economic and social lives. Venice and Genoa built ships and cast cannon to further their commercial enterprises. Renaissance princes saw a well-filled artillery park as important to their prestige as well-filled coffers and priceless art. The reconquest of Moorish Spain, the defence of Christendom against the Ottoman threat and the endless wars to establish dominance in Italy between France and the Empire, all produced a massive demand for the latest weaponry. It is therefore hardly surprising that it was on the continent of Europe that gunfounders would find a ready demand for their products.

The author of *Pirotechnia*, Vannocio Vincenzo Austino Lucca Biringuccio, was a scion of Sienna, born in 1480. His father was rather surprisingly an architect not a gunfounder. His family were supporters of the powerful Petrucci family, and their fate and fortune were closely entwined. Under their patronage the young Vanoccio travelled widely in Italy visiting the bronze foundry in Milan and industrial enterprises in Germany before being sent to manage the Sienese iron mines. In 1513 he was given an appointment in the armoury. In 1515 the Petrucci were expelled from Siena and Vanoccio was accused of debasing the coinage, possibly a trumped-up charge but he certainly possessed the expertise to have done it. In exile he travelled further in Italy visiting Rome, Naples and Sicily. The Pope supported the return of the Petrucci and back with them came Vanoccio. He was restored to his property and his offices, now being granted the monopoly of the production of saltpetre. His return was short-lived and the Petrucci and the unfortunate Vanoccio were exiled and condemned as rebels. Despite the support of the Pope and Florence, an attempt to re-take Sienna failed despite his active participation.

He now had the opportunity to work in the foundry of Duke Alfonso I d'Este of Ferrara during the Imperial siege of 1529. It was here he cast a large double culverin weighing an enormous 18,000 pounds. It was about 26 feet long and was embellished with an elephant's head. Peace and Vanoccio returned to Sienna in 1530 and he became a Senator of the Republic. He now worked as an architect, gunfounder and military engineer for the Farnese in Parma, the Duke of Ferrara, the Venetian republic as well as of course Sienna. His final appointment in 1538 was as head of the Papal foundry and administrator of the Papal munitions. Unfortunately, he died soon after in 1539. There can have been few who had such varied experiences in their lives. His career was perhaps more disordered than most, but he seems to have set out to discover as much as he could about the practicalities of industry wherever he went. He was not only an observer, but a skilled practitioner and his account is highly informative, filled with practical advice.

If Sienna was well served by Biringuccio, Venice benefited from the services of the Alberghetti. They were a remarkable dynasty of Italian gunfounders working primarily on behalf of the Serene Republic. They originated from the Valle di Comacchio near Ferrara. Sigismondo I served there, when his father was chief overseer in the foundry, then moved to Florence and finally to Venice in 1487. He was appointed Chief Founder in the Arsenal by 1514. This office was held by several of the family well into the eighteenth century. Alfonso who succeeded Giulio in 1568 as chief founder, cast two beautiful bronze well-heads in the courtyard of the Doge's Palace, an indication of the range of complex castings of which he was the master. This ability to cast works of art, bells and cannon was an indication of the breadth of their skills.

If the Alberghetti were the gunfounders of the Serene Republic the Gioardi were casting the guns of their great rival Genoa. Luchinio I was the first member of the family established as a gunfounder in the mid fifteenth century. It was Gregorio I who was to build a foundry on the Molo, which was the sea mole in Genoa's harbour, the site of the State foundry. He cast guns for the important fortress of Sarzanello. His son was Vincenzo, who after an apprenticeship in Genoa eventually succeeded Biringuccio in Rome, reorganised the Papal artillery as well as cast a bell for Castel St Angelo. His son Gregorio II, after learning his craft in Genoa, succeeded his father in Rome and was busy casting guns in Citavecchia, Perugia and Ancona which then belonged to the Papal States. There were other Gioardi, fathers, sons and brothers, although it is often difficult to accurately establish relationships that existed between them. Stefano and Battista both cast guns for Andrea Doria but Battista eventually returned to Genoa, Stefano succeeded to the office of Papal Gunfounder and was still alive in 1582.

One of the strangest families was that of Hedwige van den Nieuwenhuisen who was married to Hans Poppenruyter. Hans Poppenruyter was one of the most famous gunfounders in this period, serving Charles V as Master of the Imperial foundry at Malines. He was probably German in origin and there are many spellings of his name, as well as the alias Hans von Neurwerk. He cast well over 100 pieces for his Imperial master's ally King Henry VIII in preparation for his first French expedition, including the famous *Twelve Apostles*. He was married to Hedwige van den Nieuwenhuisen who on his death in 1534, chose Remigy de Halut to succeed him in her bed and in the Mastership of the Imperial foundry. He produced some of the finest bronze guns of his generation and was appointed Vicomte de Bergues St Winoc and a Captain in the Regiment of Philip de Montmorency, an indication of his high status. He was succeeded, probably in 1568, by Gerard Nieuwenhuisen who was his wife's nephew, who continued in office until his death in 1588.

The Löfflers were a Swiss family who cast guns, bells and statuary in Innsbruck throughout the sixteenth century. The name Löffler means 'spoon maker' which was the specialist product of the area they had originally come from. The founder of the family was Christof I, but it was his son Peter who would become ennobled by Emperor Frederick II for his service as gunfounder, many of his pieces are recorded in the Imperial *Zeugbuch*. Peter had three sons Alexander, Gregor, Wenzel and Phillip. Alexander like many founders was also a master gunner and served at the battle of Pavia in 1525 and at the siege of Vienna in 1529. He was finally killed at the battle of Offen in 1542 where he and all the Austrian artillery were lost. Gregor although also a master gunner, concentrated on foundry work and produced well over 1,000 pieces at Innsbruck. He was perhaps the most important gunfounder of his generation. He first served Ferdinand I of Austria before being appointed Imperial Master Gunner in 1523 by Emperor Charles V, the brother of Ferdinand who he continued to serve. An indication of the scale of his undertakings relate to a contract given in 1541 and completed in 1543.

In less than 18 months he cast 104 guns, including 24 cannons and the same number of demi-cannons, as well as 18 quarter cannons and 38 culverins and sakers. His brother Wenzel was a gunfounder and master gunner and was killed in action with the army in Hungary in 1527. Phillip was apparently the most difficult of the brothers moving from Worms to Innsbruck, Feldkirch and Zengg. He fell out with the council at Feldkirch to such an extent that he was imprisoned by them and only released by royal command.

Gregor provided two sons to continue in the family business, Elias and Hans. Elias served as gunfounder to the Duke Ferdinand of Tyrol but later retired to Innsbruck. Hans was more active and although younger than Elias succeeded to his father's foundry in 1563. He was Master founder for the Dukes of Tyrol as well as for Emperor Maximilian II. His service was so well appreciated that he was granted a patent of nobility in 1591. Christoff II was the son of Hans and the last of the great family. He moved the business to Vienna and continued in the Imperial service, with his father's noble rank confirmed by Emperor Ferdinand III. On his death in 1623 a dynasty of gunfounders stretching back almost 150 years came to an end.

As the name would suggest the Borgognone were a French family of gunfounders that originally from Burgundy, but the family worked in Italy. Pitero Borgogne, to use his Italian name, ran a foundry in Trento serving the Emperor Maximilian and then Archduke Ferdinand. Like many gunfounders he was a master gunner as well and served and died for the Imperial cause at Milan in 1528. Annibale was his son and benefited from having served as a pupil of Georg Löffler. He served many masters including the king of Hungary and Bohemia (later emperor), as well as the Duke of Ferrara. In 1550 on discovering the apparent adultery of his wife he murdered her and was forced to flee to Mantua where his wickedness did not stop him from casting guns for the King of France. He returned to Ferrara (apparently forgiven) and the service of Duke Ercole in 1553. His guns were recognised as being outstanding both in performance and beauty. His equally murderous brother Odorico served his apprenticeship with Hans Turing in Vienna but moved to Trento where he killed master gunner Sebastian Glaser in a quarrel. Further trouble led to his imprisonment by the city, and he was only released because of the intervention of the Bishop Cardinal von Madrutz. After this he seems to have assisted Annibale and not murdered anyone else that we know of.

Cast Iron Gunfounding

The first blast furnaces in Europe appeared in the second quarter of the fourteenth century. The first date given for the manufacture of cast iron ordnance in Europe was 1377 in Erfurth, although its impact appears minimal, with wrought iron and cast bronze guns still dominating European production. There is evidence for the existence of cast iron guns in Burgundian hands in 1414–1415, while in 1445 30 guns were cast in Siegen in Germany. A cast iron fowler with three chambers was made in 1461 by Jean Pothier with a total weight of 500 kg (1,075 lb). These seem to have been copies of the wrought iron system where barrel and breech were separate, not taking advantage of the obvious opportunity to make a muzzle loading piece. There is reference to a single muzzle-loading gun weighing 375 kg (825 lb), but this seems to have been unusual if not unique. Cast iron was used for the manufacture of small breech-loading guns in France as well during the mid sixteenth century.

Bronze guns continued to predominate despite their high cost. In the 1530s bronze guns in England cost 10s a hundredweight for workmanship and 20s for the metal or £30 a ton, while cast iron was sold at £10 a ton in total. In addition, gunfounders were allowed 12 lb per hundredweight (112 lb) as 'waste', a more than generous amount that enabled the Owens to manufacture in secret a complete cannon cast from saved 'waste', a piece of cheek they eventually had to pay for. Although the bores of bronze guns were liable to greater wear the gun could be melted down and recast, a significant saving and an admirable example of recycling. Even with its high price bronze was far preferred as the figures for English royal purchases in the period 1550–1558 shown below.[2]

Table 20.

Type	Cast Iron	Bronze
Cannons	–	2
Demi-cannons	–	5
Culverins	–	1
Demi-culverins	6	26
Sakers	8	69
Falcons	3	77
Falconets	–	15
Bases	–	2
Total cost	£164. 4. 8.	£1,543 2. 8.

Casting in iron was taking place elsewhere in Europe. Falconets were being cast in 1539–1540 in the Ardennes and contemporaneously guns ranging from falconets to bastard culverin were being cast in the royal foundry at Breteuil in Normandy. Francis I seems to have taken considerable interest in the Breteuil foundry and in 1542 an order for 126 pieces was made for bastard culverin, culverins, falcons and falconets. The largest weighed 1,740 kg and was some 3.45 m in length with a bore of 75 mm. This sudden burst of activity was not sustained, this could have been because of a weakness or distrust of the new materials or quite simply because Francis had more than enough money to pay for superior bronze guns.

2 Ordnance Office records E351/2614.

Casting in England

England in the late Middle Ages possessed neither the skills, the materials or the market to match the technological achievements and volume of gun production to be found on the continent of Europe. Cornwall was the principal supplier of raw tin and lead in Europe but there was no great tradition of metal working and especially casting such as was to be found in the Low Countries, Northern Italy and Germany. England was fundamentally poor and sparsely populated but was fortunate in that her close neighbours, Scotland and Ireland, were poorer still.

There were certainly individuals capable of making wrought iron pieces and casting bronze guns as well, but the volume required was relatively small. Although artillery was purchased by towns and cities, merchants and the aristocracy, the only major purchaser was the crown. The consequences of the Wars of the Roses and the accession of Henry VII meant that England experienced a period of prolonged domestic peace and avoided expensive foreign wars. Humphrey Walker was a fine English foundryman and servant to the first Tudor king. He cast the magnificent railings for the tomb of Henry VII and a *basilisk*, an impressively large but ill-described piece, which was cast in the Tower of London. He demonstrated skills equal to those of his continental colleagues, but England could not match the volume of production demanded by an ambitious and aggressive monarch. In preparation for a succession of wars with the French, Henry VIII was forced to purchase very large numbers of the most up to date bronze pieces from Hans Poppenruyter, the Imperial gunfounder working in Malines. In 1509 Henry purchased 24 courtaulx and 24 serpentines and later he gave Poppenruyter a contract for 48 guns with each averaging a little over one ton. Simon Giles of Malines may also have come to England to cast guns for the King as in the 1513 Tower inventory there are references to guns 'of Simon's making'. Cornelius Johnson, the King's Smith, supplied 91 wrought iron guns for Henry between 1512 and 1513, highlighting the continued use of these effective and comparatively cheap weapons.

Henry sought the services of the Frenchman Peter Baude in 1514. Baude operated a foundry in Salisbury Court and was soon well established being granted a daily pension of 16 d as a Tower gunner. He received a contract in 1538 worth £210 for guns and acquired metal and broken guns for recasting. He was so much in favour that in 1542 he was rewarded with the granting of the status of a denizen or citizenship. Baude was joined by Francesco Degli Arcano from Cesena with his two sons Archangelo, and Rafaello (probably joining later). Francesco was the 'master of the Mynes' in the French expedition of 1523. Francesco went on to build a gun foundry in Salisbury Place on the south side of Fleet Street. Although England may appear particularly backward because of the employment of 'stranger smelters', this transfer of skill was a pan-European phenomenon. The exchange of technology via masters in their craft had been essential for the 'modernisation' of Italian ordnance after 1494–1495 and was a feature of the careers of expert craftsmen in all fields.

With the creation of a strong domestic manufacturing base, even if using the skills of 'strangers', very little ordnance seems to have been purchased from abroad after 1523. The exception to this was the purchase by Henry of 19 guns from Italian gunfounder Alessandro Gioardi. These were cast in Sicily in 1528 for the knights of St John and include a fine octagonal couleuvrine moyenne now to be found in the museum of the Order in London (mistakenly, the author believes), attributed to Francesco Arcano.

Peter Baude did not establish a dynasty of gunfounders, but he did train the four Owen brothers (Robert, John and John junior and Thomas) in the craft, in the Houndsditch foundry. They were casting large-calibre guns by 1531. In 1540 they were granted the tenement known as the Bellfounders House in Houndsditch and in 1546 succeeded to the role of Master Gunner after Baude's death. Robert's son Samuel followed his father as a royal gunfounder in Houndsditch until his death in 1603, leaving his son Israel to continue the trade.

Casting in iron was one area in which England would eventually become recognised as being expert, although initial progress was faltering. The first recorded blast furnace in England was built, probably with royal patronage, at Newbridge in 1496. In England the first reference to a cast iron gun dates from 1508. Pauncelett Symart of Newbridge iron works manufactured an iron gun in that year, but it burst on firing. In 1509–1510 he successfully manufactured 'iii caste gonnes of iron with vii chambers of iron', which were sent to equip the royal ship the *Sovereign*. However, this small-scale manufacture does not seem to have resulted in the 'take-off' of the Wealden gun industry.

It appears likely that cast iron gun founding began again in 1543 at Buxted. This was in response to the desperate demand for guns caused by European hostility to Henry VIII, culminating in the threat of invasion in 1545. Parson William Levett, his 'man' Ralph Hogge and Peter Baude the famed gunfounder, worked together to produce the first successful cast iron ordnance. In the 1547 Inventory there are a number of references to cast iron guns including three in the South Castle of Portsmouth:

> 4428 Demi Culveryne of yrone of petir bondes [Baude?] makinge oone
> 4429 Sacre of yron of parsone levettes making ij
> 4430 Fawcon of yron of parsone levettes making oone[3]

One of the Parson's products a demi-culverin was sent to Pevensey Castle in 1547 where it remains, mounted on an impressive field carriage. Levett's enterprise did not result in the development of a large industry, as domestic demand was poor, and the disadvantages of cast iron were very apparent. In England Ralph Hogge became a stalwart of the Wealden gun founding industry until his death in 1585, becoming *le gonnestone-maker* for the Tower in 1559 and later the *Queen's Gunfounder of iron*. These titles did not signify a great increase in demand, as bronze ordnance was strongly preferred and there were only three gunfounders in 1569 rising to seven by 1578. After this

3 Starkey, *The Inventory of King Henry VIII*, p.112.

date the numbers remained more or less static with six furnaces in operation in 1585 but with only five in 1600.[4] The output for each furnace was between 100 and 150 tons per annum, which meant that a likely annual production was from 500 to 800 tons per annum. The Crown's demand for iron ordnance was very small, perhaps 50 tons per annum in the period 1585–1600. In 1574, 620 bronze guns were sent to arm royal ships as opposed to only 46 cast iron guns, and even these were smaller pieces.

Cast iron English pieces did however find a ready market amongst merchant vessels, where their comparative cheapness recommended them, as well as to overseas purchasers desperate for ordnance of any sort. In the 1570s the demand for ordnance in the Dutch revolt led to such an increase in demand that prices doubled. So concerned was Elizabeth's government, that in 1574 it passed legislation to control sales by licence, with founders and their landlords required to pay large deposits of cash. These would be forfeit if they were found to have broken the law. All cast iron guns were to be delivered to the Tower wharf to be sold to legitimate purchasers and records of merchant ship armaments were taken so that their guns could not be sold overseas.

Although the export of bronze ordnance was banned from England, cast iron was exported or smuggled overseas in very large quantities, perhaps involving three quarters of its total output. Guns larger than demi-culverin were banned from export but to get around this, guns were cast with a reduced bore and the customer would then bore them out to the larger size. Cast iron guns found their way by various channels to Spain, although from 1585–1604 she was at war with England. The Spanish were prepared to pay £19–22 per ton for English guns in 1587, twice the normal price. In 1589 Lord Buckhurst complained to the Lewes justices about their inaction in preventing smuggling. He noted, more in anger than sorrow that: 'Their lordships do see the little regard the owners of furnaces and the makers of these pieces have of their bonds, and how it importeth the state that the enemy of her Majesty should not be furnished out of the land with ordnance to annoy us'.[5] Sir Walter Raleigh complained that so widespread was the sale of English pieces to the Spanish that:

> Without which … the Spanish King durst not have dismounted so many pieces of brass in Naples and elsewhere, therewith to arm his great fleet in '88. But it was directly proved in the lower house of parliament of Queen Elizabeth, that there were landed in Naples above 140 culverins English … It is lamentable that so many have been transported into Spain'.[6]

To counterbalance illicit sales to Spain the industry sold large numbers of guns to Holland, England's ally and sworn enemy of Spain. The Dutch provided good customers for English guns. Between 1615 and 1618 they

4 Employment in the Wealden gunfounding industry was still very small, at perhaps a few hundred men.

5 Frederick Leslie Robertson, *The Evolution of Naval Armament* (London: Constable & company Ltd, 1921), p.54.

6 Robertson, *The Evolution of Naval Armament*, p.54.

purchased 300 guns, in practice half the total production, from the foundry of the King's Gunfounder Sir John Browne at Horsmonden in the Weald. The availability of 'cheap' English guns damaged the European bronze gun industry and encouraged the Swedes to develop their own iron gun industry. Sweden had always been a source of high-quality iron. The Finspång foundry was founded by the Crown in 1580 and produced the famous Finbankers.

The advantage that English cast iron pieces had over their European counterparts has been difficult to establish. It was asserted that Wealden iron was less brittle than Spanish, Swedish or Danish, and in a speech to Parliament in 1601 Edward Peake declared that 'there was no other iron of the toughness and validitie to make such iron ordnance'. The key to the iron used was that it was free of slag. The ore was weathered for several months and was crushed and washed to remove impurities. The fossilised grey shells to be found in the ore acted as a flux during smelting, a role that limestone would play in the seventeenth. The moulds were heated, and the molten metal was kept at an even temperature. The castings were allowed to cool slowly and not quenched with water, as was the case in Spanish production at the time of the Armada, where speed of production was of the essence. The cause of their good reputation was therefore partly in the materials but also in the craftsmanship involved. Only in the 1620s did competition from the Low Countries and Sweden begin to supplant English cast iron ordnance from its prime position.

9

Modern Reconstructions

Introduction

There are few institutions that have the imagination, energy, expertise and funding to promote large scale reconstructions. In the past, information on historical firearms of the late Middle Ages and early modern period has either been extrapolated from reliable nineteenth-century data or by informed and sometimes uninformed guesswork. In recent decades the Middelaldercentret (Danish Middle Ages Centre) and the HO group, the *Vasa* Museum in Sweden and in the UK the Royal Armouries and Mary Rose Trust have created and tested historical ordnance. These have provided definitive data which have established the capabilities of a range of representative weapons from this period.

The difficulty facing any attempt to convincingly reconstruct the performance of early weapons goes far beyond merely the gun itself. The ammunition and powder should hopefully match what was originally used. Modern 'black' powder is generally thought to produce 60 percent more 'push' (total energy released) than powders made before about 1750. The method of loading would also have an effect on the performance of the gun. The context in which a gun was used is important when considering its capability. Handling a heavy gun aboard a ship in high seas, or inside a castle casemate provides very different conditions from those found on a modern testing range. To replicate every factor is nearly impossible, although the Danish trials of a late medieval gun come close and was a very remarkable achievement.

The different groups testing historic reproductions varied considerably in how they approached their projects. As Dr Hocker has pointed out:

> it is not only not necessary to replicate every component of the system in order to have a valid test, it is preferable not to do so. Instead, the choice depends on what questions are being asked and what variables are being tested. It is not possible to test the system as a whole until after evaluating all of its components individually, which is why a proper table of fire needs to be somewhere around

35 rounds or more. Firing five or nine rounds provides useful qualitative data, but not meaningful quantitative data.[1]

Despite these serious challenges there have been four important attempts to recreate and test such weapons and each has produced valuable and often surprising results.

Late Medieval Bronze Cannon or Pesterau[2]

The most recent tests were organised by the HO Group – a group of international scholars, at the Middelaldercentret in Denmark. They set themselves the task of designing, casting and test firing a fifteenth-century bronze cannon. They chose a bronze cannon from the Historische Museum, Berlin (Steinbüchse W 437) which weighs approximately 120 kg.

The project consisted of four distinct phases:

1. Making the clay mould for the cannon, firing it and preparing it for casting
2. Reconstruct a small-scale fifteenth century furnace to melt the bronze and cast the cannon
3. Making a form of gunpowder which would be appropriate for the period
4. Test firing the cannon using a variety of projectiles and loading techniques.

Preparatory visits were made to Southern India where the traditional lost wax casting process is still used for craft production as well as to the well-preserved cannon foundry at Jaigahr Fort, where it was possible to study an unused cannon mould.

The techniques of mould-making were very similar to those described by Biringuccio. On a wooden spindle wound with rope, a clay model of the exterior of the cannon was slowly built up. On this was placed a layer of wax. Onto this was built a further clay layer 10 cm thick. Using the clay/wax/clay mould as a chimney, a fire was lit so that the wax was melted out. The inner clay layer was removed leaving a clay model of the exterior of the cannon. A separate clay mould of the sub-calibre powder chamber and barrel was made and fixed to an iron plate. This was placed inside the mould of the exterior of the barrel, as the gun would be cast muzzle down in the same way as a bell.

The moulds have to be carefully dried to drive off any moisture. The exterior mould was reinforced with steel bands and more layers of clay to resist the force of the weight of bronze as it was poured into the mould. This process closely followed that described by Cristóbal Lechuga (*Discurso en el que trata de la Artilleria con uno de Fortificación*, Milan, 1611). It is thought that the change to casting muzzle up took place about 1500 when it is known

1 Correspondence with Dr Hocker. 26 July 2021.

2 A short-barrelled muzzle-loaded weapon similar to a mortar, usually of wrought or cast iron from the very late fifteenth century, described in Smith and DeVries, *The Artillery of the Dukes of Burgundy*, pp.225–226.

The *Zeugbuch* from the beginning of the sixteenth century contains a large number of what might be considered obsolete types of artillery, including large bombards. In the collection are a number of pieces very similar to the type trialled by the HOGroup. It is of cast bronze, has a sub-calibre chamber and fired stone shot. (*Zeugbuch Kaiser Maximilians I*. BSB Cod. icon. 222, p. 0038. Courtesy of the Bavarian State Library)

that both methods were in use. The box was packed with sand in order to ensure that the mould did not shift.

A reverberatory furnace was built according to plans by Dr Bastian Asmus, a massive construction effort in its own right. Managing the temperature of the furnace was also an important learning experience. There were three castings poured, in one the barrel/breech core moved and in another the molten bronze did not completely fill the mould. In order to complete firing trials on range facilities provided by the Danish Artillery Regiment, a flawless example was required by the army. Thus, a copy of this cannon was made from a modern solid casting which was machined.

The HO Group's replica gun was mounted on a solid wooden bed and held in place with iron strapwork and then mounted on a wheeled carriage for ease of transport. Sensors were fitted into the cannon to measure the bore and breech pressures. 'Though she be but little she is fierce'. (Courtesy of Dr Peter Hansen)

The HO Group has over a decade developed an unequalled expertise in the manufacture of medieval gunpowder and incendiaries. The source for the sulphur rich mixtures was the *Fireworkbook* inventory number I. 34 47r 47v held at the Royal Armouries in Leeds.

Table 21.

Powder	Saltpetre	Sulphur	Charcoal	% Saltpetre
Common	4	2	1	57 %
Better	5	2	1	62.5 %
Stronger	6	2	1	66 %

The powder was made by combining the dry ingredients thoroughly, then wetting the mixture forming paste balls 10cm in diameter. After being dried these were then crushed into a fine powder. In these trials the 'better' powder was used although the intention is to use the other grades in future trials.

Two issues arose when considering the loading of the powder. Firstly, whether a wooden plug or tampion should be used to close the mouth of the powder chamber and if one was what size and type of wood would be best? Should the wood be green or seasoned and whether hard or soft. The second question was whether the powder chamber should be completely filled? It has been argued that space had to be left in the chamber so that when the fine powder was ignited it had space to be 'lofted' and burnt more effectively, in the same ways that many fine powders can be ignited. Custard powder shaken in a tin can be satisfactorily ignited. Loading the ball was not a matter of simply rolling it down the barrel. Wedges and waxed cloth were used to centre the shot and reduce the windage, a procedure which produced some surprising results.

The powder chamber was loaded ⅗ or ⅘ full (324 g or 432 g) with a tampion hammered hard in, taking up ⅕ of the space. The ball could then be wrapped in waxed cloth or seated firmly with three wedges, in a variety of combinations. As the army insisted that no one should be in front of the barrel, a unique set of loading tools including a selfie-stick mirror were made to avoid this danger.

The firing trials were conducted on the Oksbøl Skyde-og Øvelseterræn military range over a period of two days. The facilities at the range permitted the following to be accurately measured: muzzle velocity, range, chamber and barrel pressure and noise level (which was about 160 decibels). Chamber pressure was tested at a point just ahead of the touch hole and just ahead of the ball. There was a team of nine involved in the trials including five members of the Danish Artillery Regiment. The barrel was mounted on an extemporary field carriage, which was chocked in order to stop any movement. On the first day the barrel was mounted level and on the second day it was shot at an angle of 20 degrees to the horizontal. A rate of fire of one round every five minutes could be achieved. There was very little evidence of wear in the barrel. The original gun had evidence of considerable use with a number of obvious striations.

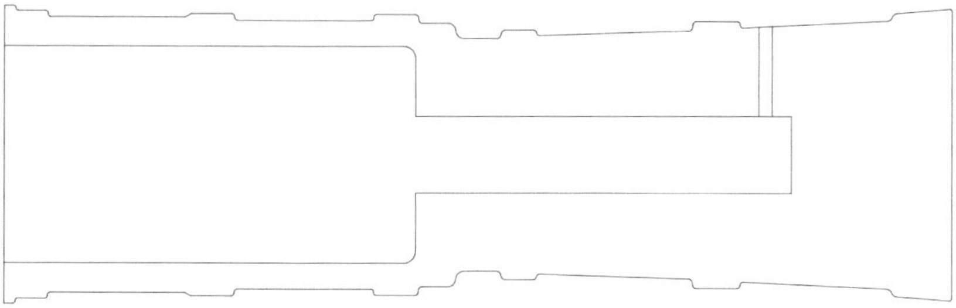

The HO Group cannon had a bore 150 mm in diameter. The gun was 645 mm overall, the powder chamber was 254 mm long and 51 mm in diameter. The thickness of the barrel wall was only 15–20mm except at the reinforcing bands. The thickness of metal around the breech varied from 55 mm to about 65 mm, while at the end of the gun there was up to 200 mm of metal. (Courtesy of Peter Hansen)

Table 22.

Shot/s No	Powder	Tampion/ wedge/cloth	Ball type/ weight *	Chamber/barrel pressures Mpa §	Muzzle velocity m/s	Range m
2	4/5	–	G 4.1	22.5/4.7	107.8	343
4	3/5	T	G 4	20.9/8.9	104.1	
3	4/5	T	G 4	36.8/9.9	135.1	
6	3/5	T C	G 4	26.1/10.8	119	
5	4/5	T C	G 4.2	39.0/12.6	145	
9	3/5	T W	C 3.7	23.5/6.4	116.5	
8	4/5	T W	G 3.9	36.8/13.4	137.2	429
11–15	4/5	T C	C 3.7	34.8/9.3 av.	148.5/6.5 ¥	680/12 ❡
16–20	4/5	T W	C 3.6	36.6/11 av.	154.2/17.8	565/20
21–25	3/5	T C W	C 3.6	24.7/8 av.	130.2/6.7	632.6/25
27–31	3/5	W	C 3.5	9.3/4.7 av.	106.1/20.2	493.7 av.

* C concrete, G granite. Weight in kg.

§ Measured in megapascals

¥ The difference between the highest and lowest velocities achieved

❡ Number of shot measured

Summary of results:

- These trials were conducted with great care and thoroughness. It is clear that the use of the tampion was essential to produce consistently high chamber pressures and velocities. Where no tampion was used the results were very wayward and disappointing; the chamber pressures were much lower as were muzzle velocities, with considerable variation between shots.
- The tampion was made from softwood and hammered into the breech chamber, hard. The wedges that were thought were primarily for centring held the ball firmly in place and led to significant pressure increase in the barrel.
- The ⅗ loading of the powder chamber gave no discernible advantage,

there was no 'custard powder' effect. Larger powder loads produced higher pressures and velocities, which is exactly what would be expected.

- It was not possible to establish the precise range achieved by every shot and the high wind clearly affected their flight. Where range is given the number of the shot is also shown. Where no tampion was used (shots 27–31) three ranges were measured varying from: 488.3/27, 565.0/28 and 427.9/31. There were curious anomalies, so that shot 25 flew 200 m further than shot eight while using less powder and achieving higher chamber pressures and velocities. A tentative conclusion would be that a consistent and reliable range of 500 m could be achieved with a muzzle velocity of about 150 m/s, as long as the same powder load and loading techniques were used.
- Barrel pressures were about a third or a quarter those of the chamber. Although the difference in chamber pressures was often comparatively small, the difference in barrel pressure varied considerably, even when the powder loading and procedures were the same. This did not seem to have a significant impact on muzzle velocity, in that a variation of up to 50 percent in barrel pressure made no discernible difference to velocity.
- It was clearly impossible to reproduce a high rate of fire, as modern safety procedures allied to the limitations placed on them by the range authorities precluded any such attempt. Loading was certainly easily manageable but would never have permitted a high rate of fire.

This project was a remarkable achievement by the HO group, providing for the first time a convincing reproduction of a late medieval weapon and producing significant results. The gun was highly portable, safe, reliable and hard-hitting, if loaded carefully and handled well. Much was discovered about the management of such weapons, especially the proportion of the chamber to be filled with powder and the key roles played by the tampion and wedges. This weapon would have been a powerful force in battle or siege.

The HO Group gun performed safely and impressively on the range. Although fitted on a wheeled carriage it was not allowed to recoil, replicating firing from a fixed bed. (Courtesy of Peter Hansen)

There was no significant wear to the barrel during the trials. The original gun showed evidence of damage, probably caused by poorly fitting and poorly finished stone balls. The comparatively small size of the powder chamber is evident in this photograph. It was felt important by early gunners to centre the projectile over the mouth of the powder chamber in order to maximise the force of the explosion, especially as it would be preceded by the driving force of the tampion. This explains the use of wedges to centre the ball. (Courtesy of Peter Hansen and the HO Group)

The *Mary Rose* Wrought Iron Port-Piece

The gun that was used is described in detail in Chapter 5 and although of new manufacture the materials and techniques used did not in any way compromise the trials. These guns certainly did not deserve the rather contemptuous opinions expressed in the past when comparing them to bronze pieces of the same era. They clearly justified their position on the main gun decks of vessels in the middle of the sixteenth century.

It would have been ideal to use 'home-made' serpentine powder. Unfortunately, this was not possible, and the closest modern equivalent was used ICI NPCG, which was the slowest burning powder that could be sourced. The other problem with modern powder was that it took up considerably less space than serpentine. John Lad, an Elizabethan apprentice gunner, recorded the following loadings for a port-piece in his notes, taken *c.* 1586,[3] 'Item a port-piece chamber of 16 inches long and three inches and a half high holdeth in powder 3 pound and a quarter, the shot of stone of 5 inches three quarters high and weigheth 9 pound.'[4] The ratio of powder to shot for safe service use was 1:3, with the powder almost certainly being serpentine. During the trials the ratio of modern powder to shot was 1:4 except in the shot which damaged the breech where it was closer to 1:3. Modern powder is clearly significantly more powerful.

3 In Caruana, *Tudor Artillery*.

4 Caruana, *Tudor Artillery*, p.25

This photo gives a good impression of the scale and power of the port-piece. The rusty fragments recovered from the ship in the nineteenth century gave the impression of a weak and dangerous gun. Little could be further from the truth. (Courtesy of the Royal Armouries)

The majority of the original stone balls were made of Kentish ragstone and one of these was used in the trials. The ball found in the excavated gun, from which the trials piece was copied, was 172 mm in diameter providing a windage of 8mm or rather more than the 6mm officially prescribed. The remaining balls in the trials were new made from Purbeck limestone in three sizes with windages varying from ¼ in, ½ in and 1 in. A lantern shot, 160 mm in diameter and 500 mm long, filled with broken flint held in a light mortar mix, was also used. Although this type of projectile was unlikely to be used in a ship-to-ship action, port-pieces were used against people. During the Prayer Book rebellions at the siege of Exeter in 1549 the guns 'were charged with bags of flint stones and hail shot and as they [the rebels] were approaching unto the gate the gate should be secretly opened and the said port-pieces discharged and so they were spoiled divers of them'.[5]

The trial crew was made up of six people. The gun captain, four 'heavy-lifters' working in pairs either side of the breech lifting it in and out of the gun and an ammunition carrier who had to place the ball in the barrel before the breech was fitted. If a spare breech was in use, which would have been common practice, then that would require at least one or two people to have the necessary materials and tools to load it. There would of course have to be others bringing powder to the gun, performing the role of the Nelsonian powder monkey.

Working in a confined space and amidst the noise and smoke of firing, let alone facing the consequences of enemy fire, conditions would have been

5 John Hooker, *The Ancient History and Description of the City of Exeter* (Exeter: Andrews and Trewman, 1765), p.70.

uncomfortable to say the least. It has been suggested that the skeletal remains of well-built men were those of gunners, because of the heavy task of moving the breeches. It seems more likely that their duties were assigned because of their size and strength and not the cause of it. The 'professional' gun captains were few and well paid, they were not employed to do the heavy work. In the Anthony Roll there were 39 large carriage mounted guns recorded but only 30 gunners. The Tudor world provided more than enough men inured to hard work to provide the labour needed; there is no evidence to suggest that Tudor gun crews were employed on a permanent basis, although some skilled gunners were.

The process of loading was a relatively slow one. The correct weight of powder was loaded, followed by sawdust and wood shavings to fill the remaining space in the chamber and a wooden tampion placed on top. Unlike the bombard trials there was no airgap left in the chamber and the tampion was not hammered home hard. This may well have had an effect on the performance of the gun. The ball was placed in the barrel and held there until the breech was in position. Almost certainly a wad of some sort would have been used in a heaving ship to hold the ball close to the chamber mouth and to avoid it rolling out. The wadding of the ball was clearly important in the loading procedures for bronze guns, to avoid an airgap opening up between ball and powder. Cyprian Lucar recommends both hard ramming the tampion and placing wadding before the shot in the barrel:

> Put into every chamber so much powder as his peece requireth for a due charge, and with a rammer beate a tampion of softe wood downe vppon the gunpowder. Moreover, put a biggewadde into the peece at that ende where the mouth of the chamber must goe in, and after the waddethruste into the peece at the sayde ende a fitte pellet, & when you have done all this, put the chamber into the lowest ende of the peece, lock them fast togeather, and cause the sayde tampion to lie harde vppon the powder in the sayde chamber, and the pellet to touch the tampion, and the wadde to lie close by the pellet.[6]

Grommets of rope and straw were used to seal the gap between the breech and barrel but had no apparent effect. The breech was held in place by a forelock and a pair of fishtail wedges, one of which inserted from the top provided a closing force on the forelock. It was found that it was best if the fit of the forelock to breech was snug rather than hard.

Firing was very sensibly done remotely and electrically especially considering the failure of the breech. During the trials it was noticeable that the staves opened slightly but the hoops and bands held well, 'in other words, it expanded'.[7] There was considerable gas escape from the breech-to-hull join, far more than from the vent. It would not have been a comfortable gun to serve in an enclosed environment. The author has experienced something similar in a breech-loaded cannon equipped war wagon; it was not pleasant.

6 Lucar, *Appendix*, p.47.

7 Hildred, *Weapons of Warre*, p.285.

Table 23. Summary of firing trials of *Mary Rose* port-piece. The intention in the trials programme was to increase ball and charge weight as the trial progressed, but the failure of the breech on the first day made this impossible.

Shot No.	Windage	Shot type/ weight	Charge weight	Target type	Distance	Muzzle velocity	Notes
1		None	1 lb 8 oz				
2	25 mm	Limestone 11 lb 15 oz	2 lb	Earth bank	*c*. 20 m	141 m/s	
3	7 mm	Limestone 15 lb 8 oz	5 lb 2 oz	Stone wall	*c*. 20 m	141 m/s	Failure of barrel/breech
4	25 mm	Limestone 12 lb	3 lb	Stone wall	17.2 m	160 m/s	
5	25 mm	Kentish rag 13 lb	3 lb 3 oz	Oak target	17.2 m	240 m/s	Original *Mary Rose* shot
6	13 mm	Limestone 13 lb 11 oz	3 lb 8 oz	Oak target	17.2 m	298 m/s	
7	25 mm	Lantern 25 lb 2 oz	3 lb	Oak target	15.7 m	n/a	
8	7 mm	Limestone 15 lb	3 lb 12 oz	Reinforced bunker	131 m	338 m/s	Damage to carriage

It is difficult to establish which was the most representative of the experimental shots. The variations in the projectile and the quantities of powder were so great as to produce muzzle velocities varying from 141 m/s to 338 m/s. Undoubtedly a loading of over 5 lb of powder was too great as it led to a disastrous breech/barrel explosion, seriously damaging the gun and bringing the first series of tests to an end. In shots four and five, balls of similar weight with identical windage produced muzzle velocities of 160 m/s and 240 m/s even though the difference in powder loading was only 3 oz. A 50 percent variation in muzzle velocity for an increase in powder loading of 8½ percent would suggest that consistency was difficult to achieve. In the next shot (number six), with an increase of powder of over 12 percent and a reduction in windage by almost half (from 25 mm to 13 mm) a muzzle velocity of 298 m/s, an improvement of over 20 percent, was achieved. In shot number five an original stone shot from the *Mary Rose* was used. The projectile broke up in flight, but even the fragments travelled at a velocity of

240 m/s. The ranges at which the gun was used were very short: under 20 yards. A longer shot (number eight) lost about a sixth of its velocity.[8]

Damage to the gun and carriage in shot three was considerable and was a classic failure that must have been not uncommon when such guns were in use. The ejected wedge is what probably caused the death of the Scots king in 1460 in a similar accident. It was caused by loading a heavy charge and a heavy ball which had a very small windage, the pressure created at the breech–barrel point was too much. The breech failed, and the trials had to be abandoned. The damage was recorded as follows:

> Plug at rear of breech chamber set back. The inner staves were pulled back. The elm forelock was broken; it and the iron wedge were pushed up and ejected. The breech chamber was partially displaced from the bed. The staves at mouth of the chamber were also damaged, with one piece of stave being broken off and thrown some distance.[9]

The damage was repairable by Chris Topps, the blacksmith responsible for making the gun, and the trials continued. This points to one of the strengths of these pieces: that they could be made and mended by the readily available skilled labour of the time. The woodwork also suffered damage to the wheels and rear of the carriage; similar damage has been identified on excavated guns. The damage also explains why stocks of spare wheels, axles and forelocks were also kept on board.

The target was a representative section of a ship's hull, using the *Mary Rose* as the model. Of course, the timber was green and not the seasoned timber of an old vessel.

> The target was of oak and consisted of an outer section of six 70 mm (2¾ in) thick edge-to edge planks 280 mm wide (original was 300 mm). The middle section was formed of six vertical oak frames of 250 mm (¾ in) with 200 mm (8 in) gaps between the frames. The inner layer was made of three horizontal stringers 70–90 mm (2¾–3¾ in) thick and 340 mm (original 380 mm) wide with gaps of 200 mm (8 in). It was held together with wooden treenails.[10]

The damage to this target, even though at very short range, was impressive. In shot number five the ball flew into two parts on leaving the barrel, but both penetrated the target. The most destructive hit was from shot number six.

> The ball caused the top of the frame to be splintered completely for the height of the top plank, displacing each frame on either side of the frame that was hit (both outwards to each side of the target), and blew off the internal stringer. This is

8 All the data on the *Mary Rose* port-piece trials comes from Hall, 'Building and Firing a Replica *Mary Rose* Port-Piece'.

9 Hall, 'Building and Firing a Replica *Mary Rose* Port-Piece', p.64.

10 Hildred, *Weapons of Warre*, p.285.

> substantial damage to all three layers of the structure, the outer planking, frames and inner stringer. It actually pushed the stringer off the treenails.[11]

It is quite clear that at short range the port-piece was capable of causing serious structural damage to any contemporary vessel.

In the trials the stone wall was significantly damaged to the depth of two to three inches, with surface damage extending about 20 inches either side. The balls themselves, it must be said, suffered even greater damage! It is not surprising that Biringuccio when considering the use of stone-firing weapons declared that: 'These are not good for breeching walls and serve only for shooting at infantry or cavalry and at armed ships at sea'.[12] This port-piece would have been of little use in a siege and its performance explains why the much larger calibre bombard was preferred for such operations. The very heavy lantern shot proved very effective with the majority of the flint in a pattern 1.3 m wide but with some extending two metres either side of the point of impact. The paper roll used as a target to record the impact was almost completely destroyed but the damage to the oak beneath was easily identified.

It would not, on the basis of these tests, be unreasonable to suggest that a muzzle velocity of about 250 m/s could be achieved safely with a 3 lb charge and a ball weighing about 13 lb. This would certainly be effective against a ship's hull at a short range of perhaps 200 yards. Against people, using lantern shot, it would have been devastating. As a siege weapon facing a 'hard' target it would be far inferior to a bronze gun.

The *Mary Rose* Bronze Culverin

The Mary Rose Trust and Royal Armouries cooperated to make and test a copy of a representative bronze gun from the *Mary Rose* excavations. The gun chosen was 81A1423, a culverin cast in 1543 by the French gunfounder Peter Baude in the service of Henry VIII. It was a long piece 12 feet in length with a 5½ inch diameter bore and weighing 3,888 lb (2,071 kg). The pattern was made using modern technology, although it was to be cast with its bore rather than solid and bored out. A casting pit some 4.3 m had to be dug out, fortunately the ceiling was high enough that the gun could be raised from the pit. It is noticeable that the Gioardi foundry on the Mole in Genoa was provided with a tower into which guns could be lifted from the casting pit.

Modern techniques were used to produce the pattern and mould and great care was made to ensure that all external decoration and details were copied precisely. The first casting took place in July 1999, but the metal cooled too quickly in the mould. Although the breech was nicely cast, the chase was cracked and the moulding at the muzzle nearly cracked off. The metal was melted down and a second gun was successfully cast. Even so, the removal of the core was a difficult process which took several days. The gun

11 Hall, 'Building and Firing a Replica *Mary Rose* Port-Piece', p.64.

12 Biringuccio, *Pirotechnia*, p.225.

The massive size of the culverin is clear from the kneeling figure aligning the barrel with the target. The loading tools were also on a monumental scale, highlighting the problems that gunners would face at sea when reloading. The carriage, if anything, looks too small to manage the weight of the gun but survived with relatively little damage. (Courtesy of the Royal Armouries)

tube had to be bored out to an extra ¼ in (6 mm) to 5¾ in which somewhat compromised the final piece, as it reduced wall thickness and was larger than the original gun. The carriage was a modified version of a reproduction carriage that had been made for the *Mary Rose* demi-cannon held by the National Maritime Museum.

The target was a reproduction of the *Mary Rose* main deck. The shot used was designed to assess the differences between shot with a variety of windages and powder loading at elevation of 0, 5, and 7 degrees. Time and finances permitted only nine shotted rounds to be fired over a two-day programme. The powder used was modern coarse-grained which was contained in silk bags. The gun was staked out with a two-inch diameter breeching rope to control recoil. The gun recoiled 600 mm on the final shot, during which the carriage was damaged with a transverse crack in the bed. The effect upon the target with every round was substantial. A clean hole was visible on the exterior but to the rear with planks, frames and stringers were smashed together with extensive splintering.

Table 24. Summary of *Mary Rose* Culverin Trials

No.	Shot diam. in/cm	Weight lb/kg	Weight Charge lb/kg	Elevation	Muzzle Velocity ft/s & m/s	
1	5/12.7	16.2/7.35	2.75/1.25	0°	472/144	1st bounce 150 m
2	5/12.7	16.3/7.4	3.3/1.5	0°	533/162.5	
3	5.25/13.35	19/8.62	3.8/1.73	Target 45m	515/157	
4	5.25/13.35	19/8.62	4.75/2.16	Target 90 m	428/130.5	
5	5.25/13.35	19/8.62	4.75/2.16	5°	NA	1st bounce 900 m
6	5.5/13.97	22.9/10.38	4.6/2.08	Target 45 m 0.5°	NA	1st bounce 1,200 m
7	5.5/13.97	22.9/10.38	4.75/2.16	Target 45 m 0.5°	653/199	
8	5.5/13.97	22.9/10.38	5.7/2.6	7°	1,492/454.9	1st bounce 1,000 m
9	5.5/13.97	22.9/10.38	7.7/3.5	7°	1,648/502.5	1st bounce 1,200 m

With only one sixth the number of rounds fired in the *Vasa* trials it is hardly surprising that the results are rather less conclusive. Clearly even at relatively low velocities the balls, irrespective of diameter, did major damage, which was as expected. There were a number of anomalous results including shot one,two and three and four, where higher powder loadings led to lower velocities. This could be explained by the larger windage of these rounds and that therefore the larger charge had no effect as so much gas escaped around the ball. This is apparent when studying slow-motion footage of the trials, as a vast plume of hot gas exits the barrel way ahead of the shot. No useful data emerged in relation to velocity and windage of the projectile as no projectiles of different windages were shot with the same load. The velocities achieved for most of the rounds were surprisingly low, little more than those achieved by the port-piece.

On the other hand, shots eight and nine produced velocities of 450–500 m/s, much greater than those achieved in the *Vasa* trials (330–350 m/s), with only the proofing round fired by the *Vasa* gun achieving a similar velocity (399 m/s). The powder charges of shots eight and nine of 2.7 kg and 3.5 kg were much larger than the standard service charge used by the *Vasa* trials of 2.2 kg. The damage to the carriage with the final shot would suggest that the combination of gun/carriage was tested to its limits. As the gun was recoiling on a damp field rather than, as in the case of the *Vasa* gun, a wooden deck this may have put more strain on the carriage?

Although these trials were beset by a number of problems, they did demonstrate the superiority of large bronze guns in achieving high velocities and causing great damage to the oak targets. The low velocity of some of the shots did not diminish their ability to penetrate the hull and still caused significant structural damage.

The *Vasa* Bronze Demi-Cannon

The *Vasa* was a massively powerful vessel when she was sailed out of harbour in 1628, a Dreadnought of her time. The *Vasa* Museum has built from scratch and test-fired the standard 24-pdr Swedish artillery piece, which was intended to be the main armament of the *Vasa* and Swedish siege artillery. The purpose of the trials was to better understand the capabilities of the *Vasa* and the tactical environment of naval warfare in the first half of the seventeenth century. In this task the trials proved a triumph, greatly adding to our understanding of these issues.

The *Vasa* had a planned total armament of 72 pieces with a main battery of 46 of the brand new 24-pdr. The *halv-kartog* (demi-cannon) was a recent design from 1620 and weighed only half that of its predecessor. They had a uniform bore of 146 mm and weighed 1.25 tonnes. The bores of 12 surviving guns varied by only 0.5 mm from 146 mm, indicating a remarkable degree of consistency. The barrel was 2.65 m long, a modest 17 calibres, far less than earlier culverin-style guns. At the breech ring it had a diameter of 420 mm, providing 137 mm of metal around the bore almost exactly the diameter of a ball. The ammunition found on the ship was far less consistent in diameter than the guns, varying from 135–142 mm. This meant a windage of 4–11 mm, while the preferred windage would have been 7.3 mm. The weight of the shot varied between 9.18–10.57 kg, and this, together with the difference in windage, would have affected long-range accuracy.

The *Vasa* cannon mounted on a firing platform at the Bofors Test Centre, on a suitably atmospheric misty Swedish morning. The combination of gun and carriage proved consistent and resilient. The 17-calibre barrel would achieve a velocity as great as that of a gun twice its calibre from earlier generations, when it was thought that barrel length determined velocity. (Photograph by Richard Belles, courtesy of the Vasa Museum)

The new gun, cast in November 2013, was a very careful copy of the original. It was cast vertically in a special pit, dug for the occasion. The bronze that was used was a median alloy of original examples with approximately 93 percent copper, four or five percent tin, and trace amounts of zinc and lead. The metal was poured into the ladle at a temperature of 1,300 degrees to ensure a pouring minimum of 1,150 degrees which was a little too hot and caused honeycombing in the chase and muzzle. The pour went well, and the mould was left to cool for a week before opening. In the spirit of the original it was cast with its bore, which was to be drawn out using a 100 tonne jack. This unfortunately failed, breaking the jack, and the core had to be bored out. The bore was then chased or honed, as was the original. There was almost a year of further preparation before firing trials, the barrel had to be cleaned of dross and planished, a noisy process that involved beating with a hammer, that provided a finer finish and a stronger metal. The substantial truck carriage weighing 350 kg and the loading tools also had to be made.

The type and quantity of powder is always a contentious issue when conducting trials. The type of powder used will have a significant effect on the peak pressure in the bore and the speed with which it is achieved. The volume of the powder will also determine the position of the ball. The gunfounder clearly took this into account when he was establishing where the breech reinforcement should extend.

The powder used was GmbH G7 3–5 mm, with the quantity calculated on the basis of muzzle velocities achieved by contemporary weapons. There is a limit to the velocities that could be achieved by muzzle loading guns because of the nature of the powder. The charge size of original guns was known, as the tubes used for transporting powder had been recovered from the ship and was 8 lb (3.33 kg) one third the weight of the ball. The trials team tested three sizes of charge to generate a muzzle velocity that contemporary shot tables suggest was typical and arrived at 2.65 kg. They then used two reduced charges of 2.2 kg and 1.1 kg to simulate impacts at longer range. Even with this low charge the target was easily penetrated. There was a variety of ammunition used including, round shot, crossbar shot (ball with a spike through it), scissor, and chain, as well as two rounds of case shot. In total 54 rounds were fired and 125 kg of powder used.

The target was a reconstruction of the lower gun deck of the *Vasa* from the waterline to just above the gundeck. It weighed five tons, was 4.5 metres long and 3.5 metres high. The hull section included three large knees and the hull itself constituted between 45 and 75 cm of green oak. It was held in place quite firmly, but it was clear that high velocity projectiles hardly ever impart much kinetic energy into the target due to inertial differences. Behind the target at 1,000 metres was a granite wall backstop and at shorter range an area of marshland.

A gun crew of seven was thought optimal in the limited space of the gundeck on the basis of experience. This consisted of, 'gunner, a loader and a dedicated rammer/sponger, assisted by four, sometimes only two, men who moved the gun into or out of battery and assisted in aiming it, as well as passing ammunition to the loader'.[13] The jobs could be shared between them as long as each clearly understood their roles. A well-drilled gun crew is a well-choreographed belligerent ballet, a badly-drilled crew are a danger to themselves and others.

The gun recoiled at 3.4 m/s, which would have been virtually impossible to avoid by a crew member who was carelessly standing behind it. The breeching rope limited recoil of the carriage to 1.5 m, which would bring the cannon inboard far enough to be reloaded. Now resting on the crown of the deck it would be much easier to roll the gun downhill to be fired again, especially if the roll of the ship was used. One surprising feature of the trials was that although the noise outside the hull was deafening, one metre behind the breech it was one sixth of that. Inside the ship with the gun and carriage filling the gunport the noise level would be even less.

There were four proofing rounds fired in order, 1.1 kg, 2.2 kg, 3.3 kg, 2.65 kg. The overproof round of 3.3 kg achieved a maximum breech pressure of 74 mPa (about 10,700 psi) and velocity of 399 m/s (Mach 1.17). Using the service charge a muzzle velocity of 360 m/s (Mach 1.07) was achieved but due to the inefficiency of the spherical projectile it was subsonic within 100 metres, with a 2.2 kg charge a muzzle velocity of 300 m/s was reached. The range with 3° of elevation was 1,000 metres, and at 40° it would have reached 4,200 metres but with little or no accuracy.

13 Fred Hocker, *Understanding the Gundeck Experience*, pp.125–140. In Johan Rönnby (ed.), *On War on Board: Archaeological and historical perspectives on early modern maritime violence and warfare* (Stockholm Södertörns högskola, 2019).

The ammunition used in the test was representative of the types available, with round shot, chain shot, crossbar and scissor shot being fired. Round shot was designed to damage the hull whilst the other types would be of use against sails and especially the standing rigging. (Photograph by Evelyn Ansel, courtesy of the *Vasa* Museum)

Accuracy at short range was impressive and would have produced a group at 20 metres of 50 cm in diameter. Extrapolated to a *Vasa*-sized target at 1,000 m the team calculated that there would be 11 hits to the hull, eight to the rigging and three misses. In practice this level of precision was unlikely from a ship at sea. From a Device fort such as St Mawes, hitting a ship in the middle of the channel to Falmouth harbour would have been a reasonable prospect, with a crew that knew its business.

Fred Hocker, who managed the programme, was sanguine when considering the practical engagement range of this gun. He has pointed out that at 1,000 m the *Vasa*-sized target would subtend only 0.2° and that the roll of the engaging vessel would be up to 6°. The point of aim for the hull would be well above the topmast of the enemy ship. The slow speed of contemporary vessels, perhaps five knots and the velocity of the projectiles, covering 1,000 m in three seconds meant that little adjustment needed to be made for the relative movement of ships. Ignition time was one to three seconds, travel time of the shot in the air to a target 1,000 m away was three seconds. A ship travelling at five knots (an average or high sailing speed under battle canvas) covers just three metres in six seconds. Even if both vessels are on opposing courses the relative speed distance only accounts for six metres of displacement. Given a spread of shot of 20 metres at that range, the relative movements of the ships were not tactically significant. It would seem that successfully engaging a target, even as large as a ship, at ranges over a few hundred yards would be a matter of sheer luck.

The *Vasa* gun was a most impressive piece and performed flawlessly. The results in terms of the performance of the projectiles and their impact on target were definitive. There would be little discernible improvement in artillery for over 200 years for the simple reason that what was a near perfect combination of material, propellant and design had been achieved.

Conclusion

It is difficult to appreciate the speed at which the projectiles fired from these very varied pieces flew. The late medieval bronze gun and iron port-piece sent their large stone projectiles at 300–400 mph, while the bronze culverins let loose their iron shot at 600–700 mph sometimes exceeding the speed of sound at the muzzle. It was hardly surprising that they did so much damage to the targets that replicated the vessels against which they would have fought.

All the guns performed well, safely and consistently when a moderate service charge was used. The late medieval piece was not used against a target but would have performed well in the field or against a ship's hull. The excellent performance of the port-piece as a ship smasher was perhaps more surprising than the excellent performance of the *Vasa* gun. If anything, the 24-pdr was unnecessarily powerful as it achieved a similar impact on the target to the port-piece but did not extend the engagement range as much as one might have expected. The *Mary Rose* culverin and *Vasa* guns both achieved an impressively long range, 1,100 metres at 7° degrees and 1,000 metres at 3° of elevation respectively.

Fred Hocker and Howard Douglas have pointed out that long range fire was almost irrelevant at sea. On the other hand, for coastal defence purposes from one of Henry VIII's Device forts, this long range could have been used effectively. A heavy gun when used in a siege would be deployed a few hundred yards at most in front of its target and again preferably far less. Field guns would have engaged targets with direct fire, which would mean usually again at only a few hundred yards, or point-blank range. What those who ordered, manufactured and served these guns appreciated from the new bronze guns was their safety, reliability and high velocity. This gave them the ability to smash ships and fortification and provided them with a flat-shooting weapon that minimised the problem of aiming.

10

Conclusion

If I may address myself to some of the criticisms of early artillery that Kay Smith highlights, that it was 'unsafe', 'leaky' and 'lacking in power'.[1] I think it is apparent that it was none of these things. Everything that could be done to ensure the guns were well-designed and 'safe' was done. 'Leaky' they certainly could be if you look at photographs of gas escaping from the breech to hull join of the port-piece but that was the consequence of the design not a flaw and the gun itself performed very respectably. 'Lacking in power' hardly describes the performance of any of these guns.

Were the guns inaccurate? Not at the ranges at which they were used. Were they ineffective? Again, in action it depended on the competence of commanders and crews and the nature of the target, but they were certainly as effective as the guns that would succeed them over the next few hundred years. Were they difficult to transport? The development of the limber made their transport massively easier and with enough horses or oxen they could be moved at a fair rate, taking into account the condition of the roads and the distance they had to travel. What mattered was whether they had the long-term logistical and technical support to both serve and maintain them. There were certainly deficiencies in standardisation, which were the inevitable consequence of the processes by which they were acquired and designed, as well as their longevity. The necessity of having different loading tools and ammunition stocks was inconvenient but it seems that this was more a cause of individual frustration than martial catastrophe.

The author's falconet (see colour plates).

It is only when we try to reproduce these weapons that we can begin to understand the skill and commitment of those who first made them. Even with modern technology it has

1 Smith, 'All Manner of Peeces', p.130.

sometimes been a challenge to reproduce what they made by hand, eye, intelligence and experience. The gunners themselves were no fools. They lacked our modern understanding of ballistics and the chemical processes that underlay the manufacture of gunpowder, but they were not ignorant of what mattered and did their best to manage their guns safely and hit their marks. I hope if my loyal reader has reached this point in the story, that they will be as in awe of my predecessors, shrouded as they are in the smoke and heat of the foundry and the battle, as I.

Appendix I

Annotated Drawings

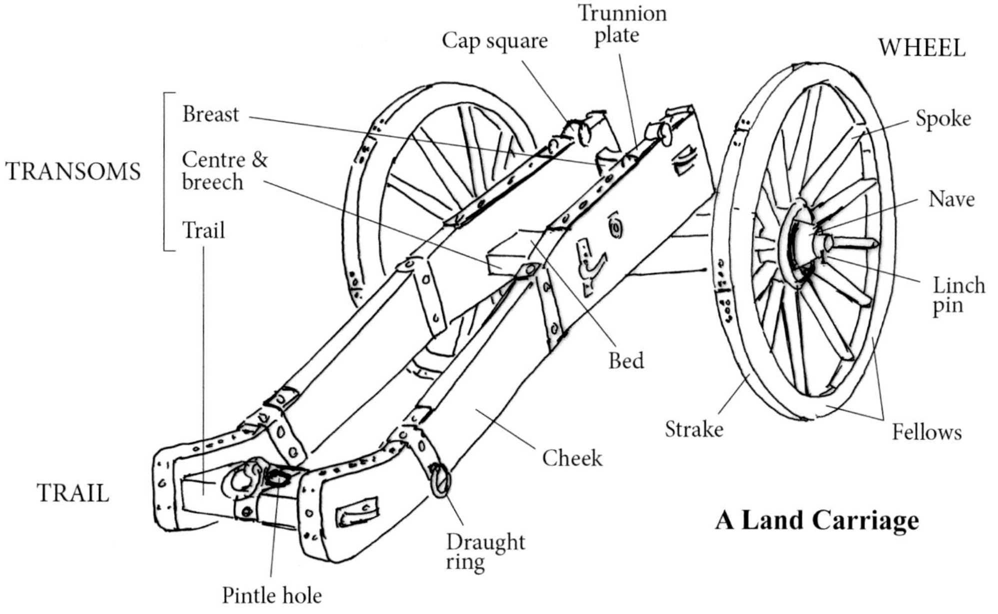

A Land Carriage

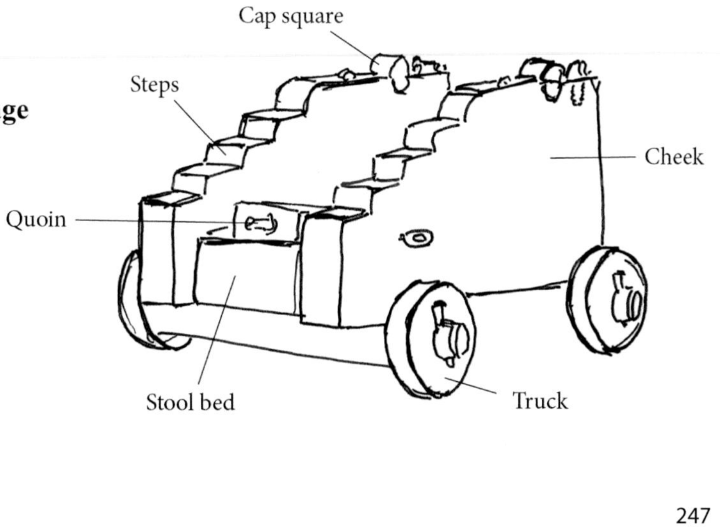

A Sea Carriage

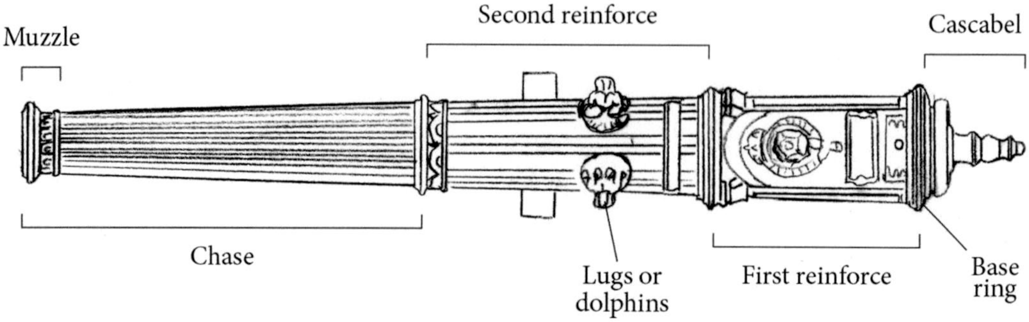
Muzzle
Second reinforce
Cascabel
Chase
Lugs or dolphins
First reinforce
Base ring

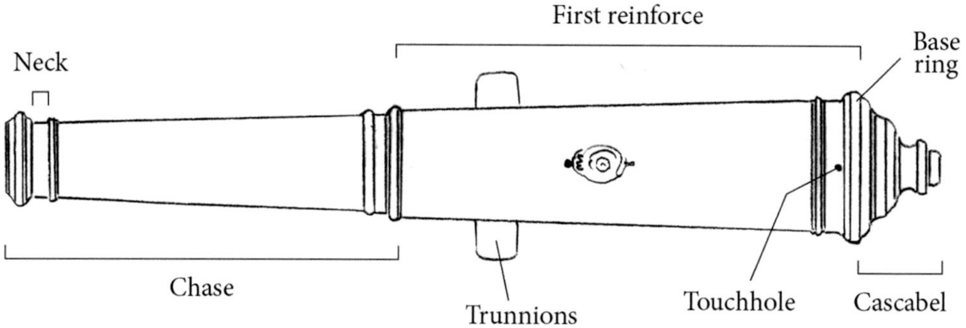
Neck
First reinforce
Base ring
Chase
Trunnions
Touchhole
Cascabel

Appendix II

List of Artillery Required to Support an English Army of 15,000 Foot and 3,000 Horse 1513

Letters and Papers, Foreign and Domestic, Henry VIII, Volume 1, March. 1513 (SP Hen. VIII, 7, f. 51, R.O. 1726 [4633] ORDNANCE)

Name	Wt of iron shot lb	Wt of powder lb	Rounds per day
Bombard	260	80	5
Curtow	60	40	40
Culverin*	20	22	36
Nuremburg piece*	20	20	30
Apostle*	20	20	30
Lizard*	12	14	37
Minion*	8	8	N/A
Pot gun	56 (stone)	40	N/A

* The equivalent weight of powder and shot would suggest culverin construction. The lower proportion of powder to shot for the curtow would suggest the use of a sub-calibre chamber or cannon construction

Appendix III

Lists of Artillery Dated 1535–63 Quoted in Caruana, *Tudor Artillery*, pp. 13–14

Name	Calibre (in)	Weight of Shot (lb)
Cannon royal	8–8¼	60–68
Cannon	7½	44
Demi-cannon	6¼–6¾	30–35
Cannon perrier	8¼	30
Culverin	5¼–5¾	18
Demi-culverin	4½	9–9½
Saker	3¼–3¾	5½–7
Minion	3¼	4
Falcon	2¼–2¾	1¾–2½
Falconet	1¾–2¼	1¼–1½

Appendix IV

European Standardisation of Ordnance

Drawn from Miller and Embleton, *Landsknechts*, p.153.

Emperor Maximilian I			Emperor Charles V		
Type	**Weight (tons)**	**Weight of shot (kg)**	**Type**	**Weight tons**	**Weight of shot kg**
Scharfmetze	5	50			
Basilicus	3¾	35			
Nachtigall	3	25	Cannon royal	2.9	18.7
Singerin	2	10	Medium piece	2.3	11.2
Large kartaune	1½	8			
Small kartaune	1¼	5	Culverin	1.38	5.6
Rothschlange	2½				
Demi-rothschangle	1	3.5	Saker	3	3
Sau (bauer, Ochse)	1	10			
Falkaune	1¼	2.5	Demi-culverin	1.23	2.8
Falkonet	¾	1	Falconet	1.4	1.4
Scharffetinnle	1½	¼			
			Mortar	2.6	46.7

Henry II of France 1552	
Type	**Weight of shot in livres**
Cannon	32
Great culverin	15
Bastarde	7
Moyenne	2
Faucon	?
Fauconneau	?

Appendix V

List of Ordnance from Bourne's *The Arte of Shooting in Great Ordnaunce* (London, 1587), pp. 65–71

Name	Cal. (in)	Lth of piece (ft)	Wt of piece (lb)	Diam shot (in)	Wt of shot (lb)	Wt serp. powder (lb)*
Old double cannon	8¼	12	8,000	8	70	46
Ordinary double cannon	8	11/12	7,500	7¾	64	42
French double cannon	7¾	11/12	7,000	7½	58	40
Old demi-cannon	6¾	11/12	6,000	6½	38	26
Ordinary demi-cannon	6½	10/11	5,500	6¼	33	24
Small demi-cannon	6¼	10/11	5,000–5,400	6	30	24
Foreign demi-cannon	6	10/11	5,000	5¾	26½	22–23
Old or Nuremburg culverin	5½	12/13	4,800	5¼	20	20
Ordinary Nuremburg culverin	5¼	12	4,500	5	17	18
Small Nuremburg culverin	5	various	4,300	4¾	15	16
Old demi-culverin	4¾	12	3,200	4½	12½	12
Ordinary demi-culverin	4½	10	2,700	4¼	10¾	11–12
Small demi-culverin	4½	9–10	2,200	4	9	10–10½
Old saker	4	10	1,800	3¾	7½	7½
Ordinary saker	3¾	8–9	1,500	3½	6	6
Small saker	3½	8	1,300–1,400	3¼	4¾–5	5–5½
Minion	3¼	8	1,000	3	3¾	3¾–4
Ordinary minion	3	8	900	2¾	3	3
Falcon	2¾	7	700–750	2½	2⅛	2½
Foreign falcon	2½	7	600–650	2¼	1¾	2
Falconet	2¼	5–6	360–400	2	1⅛	1¼

Note: Powder weights are calculated for serpentine not corned powder. Almost 50 percent more serpentine powder was required than corned powder if the figures for these two tables are calculated against the weight of shot.

Appendix VI

English Ordnance *c.* 1590: John Sheriffe SP 12/242

Name	Calibre (in)	Weight (lb)	Weight shot (lb)	*Length in calibres*	Weight Powder §	Point Blank	Extreme range
Cannon royal*	8.5	7,000	66		30	320	1,930
Cannon*	8	6,000	60	*15/16*	27	340	2,000
Cannon serpentine*	7.5	5,500	53½		25	400	2,000
Bastard cannon*	7	4,500	41¼		20	360	1,800
Demi-cannon*	6.5	4,000	30¼		18	340	1,700
Cannon perrier*	6	3,000	24¼		14	320	1,600
Culverin**	5.5	4,500	17½	*28*	12	400	2,500
Demi-basilisk**	5	4,000	15¼		10	–	–
Demi-culverin**	4.5	3,400	9½	*32*	8	400	2,500
Bastard culverin**	4	3,000	7		6¼	360	1,800
Saker***	3½	1,400	5½	*36*	5½	340	1,700
Minion***	3¼	1,000	4	*30*	4	320	1,600
Falcon of 2½***	2½	800	3	*42*	3	300	1,500
Falconet***	2	500	1¼	*48*	1¼	280	1,800
Robinet	*1.25*	*200*	*1*	*50*			
Basilisk	*8.75*	*9,000*	*60*	*14*			

* These pieces be most serviceable for battery being within 80 paces of their mark which is the chief of their force. ** Theses pieces be good and serviceable to be mixed with the above ordnance for battery to pierce being crossed with the rest as also fit for castles, forts and walls to be planted for defence. ***These pieces are good and serviceable for the field and most ready for defence.

§ Almost certainly corned powder

Note: The data in italics is drawn from the table of guns contained in Hogg, Artillery: *Its origins, heyday and decline (London: C. Hurst and Co., 1970)* p. 266

Note: Diameter of shot would be ¼ inch less for windage

Appendix VII

A Table of Proportion for the Weight and Shooting of Great Ordnance

John Smith, *A Sea Grammar* (London: Printed by John Haviland, 1627), after p. 69.

'These Peeces be most seruiceable for battery being within 80 paces to their marke, which is the chiefe of their forces'.

The names of the great Peeces.	**The height of the peeces.**	**The weight of the peeces.**	**The weight of the shot.**	**The weight of the powder**	**The bredth of the Ladle.**	**The length of the Ladle.**	**2,400 lb of powder makes of shot in a**	**Shot point blanke in**	**Shot randome in**
	Inches.	Pound.	Pound.	Pound.	Inches.	Inches.	Piece.	Pace.	Pace.
A Canon Royall.	8½	8,000	66	30	13¼	24½	80	16	1,930
A Canon.	8	6,000	60	27	12	24	85	17	2,000
A Canon Sarpentine.	7½	5,500	53½	25	10½	23⅓	96	20	2,000
A Bastard Canon.	7	4,500	41¼	20	10	23⅓	120	18	1,800
A demy Canon.	6½	4,000	30¼	18	9⅓	23¼	133	17	1,700
A Canon Petro.	6	3,000	24¼	14	9	23	171	16	1,600
A Culuering.	5½	4,500	17⅓	12	8½	22⅓	200	20	2,500

'These Peeces be good and also serviceable to be mixt with the above Ordnance for battery to peeces being crost with the rest, as also fit for castles, Forts, and Walls to be planted, and for defence'

The names of the great Peeces.	**The height of the peeces.**	**The weight of the peeces.**	**The weight of the shot.**	**The weight of the powder**	**The bredth of the Ladle.**	**The length of the Ladle.**	**2,400 lb of powder makes of shot in a**	**Shot point blanke in**	**Shot randome in**
	Inches.	Pound.	Pound.	Pound.	Inches.	Inches.	Piece.	Pace.	Pace.
A Basilisco	5	4,000	15¼	10	7½	22	240	25	3,000
A demy Culuering	4½	3,400	9⅓	8	6⅕	21	300	20	2,500
A bastard Culuering	4	3,000	7	6¼	6	20	288	18	1,800
A Sacre	3½	1,400	5⅓	5⅓	5½	18	490	17	1,700
A Minion	3¼	1,000	4	4	4½	17	600	16	600
A Faulcon	2½	660	2¼	2¼	4½	15	1,087	15	1,500
A Faulcon	2⅓	800	3	3	4¼	15	800	15	1,500

'These Peeces are good and seruiceable for the field, and most ready for defence'

The names of the great Peeces.	**The height of the peeces.**	**The weight of the peeces.**	**The weight of the shot.**	**The weight of the powder**	**The bredth of the Ladle.**	**The length of the Ladle.**	**2,400 lb of powder makes of shot in a**	**Shot point blanke in**	**Shot randome in**
	Inches.	Pound.	Pound.	Pound.	Inches.	Inches.	Piece.	Pace.	Pace.
A Faulconet	2	500	1¼	1¼	3¼	11¼	1,950	14	1,300
A Sarpentine	1½	400	½	⅓	2½	10	7,200	13	1,300
A Rabonet	1	300	½	½	1½	6	4,800	12	1,000

Appendix VIII

List of Ordnance from *The Gunners Dialogue* by Robert Norton (1628), p.7

Names of the Peeces	Height of Bore, Inches	Length in Diametres	Weight in Metal, Pounds	Weight of Powder	Length of the Ladles Inches
Cannon of 8	8	15	8,000	40 li	24
Cannon of 7	7	16	7,000	25 li	22
Demi Cannon	6½	18	6,000	20 li	21
Culuering	5½	28	4,500	15 li	20
Demi Culuering	4½	32	2,500	9 li	18
Saker	3½	36	1,500	5li ¼	16
Minion	3¼	30	1,200	3li ¾	15
Falcon	2¾	42	700	2li ½	14
Falconet	2¼	48	500	1li ¼	12
Cannon Perior	9.10.12	8	3,500	3 li 3 li ⅓ 4 li	3
Demi Can Drake	6½	16	3,000	9 li	4½
Culuering Drake	5½	16	2,000	5 li	4½
Demi Cul. Drake	4½	16	1,500	3 li ½	4½
Saker Drake	3½	18	1,200	2 li	4½

Appendix IX

French Ordnance from *The Art of Warre*, by the Lord of Praissac. Cambridge, 1639.

Of French Artillerie. Of foundings, mixtures, and measures of ordinance. CHAP. XIII pp.107–121.

The measures of the parts of a canon
The diameter of the bore of the canon is six inches and one sixt part of an inch.
The bullet weigheth 33 pound and one third part
The cilinder or concave is nine foot long, and equally wide in each part of it.
The metall of the piece weigheth about 5600 pound.
The canon may be laden ten times in an houre, and 120 times a day.
She shoots point blank almost six hundred paces. (Observe that paces are here understood to be onely steps [30 inches]: otherwise it were impossible).
There must be twenty horses to draw a cannon
For the service of a canon imployed for batterie, there must be two canoniers, three chargers, and thirty pioners.

Of the great culvering.
The diameter of her bore is foure inches and five sixt parts.
The bullet weigheth 15 pound and a quarter.
Her cilinder is nine foot & foure inches long.
She shoots point blank 800 paces. (Understand paces as hath been observed before for the canon.)
She may be laden ten times in an houre, and 120 times in a day.
Her metall weighs about 4000 pound weight.
She requireth 17 horses to draw her.
24 pioners, two ordinarie canoniers, and two extraordinarie.

Of the bastard culvering.
The diameter of her bore is three inches and three fourth parts.
Her cilinder is eight foot seaven inches in length.
Her bullet weighs seaven pound and a quarter, and her powder five pound.
She shoots point blanck 1000 paces.
She may be laden 140 times in a day, which is about twelve bullets in an houre.
Her metall weigheth about 2500 pound.
She must have thirteen horses to draw her.
twelve pioners, and two canoniers.

Of the minion.
The diameter of her bore is two inches and three fourth parts.
The diameter of her bullet is two inches and two third parts.
Her cilinder is eight foot long.
She shoots point blanck 1400 paces.
She shoots 160 bullets a day, which is about thirteen each houre.
Her metall weighs about 1350 pound.
She must have nine horses to draw her.
She is served with six pioners, and two canoniers.

Of the faulcon.
The faulcon may be laden fifteen times in an houre, which is 180 times a day.
Her bullet weighs one pound and an half,
her metall eight hundred pound.
She must have five horses to draw her.
She is served with foure pioners, and one canonier.

Of the faulconet.
The faulconet may be laden about 250 times a day, which is about twenty times each houre.
Her bullet weigheth three quarters of a pound,
her metall five hundred pound.
One horse draweth her.
She is served with 4 pioners and one canonier.

Appendix X

Details of the Ordnance Supplies for the Invasion of France 1544

Letters and Papers, Foreign and Domestic, Henry VIII, Volume 19 Part 1, 31 March R O 13.

The proporcyon off ordenaunce, artylery and habyllymentes of warr for the furnyture of an armye ryall wythe all maner of munycyone necessarye for the same.
Fyrst for the forewarde:—
Cannons 2; lymmeners 70; powder 9 last, caryages 12; shott 720 caryages 21.
Demi canons 4; lymmeners 100; powder 12 last, caryages 16; shot 1,440, caryages 21.
And so on, viz.—Culveryns 2, 1. 50, p. 4½ last 725 lb., c. 6, shot 720, c.8
Demi-culveryns 4, 1. 64, p. 5 last 4½ barrels 60 lb., c.6 , shot 1,440, c. 7.
Sakers 10, l. 100, p. 2 last I bar, c. 3, shot 1,000, c. 2. Fawconnes 8.
1. 48, p. 10 barrels, c.I, shot 500, c. 1. Mortars —— [*blank, with blank spaces for lymmeners &c.*] Prevy wagons 15,1. 80, corn powder 12 demi-barrels, c. ... [*mutilated*], shot of lead 17 cwt., C.I. Spare lymmeners 30-Demi-hakes 250, c. 2. Morisspykes 1,000, c. 6. Bows 3,000, eight chests to a load, c.8. Bowstrynges, 20 barrels 200 gross, c. 3. Levery arrowes, 6,000 sheaf in 160 chests, c. 16. Demi-lawnces 200 and northern staves 300, c.5,

Bylles 1,000 (200 to a load), c. 5. Pykeaxys and mattokes 500, c.3. Mynars toolys, c.2. Shovelles and spades 1,000, c. 4. Skowpes 100, 0. 1. Crowes of yron 50, c. 1. Hedgyng bylies 300, c. 2. Fellyng axys 200, c. 1.

Horse shewes 6,000, c. 8. Horse shew nayles 300,000 and nayles of sundry sortes, I barrell, c. 1. Tonnes of yron 3 and in steele 20 sheves, c. 4. See cole[1] 6

1 Sea coal – the contemporary reference to coal which was often collected from exposed seams to be found on sea coasts.

chalder, c. 7. Smythes forges I, c. 1. Ferrars forges I, c. 1.

Cressettes 50, cressett lyght 2,500, c. 3. Lanterns 150, c. 2. Ropes of hemp of all sortes 6,0001bs. weight, and bast ropes 30 pieces, c. 4. Cotton matches or lyntes 2 cwt., ballances and weightes 2 pair, canvas for carte-wyches 30 ells, partchement 30 skins, paper ryall I roam, shearys 3 pair
twyne 301b. weight, pakknedles 30, mowthe glew I oz., c ... [*mutilated*] Tallow candelles 5 barrels, tallow 3 barrels and blak sope 2 barrels, c. 2. Spare lymmers 20 pair, c. 1. Spare exultres [*axle trees*] with barres 10, c. 1. Spare exultres for small ordynaunce 50, c. 1. Spare wheles for battry peces 12 pair, c. 6. Spare wheles for small ordynaunce 12 pair, lynce-pynnes wasshers, tuk pynnes and shettle-pynnes I barrel, c. 5. ferons alias gynnes to mownte ordynaunce 8, and crekes 2, c. 1. Tall pressys 6 and tresselles or bearers 6, c. 1. Hande spekes 100, carte clowtes 120, and brodes and clowte nayles 6,000, c. 1. Chargyng ladles 30, moldesbrasse to shott of all sortes 8, castyng ladles 8, dyce of yron I demi barrel, demi barrelles and fyrkyns with purses for gunpowder 10,c. 1 Peyles [*pails*] of lether 150, sythes 10 doz., syckellles 10 doz., and vyneger 1 hogshead, c 1. horse hydes tanned 60, carte saddles 100, trasys in payeres 150 belly bandes 200, hames 50 payre spare chevyn boltes 10, grynde stones 30, c.4.

Appendix XI

The Artillerie more or lesse, that is used to be conducted with an Army

From *The Theorike and Practike of Moderne Warres* by Robert Barrett, 1598.

30 Cannons for battery, that shoote bullet from 44, 50, 60, and some 70 pound weight; from 6, to 10 inches in height.
20 demy Cannons, from 24 pound bullet, to 30 pound.
20 Culuerins, from 16 pound, to 20 pound bullet.
15 demy Culuerines; 20 Falcons: 25 Falconets; and 12 Sakers.

5000 Quintals or Centenaires of Cannon powder, to remaine in the place of garrison of most respect, oxe hide to couer the powder in cariage; 600 quintals of hargubuze powder, and a number of lether bagges or satchels, to cary powder behind men on horsebacke, and when any battery is set.

9 or 10 thousand Cannon shot, 7 or 8 thousand demy Cannon shot, 3 or 4 thousand Culverine shot, 3 thousand demy Culverine shot, 2 thousand Falcon shot, 2 or 3 thousand bullets of lead for Falconets, 6000 shot of diverse sizes. All these shot and bullets must be carried in coffers, and strong maunds made of oziers or mats, 300 Centenaires of lead, 500 Centenaires of match, 2000 iron shovels, 7 or 8 thousand mattocks, 4 or 5 thousand pickaxes, 2000 axes and hatchets to cut wood, 1500 wood hookes, and tooles to make faggots: the greatest part of all these iron tooles are to have handles, and healves. There must also bee tables, plankes, & boordes, & good store of such other timber for many necessary causes.[1]

1 Barret, *The Theorike and Practike of Moderne Warres* (London:1598), p.104.

Appendix XII

The Properties, Office and Duetie of A Gunner – Cyprian Lucar, *Appendix*, p. 1

A Gunner ought to be a sober, wakefull, lustie, hardie, patient, prudent, and quick sprited man, he ought also to have a good eysight, a good judgement, and perfect knowledge to select a convenient place in the day of service, to plant his Ordinance where he may doe most hurt unto the enemies, and be least anoyed by them, and where his Ordinance may not be surprised by the enemie.

A Gunner ought to be skilfull in Arithmeticke, and Geometrie, to the ende he may be able by his knowledge in those artes to measure heights, depthes, breadthes, and lengthes, and to drawe the plat of any peece of ground, and to make mines, countermines, artificiall firewoorkes, rampiars, gabbions or baskets of earth, and such like things which are used in time of warre to be made for offensive and defensive service.

A Gunner ought also to procure with all his power the frendship and love of every person, and to be carefull for his owne safetie, and for the preservation of all those that shal be about him. Also he ought to be no surfeter nor a great or sluggish sleeper, but he must governe him selfe in al times as a wise, modest, sober, honest, and skilfull man ought to doe, that through want of understanding he may never leese his credite, nor an universall victorie which oftentimes by the meanes of good Gunners well managing their peeces is gotten.

Also a Gunner ought at the receite of his charge to make an Invitorie of al such things as shall be committed to his charge as well to render an account, as to consider the want of such necessaries as to the Artillerie apperteineth.

And when a Gunner shall be appoynted to doe an exployte, he ought to want neither a fire stone, nor a tyndar box with a good steele, nor flintstones, nor tindar, nor gunmatches, nor a flaske full of good touchpowder to kindle his gunmatch and fire, when neede shall require.

Alessandro Capobianco in his study, from the frontispiece of his book *Corona e Palma Militare di Arteglieria*, printed in Venice in 1598. This volume dealt with the invention and employment of artillery on land and sea, pyrotechnics for war and peace, as well as 'new' instruments to measure distances. It also pointed to the 'errors discovered in ancient fortresses' and the superiority of the new designs, as represented in the diagram of fortifications on the table. Fig. 86 in Guttmann's *Monumenta*.

Also when a Gunner shall be appoynted to doe an exployte, he must lay his powder twenty paces from his Peeces in such a place where no fire, water, or hurt may come unto it through any person, or by reason of any winde, weather, or otherwise, and keeping his powder alwayes covered, he must not be unmindfull of this, that it is a very dangerous thing for a Gunner to trust many, because a generall hurt and death may thereuppon followe.

Also a Gunner that hath a charge ought to have alwayes in a readinesse all necessarie things for his artillerie: that is to say, wheeles, axeltrees, ladles, rammers, spunges, gunpowder, pellettes, tampions, chaineshot, crossebarres, fustian, canvas, or paper for cartredges and firewoorkes, fourmes for ladles and cartredges, needles, threed to sowe and binde the cartredges and firewoorkes, artificiall torches, candles, lantornes, mattockes, shovels, crowes of yron, handaxes, leuers, enguines for the mounting and imbasing of ordinance, ropes, little handbaskets, glwe or payst, horsecollers, horses or oxen to drawe his peeces, all manner of cartware, carters to guyde and keepe his horses and oxen, and a sufficient number of Gunners and assistants to charge, discharge, mount, imbase, wadde ramme, make cleane, scoure, and coole his peeces when they are over heated, and to have for this purpose vineger and faire colde water.

Also a Gunner ought alwayes to have a gunners staffe, or a partisant, or a halbert sticking by him for a part of his defence, and he ought to put into the cocke of his Gunners staffe a gunmatch, or wrappe about the lower end of his staffe, partisant and halbert a good gunmatch, which may geve fire unto his peeces of artillerie when neede shall require.

Also a Gunner ought not to sleepe much at any time of the day, or night, when he is appoynted to serve in the fielde, or in any other place, nor to eate or drinke in any other roome than where his peeces of artillerie are planted, because in his absence the same peeces may be choked, poysoned, and harmed by divers wayes; and that he may many times vppon a sudden have good occasion to discharge all his peeces.

Appendix XIII

Useful Data and Formulae

Comparative Weights of Materials

Material	Quantity	Weight
Water	Gallon	10 lb
Water	Cubic foot	62½ lb
Granite	Cubic inch	1.33 oz
Granite	Cubic foot	144 lb
Iron	Cubic inch	4 oz
Iron	Cubic foot	450 lb
Bronze	Cubic inch	4.7 oz
Bronze	Cubic foot	520 lb

Relative Densities of Materials

Water	1	Cast-iron	7.1–7.7	Bronze	8.7–8.9
Granite	2.5–3	Wrought iron	7.8–7.9	Lead	11.37
Marble	2.5–2.8	Steel	7.7–7.9	Gunmetal	8–8.4

Formulae

Volume of a cylinder = $\pi r^2 l$ where r=radius l=length
Volume of a sphere $\pi d^3/6$ where d=diameter

To calculate the weight of a gun, employ the following formulae:

For a bronze gun $3(D^2L - 5/6\, d^2 l)$
For an iron gun $2.5(D^2L - 5/6\, d^2 l)$

D = mean external diameter in inches; d = calibre in inches; L = external length in feet
L = internal length in feet

Weight of Shot

Diameter in inches	Iron	Lead	Stone
1	2 oz	3 oz	½ oz
1.5	6½ oz	9 oz	2¼ oz
2	1 lb 2 oz	1 lb 9 oz	6¼ oz
2.45	2 lb	2 lb 8 oz	11 oz
3	3 lb 6 oz	4 lb 13 oz	1 lb 3¼ oz
4	8 lb 15 oz	11 lb 4 oz	2 lb 13 oz
5	18 lb 10 oz	23 lb 5 oz	5 lb 13 oz
6	30 lb 10 oz	47 lb 8 oz	11 lb 14 oz
7	42 lb 14 oz	64 lb 4 oz	16 lb 1oz
8	72 lb 10 oz	107 lb	26 lb 1oz
9	91 lb 2 oz	136 lb 12 oz	34 lb 3 oz
10	138 lb	206 lb 8 oz	51 lb 10 oz
11	187 lb	290 lb 4 oz	72 lb 9 oz
12	24 lb	380 lb	95 lb

1oz (troy) = 480 grains = 31.1 g
1oz (avoirdupois) = 437.5 grains = 28.3 g

Appendix XIV

Engravings on the Art of Casting Cannon, from Diderot's *Encyclopédie* (France 1765)

'Cannon casting – [3] [Plates XI – XVII].' *The Encyclopedia of Diderot & d'Alembert Collaborative Translation Project*, Translated by Steven A. Walton (Ann Arbor: Michigan Publishing, University of Michigan Library, 2010, web, accessed 30 January 2021 <http://hdl.handle.net/2027/spo.did2222.0001.493>); translation of 'Fonderie des canons – [3] [Planches XI–XVII],' *Supplément à l'Encyclopédie ou Dictionnaire raisonné des sciences, des arts et des métiers*, vol. 5 (plates). Paris, 1765.

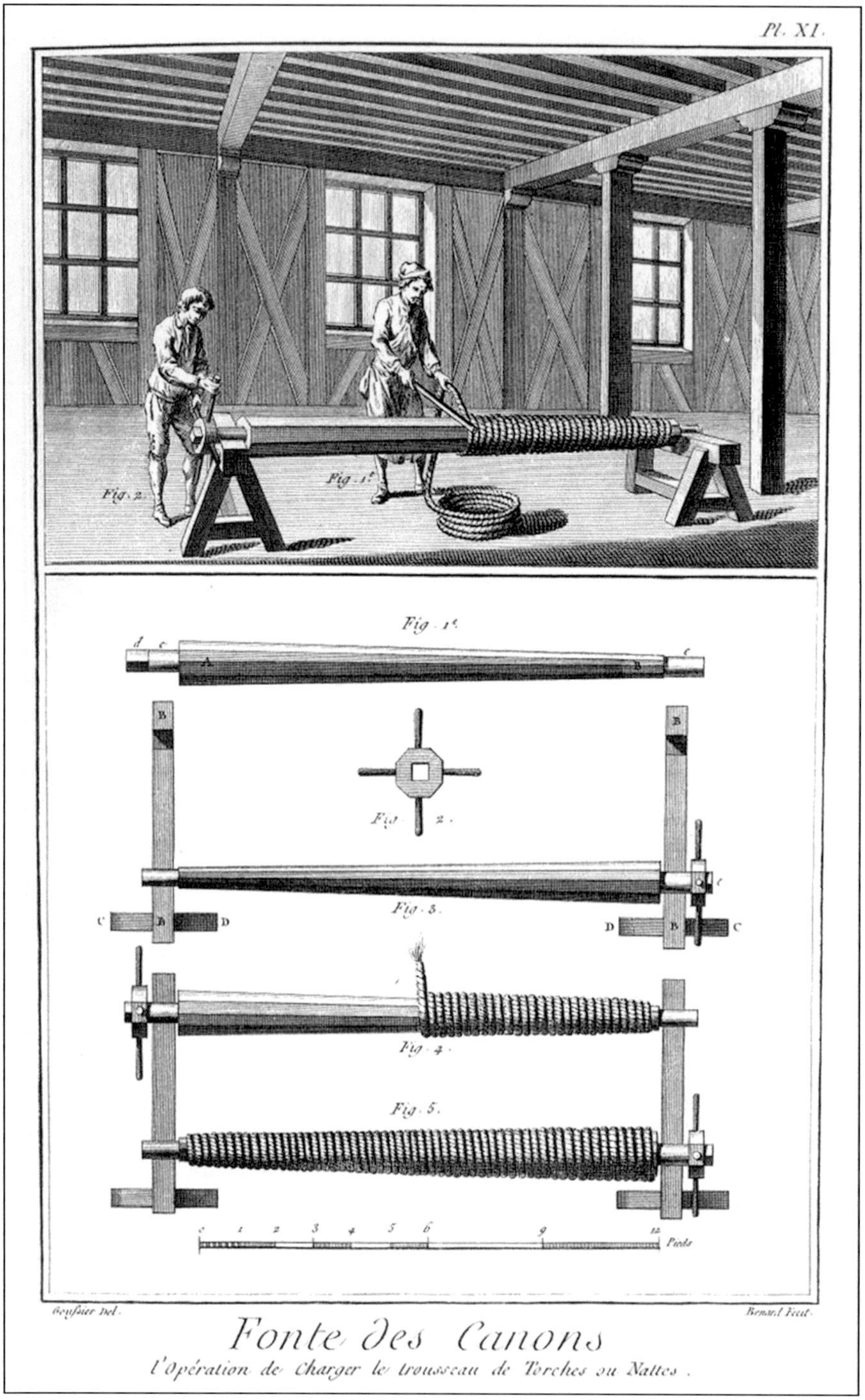

The process of making the pattern began with a wooden form around which rope was bound. The rope provided a rough surface to which the layers of clay could more readily adhere. The wooden form is faceted rather than round, this is to ensure a snugger fit for the rope. Note the use of a 'beetle' or club to make a tight fit. Two layers of rope were required to avoid too angular a surface.

The outer clay model of the barrel, but not the breech, has been completed and the decoration, dolphins and trunnions are being added. The use of wooden patterns speeded the process of decoration significantly. Even so, in preparation for the Armada Spanish gunfounders requested permission from King Phillip to not apply the royal coat of arms in order to speed the process of manufacture, a request which they were granted. Note the position of the trunnions, that had migrated to the bottom from the midpoint of the barrel at the beginning of the sixteenth century.

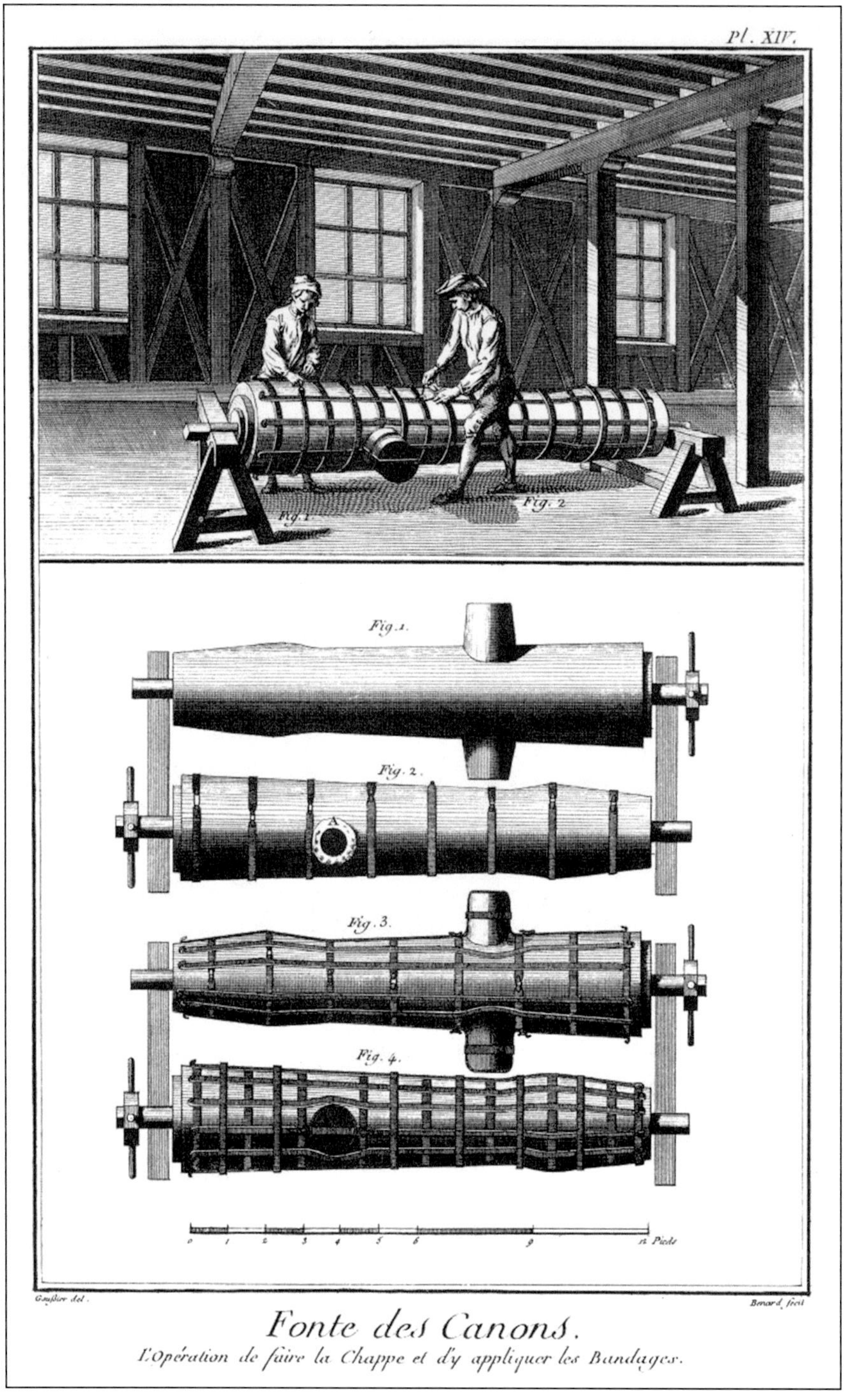

The pattern has been covered with a parting layer and another coating of clay, as described in detail by Biringuccio. Note the very swollen 'head' which was essential in providing a reservoir of molten metal and adding to the weight of bronze that would ensure a stronger breech. The clay mould has been completed and the iron bands placed around it, to which have been attached iron rods, the reverse order which Biringuccio suggests but which achieved the same result. It was essential that the mould be strong enough to stand the weight of the molten metal, or disaster would ensue.

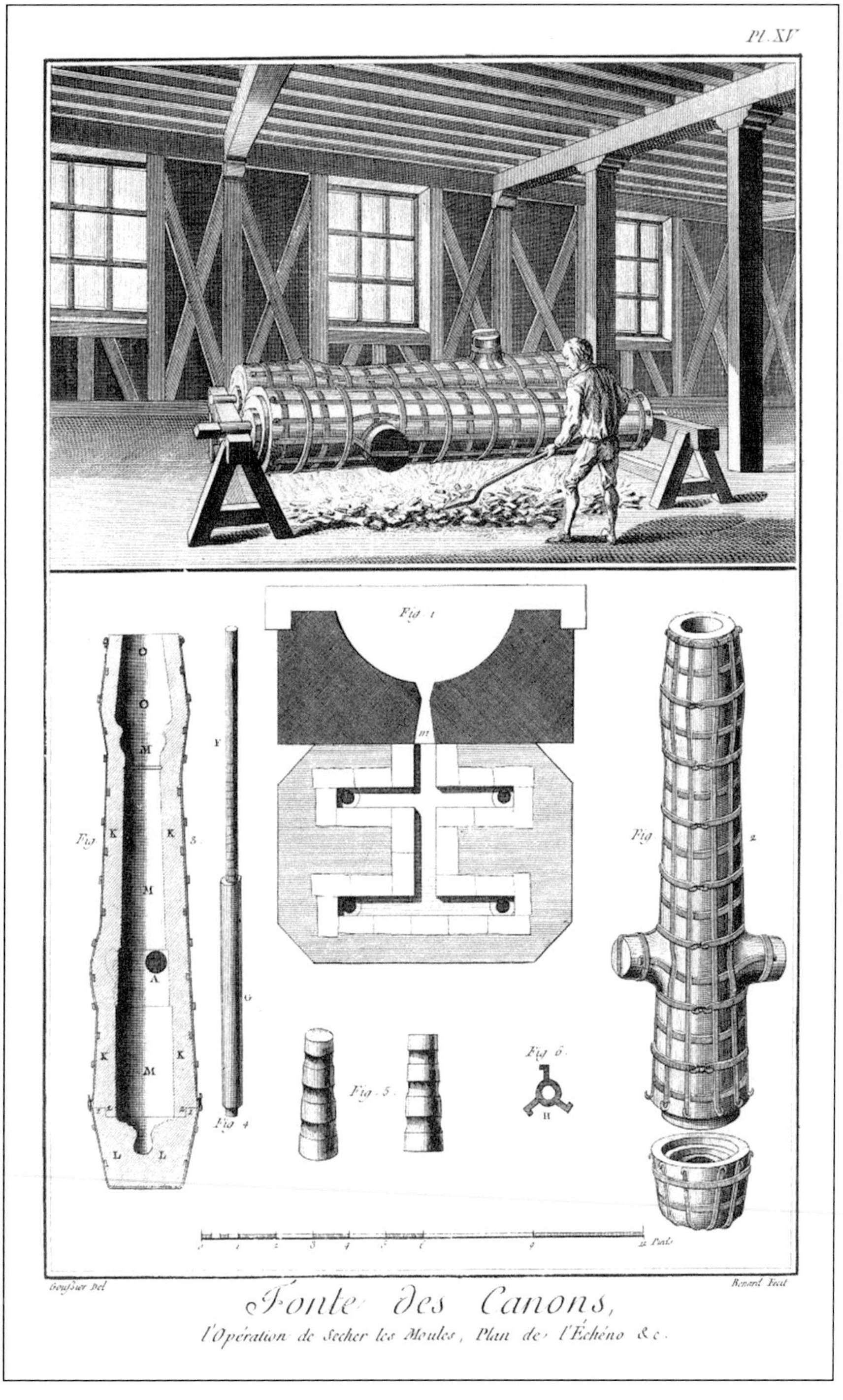

The pattern has been removed and the mould is being 'baked' carefully, to make sure that it is completely dry. Below that in the centre can be seen the arrangement of a furnace and four casting pits for guns. Any moisture in the mould could produce cracking, or even cause a volcano of molten metal to erupt, driven out by the steam generated in a damp mould. Below that on the far left is an assembled mould ready to be cast with only the core needing to be inserted. The core bar (covered and uncovered) is to its left, together with the chaplets to hold it into place. On the right the breech mould is about to be attached to the main mould and the armature, that connects the two, joined.

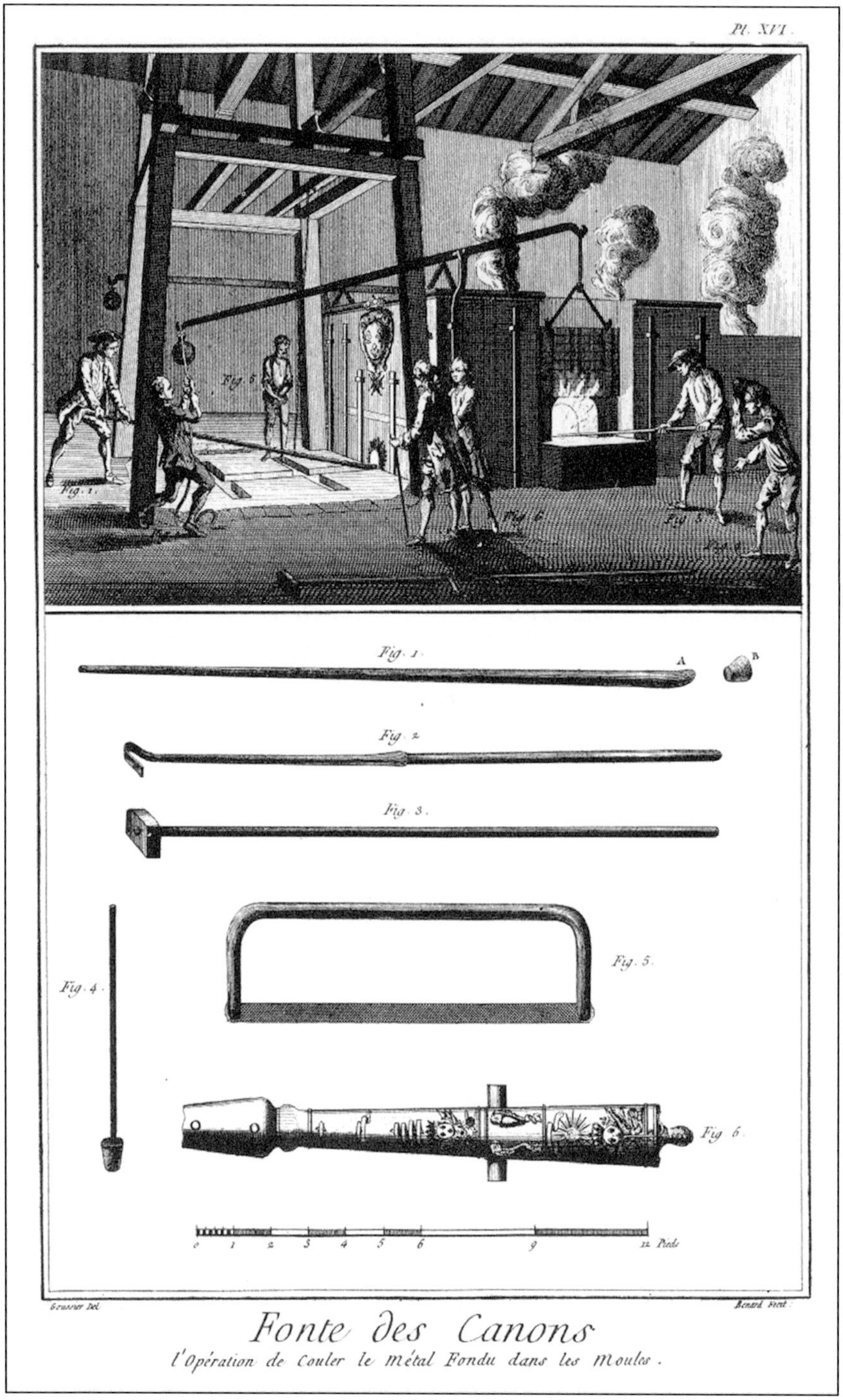

The foundry is ready to be tapped into the four waiting moulds. The door to the charging port has been raised to allow the workers to confirm that the melt is ready for pouring. The metal should not be turbulent and any surface dross should be skimmed off, which was the purpose of the numerous long-handled tools depicted. The top-most tool is designed to push the conical clay plug holding the molten metal in. The cannon with its large head of metal, now released from the mould, can be seen at the bottom, the large saw was used to cut away this head, which could then be recycled.

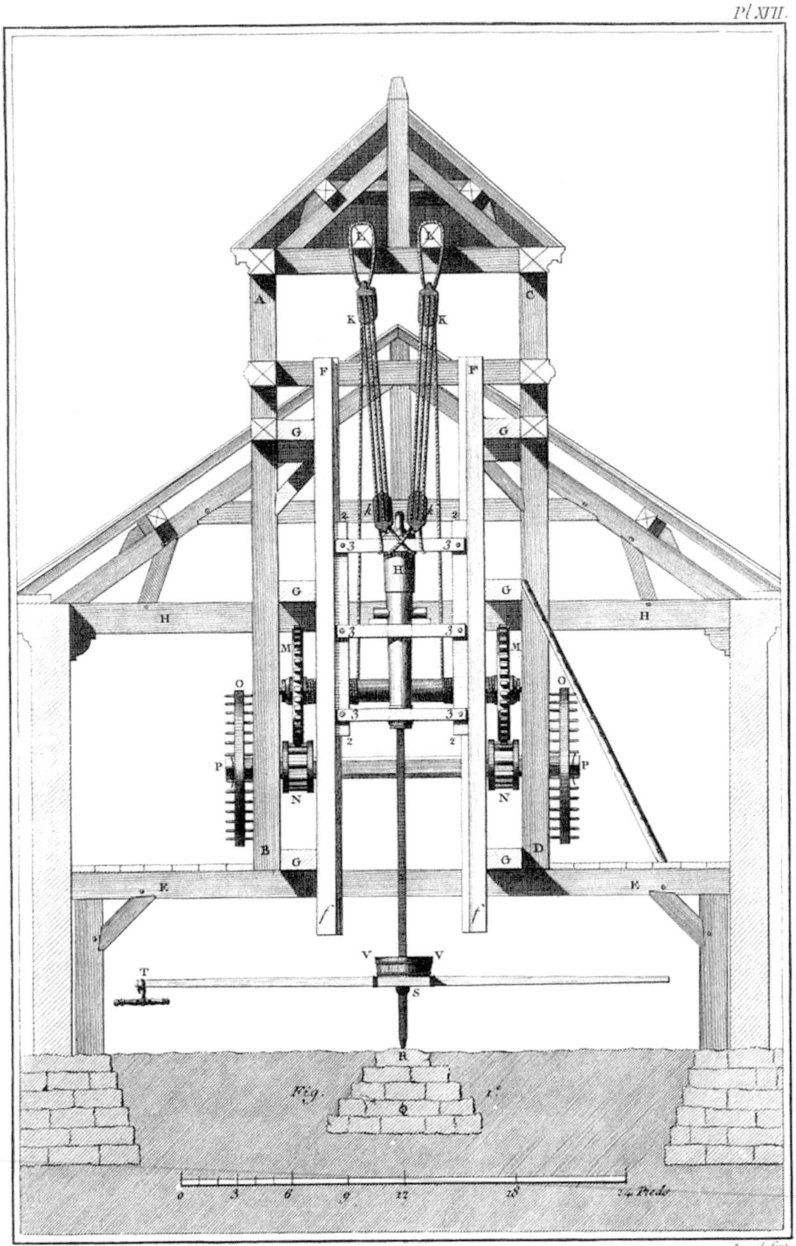

Fonte des Canons.

Élévation de l'Alézoir pour Forer et Alézer les Pieces.

This engraving should be compared with the image on p. 169 from almost three centuries before. The principle of using the weight of the gun to assist the process, and the use of horsepower, remain the same. The building is a fascinating complex of blocks, pulleys and gearing designed to achieve a precise central bore. In practice the processes, tools and skills depicted in the *Encyclopédie* are very little different from those described by Biringuccio over two centuries before. The guns themselves, less flamboyant and far less individual, were in principle identical to the guns that emerged as part of the 'new artillery' at the end of the fifteenth and beginning of the sixteenth centuries. It would take a further industrial revolution, one where the material was steel and when engineering precision had been achieved, to transform artillery. Men such as Armstrong, Whitworth and Krupp would make the new breech-loading rifled pieces, which were far uglier and deadlier than their predecessors.

Appendix XV

The Field Guns and Carriages of the Musée d'Art et d'Histoire, La Neuveville

The following pieces are all extraordinary survivals from the Swiss-Burgundian wars of 1476–1477 and specifically the battles of Grandson and Murten (Morat).They are displayed in the Musée de Neuveville as trophies of that conflict. It is difficult if not impossible to date the guns and carriages other than prior to the battles themselves. They are a unique collection of early gun carriages and demonstrate examples of both the old and new. The author is indebted to the museum for providing these photographs of these remarkable pieces of ordnance. Courtesy of the Musée d'Art et d'Histoire, La Neuveville, photos C. Ballif.

Overleaf: This wrought iron gun was made from 10 staves, bound by the usual hoops and bands. It has been suggested that the barrel was originally a breech-loader which has now had the breech fixed permanently to the barrel. The elevating arc has been strongly reinforced and the diagonal reinforcement bars would help the gun bed absorb the recoil of the piece. A gun chest has been rather clumsily fixed to the trail, and a rather fine wad hook can be seen slung alongside the two trails that are placed in the forward position. These are fully removable, but they would not be a serious inconvenience to the crew when in action even in this position. There is some doubt over the date of the carriage. A variety of stone and lead ammunition can be seen on the ground; stone would be used for the larger pieces and cast lead, or more usually lead with an iron or stone core, for the smaller.

Bore 2.8 in (72 mm). Length 63 in (1.595 m). Museum Inventory MAHLN1476.001.001.

The barrel of this gun was made rather unusually from only two staves but was held together by the usual hoops and bands. It is very noticeable that the diameter of the bands increased significantly after the trunnions and towards the breech. In this respect the barrel profile shows a trend that would be characteristic of the cast bronze pieces. The carriage is very similar in many respects to No 10 with similar ironwork reinforcing the trunnions and with the trunnions set fully into the cheeks. This photo also provides an excellent view of the trails and the way in which they were attached when ready for the draught horse to be harnessed.

Bore 2.4 in (60 mm). Barrel length 44.5 in (1.13 m). Museum Inventory MAHLN1476.001.004

Facing page: The barrel of this piece was made from four staves and held with hoops and bands in the traditional form and was built as a muzzle-loading piece. What is unusual is that the hoops and bands are not spaced evenly, with the interval increasing from breech to muzzle. The barrel is exceptionally long, with a comparatively narrow bore and a large calibre which equates to the type usually referred to as a sling. This would most likely have shot an iron projectile at higher velocity than its stone-firing brethren, explaining its reinforced breech. The lifting rings are so positioned that they can be used as a means of sighting. It is mounted into its bed, and it has the usual elevating arc.

Bore 2.4 in (60 mm). Length overall 115 in (2.925 m). Museum Inventory MAHLN1476.001.002.

Above: The barrel of this piece is made from 10 staves. It was probably originally a breech-loaded gun that had the rear of the barrel plugged and trunnions added. The trunnions are rather badly bent pins and a poor modification; this was clearly an attempt to update or upgrade an earlier piece. The two trails into which the horse was led are clearly visible, a simple and rather elegant solution to the problem of drawing the gun. The large gun chest fitted between the cheek pieces was both a practical and efficient use of the new form. Previously a chest could be carried on the trail or even over the breech of the gun. This new system was a far better solution to the problems of carrying the many smaller items needed by the gunner.

Bore 2.75 in (70 mm). Length overall 60 in (1.525 m). Barrel length 55 in (397 m). Museum Inventory MAHLN1476.001.004

Above: The barrel of this gun is a very unusual example of a single forging and is not a cast barrel as one might think at first sight. The wheels are four feet in diameter and the trail is well constructed and proportioned. It was this carriage that was used as the basis for the author's falconet. The design is pleasing and practical. The ironwork is modest but as can be seen was designed to protect the carriage from the recoil of the gun. The trunnions are completely set into the carriage, and it is clear from their diameter and width that the carriage was designed for this barrel. There is no capsquare fitted and the carriage would have to be disassembled to remove the barrel. The carriage is a design that would, with little modification, see service for the next 400 years.

Bore 2.3 in (58 mm). Length overall 87 in (2.213 m). Length of barrel 83 in (2.108 m). Museum Inventory MAHLN1476.001.006.

Below: This is a breech-loading gun converted to muzzle-loading, with the barrel also being turned almost upside down, probably in the process of conversion. The obvious advantage of muzzle-loading was a stronger piece but losing at the same time the increased speed of reloading offered by the availability of several spare breeches. The wheels are dished, and the prominent wheel hub is well displayed. The felloes have been carefully scalloped and the five iron strakes have the usual prominent nails. As was common, there are five felloes and 10 spokes. The condition of the carriage is remarkable, demonstrating little hard use, which is unsurprising considering that it was a trophy piece.

Bore 2.9 in (73 mm). Length overall 86 in (2.19 m). Barrel length 82 in (2.075 m). Museum Inventory MAHLN1476.001.005

Colour Plate Commentaries

Plate A

Two late medieval field guns illustrated in the Swiss Schilling chronicle
Both have cast bronze barley twist and round barrels. One is mounted in a wooden bed with a tiller attached, the other is fitted with trunnions and has a cast tiller. It is easy to see how the tiller fitted to the wooden bed on the right-hand gun could have led to the development of the barrel with a cast tiller on the left-hand piece. Both guns seem to be using cartridges rather than loose powder, indicative of the early adoption of this practice. This was especially sensible for field pieces that may have to go into action quickly. The gunner on the right is using a pricker to clear the vent and prick the cartridge. Both guns have a towing ring. (Diebold Schilling, *Berner Chronik*, vol. 3, p. 347. 1474–1484 (<www.e-codices.ch>, courtesy of Burgerbibliothek Bern)

The Norwich Great Hospital falconet
This is a very fine early piece of artillery, probably Flemish in origin, which Kay Smith dates to the late fifteenth or early sixteenth centuries. It has a faceted octagonal barrel which turns by half of one flat about halfway along the barrel. It is provided with trunnions placed midway in the gun's diameter and has a cast tiller extending from the flat face of the breech. The barrel feels heavy to elevate, as the trunnions are too far forward, according to Biringuccio as well as later experts. The tiller would have been used with an elevating arc and the gun would probably have originally been mounted on a carriage similar to the left-hand gun in the previous image. It was possibly the gun bought in Flanders by John Marsham Mayor of Norwich in 1518, although it may have been an old piece even when he bought it. 'Norwych' is incised in gothic script in the rear of the barrel, again an indication of an early date. There is no decoration on the cast shield, which itself is not unusual if a gun was cast 'on spec' rather than for a known buyer. The carriage is clearly recent. (Photograph courtesy of Thomas Davies)

Barrel length 53.25 inches (135 cm), muzzle to the base ring. Overall length 57.75 inches (147 cm). Bore 2⅛ inches (5.5 cm)

Plate B

A transitional bronze gun from the *Zeugbuch* of Emperor Maximilian I
The arsenal list or *Zeugbuch* of Emperor Maximilian I was compiled by Bartholomaeus Freysleben, the Hauszeugmeister (Imperial master of armaments) and is an incomparable source for those interested in the early history of artillery; it was produced in the first decade of the sixteenth century. The piece illustrated here reflects both the traditional and new approach to artillery. The barrel is mounted in a carriage and elevated using trunnions. The breech has an iron ring fitted to aid lifting the gun. The carriage itself is a single piece of iron-bound timber. All these are features of the new style; older features include the 'reinforcing' rings around the barrel, like a wrought iron piece, although these perform no useful purpose. The breech has a sub-calibre powder chamber, suggested by the different diameters of the loading ladle and rammer, and its ammunition is clearly stone rather than iron. The midriff of the piece is bulged, in the same way that the junction between the powder chamber and the barrel on wrought iron guns was 'thickened', as this was the weak spot in the design. These features are redolent of medieval wrought iron guns. No doubt the gun was an effective stone thrower and it would take little to transform it into the bronze perrier that would see service until at least the end of the sixteenth century in Europe. Although obviously completed, the carriage is 'unpainted'. Oak and elm would not really need much in the way of preservatives. (*Zeugbuch Kaiser Maximilian I* – BSB Cod. icon. 222 p. 0115. Courtesy of the Bavarian State Library)

An early depiction of a gun with a limber
This very early limber is clearly a functional item: the trail of the gun has a reinforcement around the hole through which the pintle from the limber projects. Its position is far more forward on the trail than one would expect, potentially causing a problem with the trail hitting the spokes of the limber's wheels at high turning angles. The gun is exactly of a form one might expect with its swollen midriff, but the early appearance of the lugs or dolphins is noteworthy. Martin Merz (1425–1501), who drew this, was in the service of Frederick I of the Palatinate as his principal gunfounder in 1469, a role he performed on the succession of Phillip the Upright, Frederick's successor. His performance was impressive: as master gunner he successfully besieged eight castles and cities in 1470–71 using 372 barrels of gunpowder. (*Feuerwerksbuch*, Martin Merz – shelfmark, fol. Cgm 599 p. 12 *c.* 1473. (Courtesy of the Bavarian State Library)

Plate C

A carriage in construction and completed, from the *Zeugbuch*
This image is valuable in showing a carriage in the process of construction as well as a completed version. It has a barrel redolent of the Neuchatel guns with facets and cascabel ring. A single tree trunk has been used, and as can be seen, the barrel will sit almost completely within it. The extensive reinforcement around the trunnions was expensive and clumsy but necessary to stop the wood being split by the recoil forces transmitted along the grain. The upper

gun has a vent cover and the cascabel is in the form of what appears to be an irritated hedgehog! The black and red paint scheme was associated with Imperial artillery. *Zeugbuch Kaiser Maximilians I.* – BSB Cod. icon. 222 p. 0100. (Courtesy of the Bavarian State Library)

A magnificent transitional piece and limber from the *Zeugbuch*
This is a fine example of the old and the new coming together. The barrel is round with a constant diameter bore and has a reinforced breech and lugs or dolphins. It is of long calibre, and the cannonballs would probably have been of iron. It is held securely with iron bands, with the flat breech transmitting the recoil to the wooden bed. The bed pivots on the lower part of the carriage and is fitted with an elevating arc. It is easy to see how with the addition of trunnions this gun could readily take on the form of the gun in the previous illustration. There is a strut attached to the hub to increase the stability and resilience of the axle. The sizeable limber helpfully illustrates the complex iron and woodwork structure designed to assist stability and reduce the shock forces on the axle and hub. (*Zeugbuch Kaiser Maximilians I* – BSB Cod. icon. 222 p. 0037. Courtesy of the Bavarian State Library)

Plate D

The bronze bombard *Die Katerie*
This piece was cast by Georg Endorfer in 1487 in his Innsbruck foundry. It bears the arms of Sigismund of Austria and the legend 'My name is Katerie, beware of my strength, I punish injustice'. It displays an unusually early pair of dolphins as well as cast 'rings' along the barrel and a swollen 'midriff'. It has a sub-calibre powder chamber. The rectangular projections are for seating it in a wooden bed. It is an impressive collection of old and new features, and is indicative of how capable gunfounders were in casting very large pieces of artillery by the end of the fifteenth century. The gun is 144 in (3.65 m) long with a bore 15.4 in (39 cm) in diameter, and it weighs a magnificent 10,135 lb (4,597 kg).

Numéro d'inventaire: 2012. 0. 458
Autre numéro d'identification: Cote: N 500, Cote: N 238 ROB
(Photo © Paris – Musée de l'Armée, Dist. RMN-Grand Palais/Emilie Cambier)

A pair of bombards found in the *Zeugbuch* of Maximilian I
They were both clearly considered to be operational and serviceable weapons as they were included alongside the most modern types. The upper bombard is probably of wrought iron construction and in two pieces. The breech section would be screwed to the barrel section, with the aid of the windlass sockets fitted to the circumference of the two pieces. The painted coat of arms is an intriguing and very unusual detail. The other bombard is clearly of cast bronze and mimics the external and internal shape of its close cousin. Lifting rings have been cast into the piece with iron rings forged into them, although they are far too far forward of the point of balance to be of use on their own. The projection on the breech of the piece was probably a gunsight

as it is too far forward to be the vent. The stone ammunition would suffice for both. The ratio of shot to powder for such guns was approximately 8:1, considerably more than for lighter weapons. It relied on the weight of its projectile, rather than its velocity, to achieve its purpose. It is worth noting that in 1807 an English warship was crippled by a shot fired by a weapon very similar to these when passing through the Dardanelles. The utility of such weapons should not be underestimated, what led to their abandonment was their weight and clumsiness in operation. (*Zeugbuch Kaiser Maximilians I* – BSB Cod. icon. 222 p. 0052. Courtesy of the Bavarian State Library)

Plate E

The paraphernalia of gunnery
Very rarely are the essential paraphernalia associated with a working gun displayed in such detail. Without a flame or the ability to strike a light or pierce a cartridge, the most magnificent and powerful piece of ordnance would be useless. (Heidelberg, Mönch 1496 cpg 126_0097. Courtesy of the University of Heidelberg Library)

Top left: a cresset light essential to provide light, and a source of flame for the match cord which burns fast and sometimes furiously and is quickly used up.

Middle top: an interesting hybrid piece. The carriage is made from two cheeks, reinforced with iron. The barrel is somewhat old fashioned, but it is equipped with trunnions and has a sophisticated elevating screw system.

Top right: a set of calipers to measure the ball, and a variety of prickers used to puncture the linen or paper cartridge and ream the vent. Two of the prickers are of iron, they would be unsafe to use in anything other than a bronze barrel.

Bottom left: this appears to be a cradle to hold the barrel when dismounted. No gunner would like to see his barrel in the mud.

Middle bottom: an interesting assemblage of a steel, flint, tinder and match cord, next to this is a simple jack useful for lifting an axle to replace a wheel.

Plate F

An early-sixteenth century falcon on a reproduction carriage
At present on display in the White Tower of the Tower of London, this fine falcon (XIX-14) was cast around 1520, probably in Flanders; it was captured off Spain by Rear Admiral Sir Frederick Lewis Maitland (1779–1839). The gun is 96 in (2.438 m) long with a bore of 2.3 in (58 mm). The calibre or ball-to-barrel length ratio is almost 40, making it a long culverin type. The octagonal barrel was quite a common form in the early decades of the century in France, Genoa and Scotland. There is very little decoration on the gun although the cascabel is quite complex, and there is some decoration in front of the base ring a design of

alternating trefoils and fleur-de-lys. It could easily have been still in use almost three centuries after its casting as it is in near perfect condition, although the touch hole has probably been bushed. The carriage is well-proportioned but not an attempt to provide a fully authentic example. Contemporary carriages are far rarer survivals than barrels, for obvious reasons. Those few that exist are often difficult to date with any precision, but those found in Neuveville are an exception, as is another falconet on a carriage on display in the Landeszeughaus in Graz which they date to about 1500. (Author's photograph)

An original Armada siege piece on a reproduction carriage

One of the most extraordinary finds on the wreck site of the Spanish Armada ship *La Trinidad Valencera* was a magnificent barrel, together with the remains of the disassembled carriage. It is one of three carriages on board, two of which have been recovered from Glenagivney Bay. The barrel was cast in 1556 in Malines in Flanders by Remigy de Halut, the Royal Founder appointed to succeed Hans Poppenruyter, and is a *canon de batir*, the heaviest siege gun used by Imperial armies. It is entirely conventional in design, retaining evidence of the four-pin iron chaplet in the breech which secured the core during casting. The relatively short calibre was deemed suitable for a short-range siege piece, in practice such a 'short' barrel was just as efficient if not more so than the much longer culverin types. It has a second pair of trunnion holes so that the barrel can be mounted more centrally when being transported, thereby reducing the strain on the wheels. Amongst the ornamentation are the arms of Spain impaling those of England, as at the time of its manufacture Philip II was married to Queen Mary of England (1553–1558). Like all Imperial guns the woodwork is painted black, possibly using pine resin and ironwork coloured with a red preservative. Although the gun is impressive, the size of the carriage required to carry it is equally if not more so. It is at present on display in the Ulster Museum.

Plate G

A reproduction iron saker and carriage from the late sixteenth century

A reproduction gun on a reproduction carriage, both of excellent quality. It is presently part of the armament of Portland Castle, one of Henry VIII's Device forts. The gun is a cast iron saker with a 3.5 in (89 mm) bore. It is based on a mid-sixteenth century example and mounted on a carriage that would have seen service about the time of the Armada in 1588. Firing a 6 lb (2.7 kg) shot this gun ought to be compared to the Alberghetti saker for its economy of material and usability aboard ship. The front wheels are of a larger diameter than the rear to take account of the camber of the deck. Although a quoin is provided to elevate the gun, at the short engagement range of a few hundred yards or less it is unlikely that it was used much. There is effectively no reinforcement to the barrel and the muzzle has little more than a plain ring, it is clearly designed for service rather than ornament. Short guns at sea were not seen as a disadvantage by English seafarers, as Bourne records. This gun was cast in 1978 and its carriages refurbished in 2019. (Author's photograph)

Bore: 7.24 in (184 mm)
Shot diameter: 6.9 in (175 mm)
Shot weight: 44 libras
Overall length: 115 in (2.92 m)
Muzzle to breech ring: 104 in (2.64 m)
Calibre/length: 15.1
Weight: 5,430 lb (2,463 kg)
(Bazonka, CC BY-SA 3.0 <https://creativecommons.org/licenses/by-sa/3.0>, via Wikimedia Commons)

Plate H

A military revolution in motion

It was the pike, and not artillery, that would be the *Wunderwaffe* (wonder weapon) of the age. The destruction of the Burgundian army on three occasions by pike-wielding Swiss infantry (1476–1477) caused a severe shock to the martial caste of Europe. This very rare picture is a remarkable demonstration of how warfare was changing. The principal force on this battlefield were the pike blocks of armoured and unarmoured men, now accompanied by arquebusiers. Lance- and crossbow-equipped cavalry provided mobile firepower and shock power. Artillery is placed between the formations and is provided with what should be a clear field of fire. The field guns have not been distributed along the line of battle but concentrated in batteries. Their limbers and teams of horses have been drawn up behind them in a thoroughly modern fashion. The ultra-long linstock in the hands of the 'master gunner' is noticeable, the purpose of the axe he holds is less clear! Gunpowder weapons were clearly being integrated in to warfare in increasingly sophisticated ways, although the main weapon of offence was to be the pike. The role of the military musician is also highlighted, an aspect of warfare often ignored. (Universitätsbibliothek Heidelberg, Cod. Pal. germ. 126-0105 *Kriegsbuch* [aka *buch der stryt vnd büchßen*], 1496, by Philipp Mönch <https://digi.ub.uni-heidelberg.de/diglit/cpg126>. Courtesy of the University of Heidelberg Library)

Plate I

A reproduction falconet and carriage

The carriage is based closely on 10 N in the Neuveville collection; the octagonal barrel was a reduced version of Falcon XIX-15 in the Fort Nelson collection of the Royal Armouries. This gun was probably the work of the Genoese founder Gregorio I Gioardi, cast at the very beginning of the sixteenth century, it is the type of early polygonal piece that was common at the very end of the fifteenth century and the first decades of the sixteenth. The barrel is 65.75 in (1.67 m) with a bore of 1.8 in (46 mm), and it has a calibre of 35.6. This is representative of surviving guns of this type and period. It would have fired a ball of one Italian *libra* weighing between .7–8 lb (301–361 g) with a point-blank range of about 300 paces. What is especially noticeable about this barrel is that unlike original examples it does not show the green patina of age but the golden hue of youth. This is also well illustrated in the *Zeugbuch*. It is painted with linseed oil paint with natural red ochre pigments. The wheels are dished and appear to have strakes, rather than a continuous iron tyre, fitted to the rim. The carriage and wheels were made by Greg Rowland and much of the ironwork by Tim Jones.

Plate J (top)

A bronze saker cast by Alberghetti on a reproduction carriage
Currently on display at St Mawes Castle, this fine bronze saker was recovered from the sea with six other bronze guns, comprising a saker, two minions and three swivel guns. They were found in 1975 in association with a Church Rocks wreck of what was probably a Venetian vessel. The saker and minions bear a shield and the Initials SA; these are associated with Sigismondo II Alberghetti (1539–*c.* 1610) who was a gunfounder in Venice and Turin and heir to the tradition and talent of his famous family. He experimented with breech-loading guns, a not uncommon practice. Several of his pieces are to be found in Istanbul and one in Woolwich. The saker was 141 in (3.58 m) long and had a 3.85 in (98 mm) diameter bore. It would have fired a 3.7 in (93 mm) cast iron roundshot weighing approximately 6–7 lb (2.7–3.2 kg). The carriage is a reproduction of a sea or garrison carriage. The solid wooden wheels are not shod with iron, which was not unusual for such types of mounting. The total length of gun and carriage would make it decidedly unwieldy across a deck but the long barrel would be less troublesome if mounted on the corsia or forward centreline of a galley. (Nilfanion, CC BY-SA 3.0 <https://creativecommons.org/licenses/by-sa/3.0>, via Wikimedia Commons)

Plate K (top)

Queen Elizabeth's Pocket Pistol
Queen Elizabeth's Pocket Pistol, as it is mistakenly known, is a magnificent very long culverin manufactured in 1544 in Utrecht and now on display in Dover Castle. It was a gift for Henry VIII from Maximilian van Egmont the Count of Buren and cast by Jan Tolhuys, the chief gunfounder of the City of Utrecht from 1541 to 1558. It has a bore of 4¾ inches (121 mm) and fired a ball of 20 lb (9 kg) with a charge of 18 lb (8.2 kg). It is a monumental 23½ feet (7.16 m) in length with an extraordinarily long calibre of 62.6. It is beautifully cast with flamboyant renaissance decoration, only marred by the following doggerel: 'As breaker of ramparts and walls am I known, O'er mountain and valley fly balls by me thrown'. It was installed in Dover Castle in 1547 where it was recorded as a 'Basillisches of brass' and provided with 150 iron shot. In 1644 it was used to breach the walls of the Royalist stronghold of Sheffield Castle, an impressive performance for a centenarian. (Public domain. Image from user Mlm42, Wikimedia Commons)

Plates J–K, L–M

Firing the *Vasa* reproduction gun
Very little can prepare you for the noise and power produced by a black powder

cannon in full-throated roar. This is a sequence of the *Vasa* cannon being fired with a full service load of powder and round shot. The high-speed filming at 4,000 fps (1,219 m/s) has captured the eruption of the gun, which took place in about half a second. In the first frame the blast from the vent is very apparent and something that any of its crew would be wary of. In the second frame the ball trailing fire can be seen a little way ahead of the leading edge of the muzzle flash. It is clear that at this point the ball is moving faster than the cloud of expanding gas behind it. The volume of smoke generated was enormous, and one can imagine how the sound and smell of burnt powder would dominate experience of the battlefield. (Photos courtesy of the Bofors Test Center)

Plate N

Accoutrements associated with the falconet. Clockwise from top left: ladle and rammer, shot box, gunsight.
The ladle and rammer are based on *Mary Rose* examples. The ladle was bound with a copper hoop over which the ladle was fixed. The turned portion was drilled through in the centre and the handle pushed through. The poles are of ash, barrelled in profile and a good length, making them easy to use. They are marked to indicate where they should be when fully inserted into the muzzle, as recommended by Norton. The worm or wad hook is a sturdy piece of ironwork. The mop is wrapped in lambswool fixed with copper nails and had to be carefully 'trimmed' to ensure a snug fit. (Author's photograph)

The shot box was based on early sixteenth-century German examples in the Weimar Ingenieurkunst-und Wunderbuch and chests in the Saint Anne Museum in Lübeck. The ironwork including the lock was made by Tim Jones and the woodwork by Nora Kalz. It would contain prepared cartridges as well as shot. It has the characteristic gabled lid and a lock to discourage the inquisitive; the trail also has a useful box that would contain items such as prickers, match cord, a tinder box and possibly some ready-use ammunition. A water bucket was essential in wetting the mop to clean the barrel. The powder horn was essential for priming the piece and would have contained finer powder than the main charge. The impedimenta required to operate even this simple piece are considerable.

The gunsight was made by Tim Jones after drawings by Tom Davies, and is a close interpretation of an early example found in the Museo Galileo in Florence. It is calibrated in inches and degrees and was designed to be mounted on the octagonal breech ring. It was specifically designed for the length of the barrel of the author's falconet. The pendulum intended to ensure that the piece was level was probably added at a later date. It took some 30 hours to make, using hand tools, and is a beautiful piece of work. The author doubts whether it was fixed to the barrel, and suggests it was placed and removed upon each shot if precision was required.

Bibliography

Printed Primary Sources

Agricola, Georgius, *De Re Metallica* (New York: Dover Publications Inc., 1950)

Anon., *The Compleat Gunner* (London: 1672)

Audley, Thomas., 'A treatise on the Art of War', *Journal of the Society for Army Historical Research*, Vol. 6, No. 24 (April–June 1927), pp. 65–78

Barret, Robert, *The Theorike and Practike of Moderne Warres* (London:1598)

Bent, J. Theodore (ed.), *Early Voyages and Travels in the Levant* (New York: Burt Franklin, 1893)

Binning, Thomas, *A light to the art of gunnery wherein is laid down the true weight of powder, both for proof and action, of all sorts of great ordnance* (London, 1676)

Biringuccio, Vannoccio, *The Pirotechnia* (New York: Dover Publications inc, 1990)

Bourne, William., *The Arte of Shooting in Great Ordnaunce* (London: Thomas Woodcoke, 1587)

Brown, Rawdon (ed.), *Calendar of State Papers and Manuscripts Relating to English Affairs, Existing in the Archives and Collections of Venice and in Other Libraries in Northern Italy*, Vol. 2, 1509–1519 (London: Longmans, Green, Reader and Dyer, 1867)

Corneweyle, Robert, *The Maner of fortifications of cities, townes, castelles and other places* (London, 1559)

Digges, Thomas, *Foure Paradoxes, or Politique Discourses, Concerning Militarie Discipline, written long since by Thomas Digges Esquire. Of the worthinesse of warre and warriors, by Dudley Digges, his sonne* (London: 1604)

Du Praissac, Sieur, *The Art of Warre, or Militarie Discourses of Leavying, Marching, Encamping* (London: 1639)

Garrard, William, *The Arte of Warre Beeing the Onely Rare Booke of Myllitarie Profession: Corrected and Finished by Captaine Hichcock* (London: 1591)

Guicciardini, Francesco, *The History of Italy* (Princeton: Princeton University Press, 1984)

Hawkins, Sir Richard, *The Observations of Sir Richard Hawkins Knight, in his voyage into the South Sea. Anno Domini 1593* (London: John Jaggard, 1622)

Hexham, Henry, *The Third Part of the Principles of the Art Military Practised in the Warres of the United Provinces* (Rotterdam, 1643)

Jordan, W. K., *The Chronicle and Political Papers of King Edward VI* (London: George Allen and Unwin,1966)

Lucar, Cyprian, *A Treatise Named Lucar Appendix, collected by Cyprian Lucar Gentleman, out of divers good authors in divers languages*. In Nicolò Tartglia, *Three Bookes of Colloquies Concerning the Arte of Shooting in Great and Small Pieces of Artillerie*, trans. Cyprian Lucar (London: 1588)

Machiavelli, Nicolò, *The Art of War*, in *Machiavelli: The Chief Works and Others*, Vol. II, translated by Allan Gilbert (Durham, NC: Duke University Press, 1989)

Nichols, J. G. (ed.), *The Chronicle of Calais: In the Reigns of Henry VII and Henry VIII to the Year 1540*, Vol. 35 (London: Camden Society 1846)

Norton, Robert, *Of The Art of Great Artillery: Written by Robert Norton Gunner* (London: Edward Allde, for John Tap. 1624)

Norton, Robert, *The Gunners Dialogue. With the Art of great Artillery, Robert Norton, Enginier and Gunner* (London: John Tap, 1628)

Norton, Robert, *The gunner shewing the whole practise of artillerie* (London, 1628)

Nye, Nathaniel., *The Art of Gunnery* (London: 1624)

Rose-Troup, Frances, *The Western Rebellion of 1549: An Account of the insurrections in Devonshire and Cornwall against religious innovations in the reign of Edward VI* (London: Smith & Elder & Co., 1913)

Smith, John, *A Sea Grammar* (London: Printed by John Haviland, 1627)

Smith, John, *The Sea-Mans Grammar and Dictionary* (London: 1691)

Smith, Thomas, *The Arte of Gunnerie* (London, 1600)

Starkey, David (ed.), *The Inventory of King Henry VIII (the transcript)* (London: Harvey Miller Publishers, 1998)

Styward, Thomas, *The Pathwaie to Martiall Discipline* (London: 1581)

Sutcliffe, Matthew, *The Practice, Proceedings, and Lawes of Armes described out of the doings of most valiant and expert captaines, and confirmed both by ancient, and moderne examples, and præcedents* (London: 1593)

Tartaglia, Nicolò, *A New Edition* (Max Planck Research Library for the History and Development of Knowledge), <https://edition-open-sources.org/sources/6/index.html>

Tartglia, Nicolò, *Three Bookes of Colloquies Concerning the Arte of Shooting in Great and Small Pieces of Artillerie*, trans. Cyprian Lucar (London: 1588)

Tartglia, Nicolò, *Nova Scientia* (np, 1558)

Ulpian, Fulwell, *The Flower of Fame: Containing the bright renowne, & moste fortunate raigne of King Henry the viii* (London: 1575)

Usherwood, Stephen, *The Great Enterprise: The History of the Spanish Armada as Revealed in Contemporary Documents* (London: Bell and Hyman, 1978)

Vauban, Sébastien Le Prestre de, *The New Method of Fortification, as practised by Monsieur de Vauban, Engineer General of France* (London: 1691)

Whitehorne, Peter, *Certaine Wayes for the Ordering of Souldiours in Battelray* (London: 1588)

Academic Theses, Dissertations and Archaeological Reports

Cooper, Jonathan Paul, 'Whitecoats and Rascals: In search of the fortifications and siege works from the siege of Haddington 1548–1549'. Submitted as part of the MLitt in Battlefield and Conflict Archaeology (8V7S) Department of Archaeology Faculty of Arts of the University of Glasgow, submitted September 2008

Flint, Joshua, 'A new reassessment of the importance of gunpowder weapons on the battlefields of the Wars of the Roses', Master's thesis, University of Huddersfield (2014)

France, Catherine Ann, 'Gunnery and the Struggle for the New Science (1537-1687)'. Submitted in accordance with the requirements for the degree of Doctor of Philosophy. The University of Leeds School of Philosophy, Religion and History of Science September 2014

Green, Jeremy N., *The Loss of the Verenigde Oostindische Compagnie Retourschip BATAVIA, Western Australia 1629: An excavation report and catalogue of artefacts*, Western Australian Maritime Museum, Fremantle, Western Australia BAR International Series 489, 1989

HO Group, *Casting and Testing a 15. Century Bronze Cannon*, The Ho Report no. 15, February 2016 (September 2020)

Hoskins, Sara Grace, '16th Century Cast-Bronze Ordnance in the Museu de Angra do Heroísmo', Texas A&M University Master of Arts (December 2003)

López-Martín, Francisco Javier, 'Historical and Technological Evolution of Artillery From its Earliest Widespread Use Until The Emergence of Mass-Production Techniques', a thesis presented in the Sir John Cass Dept. of Art, Media & Design, for the degree of PhD, London Metropolitan University (June 2007)

Martin, Colin John Mackenzie, *The Equipment and Fighting Potential of the Spanish Armada*, A Thesis Submitted for the Degree of PhD at the University of St Andrew (1984)

Van Duivenvoorde, Wendy, *The Armament of Australia's VOC Ships Report*, Department of Maritime Archaeology, Western Australian Museum, No. 258 (2010)

Walton, Steven Ashton, 'The Art of Gunnery in Renaissance England', Institute for the History & Philosophy of Science & Technology, University of Toronto PhD, (1999)

Waterhouse, Robert, *Tudor bronze guns of the Channel Islands* (2015), pp. 82–109, <http://www.christopherlong.co.uk/aanna/papers/WaterhouseR/TudorBronzeGunsOfTheChannelIslands.pdf>

Books

Adams, Nicholas, and Simon Pepper, *Firearms & Fortifications: Military architecture and siege warfare in sixteenth-century Siena* (Chicago: University of Chicago Press, 1986)

Beltrame, Carlo and Renato Gianni Ridella, *Ships and Guns: The Sea Ordnance in Venice and in Europe Between the 15th and the 17th Century* (Oxford: Oxbow Books, 2011)

Bennett, Jim and Stephen Johnson, *The Geometry of War 1500–1750* (Oxford: Ashmolean Museum, 1995)

Blackmore, H. L., *The Armouries of the Tower of London, Vol. 1, Ordnance* (Her Majesty's Stationery Office, 1976)

Brown, R. R. and R. D. Smith, *Bombards: Mons Meg and Her Sisters* (Leeds: Royal Armouries, 1989)

Bull, Stephen, *The Furie of the Ordnance* (Woodbridge: The Boydell Press, 2008)

Carman, W. Y., *A History of Firearms* (London: Routledge Kegan Paul, 1955)

Carpenter, Austin C., *Cannon: The conservation, reconstruction and presentation of historic artillery* (Exeter: Halsgrove Press, 1993)

Caruana, Adrian B., *The Pocket Artillerist: Or, The Art of Coarse Gunfiring: A handbook for the safe firing of muzzle loading ordnance* (Rotherfield: Jean Boudriot Publications, 1992)

Caruana, Adrian B., *Tudor Artillery 1485-1603* (Bloomfield: Museum Restoration Service of Canada, 1992)

Cressy, David, *Saltpeter: The Mother of Gunpowder* (Oxford University Press, 2013)

Cruickshank, C. G., *Army Royal: Henry VIII`s Invasion of France 1513* (Oxford: Clarendon Press, 1969)

Cruickshank, C. G., *Elizabeth's Army* (Wotton-under-Edge: Clarendon Press, 1966)

Davies, Jonathan P., *The Medieval Cannon 1326–1494* (Oxford: Osprey, 2019)

Davies, Jonathan P. (text transcribed by M Bryn Davies), *Thomas Audley and the Tudor 'Arte of Warre'* (Sunderland: Pike and Shot Society, 2002)

Douglas, Sir Howard, *A Treatise on Naval Gunnery* (London: John Murray, 1855)

Duffy, Christopher, *Siege Warfare* (London: Routledge Kegan Paul, 1979)

Embleton, G., and D. Miller, *The Landsknechts* (Oxford: Osprey,1976)

Firth, Charles Harding, *Cromwell's Army: A history of the English Soldier During the Civil Wars, the Commonwealth and the Protectorate* (London: Methuen and Co., 1902)

Ffoulkes, Charles, *The Gun-Founders* (Cambridge: Cambridge University Press, 2010)

Gille Bertrand, *The Renaissance Engineers* (London: Lund Humphries, 1966)

Guilmartin, J. F., *Galleons and Galleys* (London: Cassell and Co., 2002)

Guilmartin, J. F., *Gunpowder & Galleys: Changing Technology & Mediterranean Warfare at Sea in the 16th Century* (Cambridge: Cambridge University Press, 1974)

Guilmartin, J. F., *The Guns of the Santíssimo Sacramento* <https://www.angelfire.com/ga4/guilmartin.com/Santissimo.html>

Guttmann, O., *Monumenta Pulveris Pyrii: Reproductions of ancient pictures concerning the history of gunpowder, with explanatory notes* (London: Artists' Press, 1906)

Hall, A. R., *Ballistics in the Seventeenth Century* (Cambridge: Cambridge University Press, 1952)

Hall, Bert S., *Weapons and Warfare in Renaissance Europe* (Baltimore: Johns Hopkins University Press, 1997)

Hale, J R., *Artists and Warfare in the Renaissance* (New Haven, Conn: Yale University Press, 1990)

Hale, J R., *Renaissance Fortification. Art or Engineering?* (London: Thames & Hudson, 1977)

Hale, J R., *Renaissance War Studies* (London: Hambledon Press, 1983)

Heath, Ian, *Armies of the Sixteenth Century* (St Peter Port: Foundry Books, 1997)

Hildred, Alexzandra, *Weapons of Warre: The Armaments of the Mary Rose*, Vol. 3 (Exeter: Short Run Press Ltd, 2011)

Hogg, O. F. G., *Artillery: Its origins, heyday and decline* (London: C. Hurst and Co., 1970)

Hogg, O. F. G., *Clubs to Cannon* (London: Gerald Duckworth & Co. Ltd, 1968)

Hogg, O. F. G., *English Artillery* (Woolwich: Royal Artillery Institution, 1963)

Hogg, O. F. G., *The Royal Arsenal* (Oxford: Oxford University Press, 1963)

Hooker, John, *The Ancient History and Description of the City of Exeter* (Exeter: Andrews and Trewman, 1765)

Kelly, Jack, *Gunpowder* (London: Atlantic Books, 2004)

Kennard, A. N., *Gunfounding and Gunfounders* (London: Arms & Armour Press, 1986)

Lewis, Michael, *Armada Guns* (London: Allen & Unwin, 1961)

Mallett, Michael Edward and Christine Shaw, *The Italian Wars 1494–1559: War, State and Society in Early Modern Europe* (London: Routledge, 2012)

Mattingly, Garret, *The Defeat of the Spanish Armada* (London: Jonathan Cape, 1960)

Murphey, Rhoads, Ottoman Warfare 1500–1700 (New Brunswick: Rutgers University Press, 1999)

Napoleon III, *Études sur le passé et l'avenir de de l'artillerie* (Paris: J. Dumaine, 1846)

O'Neil, Brian Hugh St John, *Castles and Cannon: A Study of Early Artillery Fortifications in England* (Westport, Conn: Greenwood Press, 1975)

Padfield, Peter, *Guns at Sea* (London: Hugh Evelyn, 1979)

Parker, Geoffrey, *The Military Revolution: Military Innovation & the Rise of the West, 1500–1800* (Cambridge: Cambridge University Press, 1996)

Partington, J. R. A., *A History of Greek fire and Gunpowder* (Cambridge: W. Heffer & sons, 1966)

Pivka, Otto von, *Armies of the Napoleonic Era* (Newton Abbot: David & Charles, 1979)

Ponting, Clive, *Gunpowder* (London: Chatto & Windus, 2005)

Raymond, James, *Henry VIII's Military Revolution: The armies of Sixteenth-Century Britain and Europe* (London: Tauris Academic Studies, 2007)

Robertson, Frederick Leslie, *The Evolution of Naval Armament* (London: Constable & Company Ltd, 1921)

Smith, Robert Douglas and Kelly DeVries, *The Artillery of the Dukes of Burgundy* (Woodbridge: The Boydell Press, 2005)

Smith, Robert Douglas and Kelly DeVries, *Rhodes Besieged: A New History* (Stroud: The History Press, 2011)

Stewart, R. W., *The English Ordnance Office* (Woodbridge: Boydell Press, 1996)

Taylor, F L., *The Art of War in Italy* 1494–1529 (London: Greenhill books, 1993)

Teesdale, Edmund B., *Gunfounding in the Weald in the Sixteenth Century* (Royal Armouries, 1991)

Valleriani, Matteo, *Metallurgy, Ballistics and Epistemic Instruments: The Nova Scientia of Nicolò Tartaglia* (Berlin: Open Access, 2013)

Williamson, J., *Sir John Hawkins* (Oxford: Oxford University Press, 1927)

Williamson, J. (ed.), *The Observations of Sir Richard Hawkins* (London; Argonaut Press, 1933)

Wood, James B., *The King's Army* (Cambridge University Press, 1996)

Articles and Chapters

Ansani, Fabrizio, 'Craftsmen, Artillery, and War Production in Renaissance Florence', *Vulcan*, 4 (2016), pp. 1–26

Ansani, Fabrizio, 'The Life of a Renaissance Gunmaker: Bonaccorso Ghiberti and the Development of Florentine Artillery in the Late Fifteenth Century', *Technology and Culture*, July 2017, Vol. 58, pp. 749–789

Ansani, Fabrizio, '"This French artillery is very good and very effective." Hypotheses on the Diffusion of a New Military Technology in Renaissance Italy', *The Journal of Military History* 83 (April 2019), pp. 347–378

Archbold, W. A. J., 'A Diary of the Expedition of 1544', *The English Historical Review*, Vol. 16, No. 63 (July 1901), pp. 503–507

Atkinson, C. T., 'The Art of War in the Sixteenth Century', *Journal of the Society for Army Historical Research*, Vol. 16, No. 63 (Autumn 1937), pp. 152–157

Awty, Brian G., 'The Breakthrough of the 1540s in the Casting of Iron Ordnance', *Journal of the Ordnance Society*, Vol. 15 (2003), pp. 19–28

Bailey, Sarah Butler, 'Attempts to Control the Export of Cast-Iron Guns and the Market on Tower Hill', *Journal of the Ordnance Society*, Vol. 15 (2003), pp. 53–70

Bailey, Sarah Butler, 'Information Relating to the Operation of the Early Cast-Iron Gun Industry from a Manuscript Account in the Xollection of the Royal Armouries', *Journal of the Ordnance Society*, Vol. 3 (1991), pp. 11–17

Barker, Richard, 'Gun Carriages in a Notebook of Portuguese Field Artillery', *Journal of the Ordnance Society*, Vol. 14 (2002), pp. 19–30

Blackmore, Howard L., 'The take-down gun', *Journal of the Ordnance Society*, Vol. 2 (1990), pp. 1–4

Brinck, Nico, 'Three cannon salvaged from the Meuse in Dinant, Belgium', *Journal of the Ordnance Society*, Vol. 28 (2021), pp. 45–53

Brown, Ruth Rhynas, 'Comment on the Tal-y-bont Gunner's Rule', *Journal of the Ordnance Society*, Vol. 2 (1990), p. 71

Brown, Ruth Rhynas, 'Guns for merchant shipping', *Journal of the Ordnance Society*, Vol. 22 (2010), pp. 23–38

Brown, Ruth Rhynas, 'Troncks, rockets and fiery balls: Military fireworks of the early modern period', *Journal of the Ordnance Society*, Vol. 17 (2005), pp. 25–38

Bruce, J. (ed.), 'Journal of a Voyage into the Mediterranean by Sir Kenelm Digby, A.D. 1628', Camden Society, Vol. 96 (London: Camden Society, 1868)

Cassidy, Ben, 'Machiavelli and the Ideology of the Offensive: Gunpowder Weapons in "The Art of War"', *The Journal of Military History*, Vol. 67, No. 2 (April 2003), pp. 381–404

Cressy, David, 'Saltpetre, State Security and Vexation in Early Modern England', *Past and Present*, no. 212 (August 2011), pp. 73–111

Cruickshank, C. G., 'King Henry VIII's Army, Part Two: Munitions', *History Today*, (January 1969), pp. 40–46

Davies, Jonathan, 'Gun Arrows: "An amusing novelty or intelligent innovation"?', *Journal of the Ordnance Society*, Vol. 27 (2020), pp. 67–68

Davies, Jonathan, 'Henry VIII and his Twelve Apostles', *Journal of the Ordnance Society*, Vol. 27 (2020), pp. 60–66

Dillon, H. A., 'Arms and Armour at Westminster, the Tower and Greenwich, 1547', *Archaeologia or Miscellaneous Tracts relating to Antiquity* (London: 1888), vol. 51

Green, J., 'Gunner's Rule, Further Information on Gunner's Rules or Tally Sticks', *Journal of the Ordnance Society*, Vol. 2 (1990), pp. 25–32

Guérout, Max, 'Sixteenth-Century French Naval Guns', in *Ships and Guns*, ed. Beltrame and Ridella, pp. 124–132

Guilmartin, J. F., 'The Cannon of the *Batavia* and the *Sacramento*: Early Modern Cannon Founding Reconsidered', *IJNA* 11, 1982, no. 2: pp. 133–144.

Hale, J. R., 'Men and Weapons: The Fighting Potential of Sixteenth-Century Venetian galleys', *War in History*, Vol. 5, No. 2 (1998), pp. 309–334

Hall, Nicholas, 'Building and Firing a Replica *Mary Rose* Port-Piece', *Royal Armouries Yearbook*, Vol. 3 (1998), pp. 57–66

Hall, Nicholas, 'Casting and Firing a Mary Rose Culverin', *Royal Armouries Yearbook*, Vol. 6 (2001)

Haskins, Roger A., 'Effect of Shot and Shell upon a Wooden Warship', *Newsletter of the Ordnance Society*, No. 135, October 2021, pp. 21–22

Hildred, Alexzandra, 'A Gunner's Rule from the *Mary Rose*', *Journal of the Ordnance Society*, Vol. 15 (2003), pp. 29–40

Hildred, Alexzandra, and David Starley, 'Technological Examination of a *Mary Rose* Hailshot Piece; New Evidence for Early Iron Gun Casting', *Royal Armouries Yearbook*, Vol. 5 (2002), pp. 139–146

Hocker, Fred, 'Ships, shot and splinters: the effect of 17th-century naval ordnance on ship structure', in J. Litwin (ed.), *Baltic and Beyond: Change and Continuity in Shipbuilding*, Gdańsk 2015. Proceedings of the 14th International Symposium on Boat and Ship Archaeology, pp. 221–228 (Gdansk, 2017)

Hocker, Fred, 'Understanding the Gundeck Experience', pp.125–140. In Johan Rönnby (ed.), *On War on Board: Archaeological and historical perspectives on early modern maritime violence and warfare* (Stockholm Södertörns högskola, 2019)

Hodgkinson, Jeremy S., 'Gunfounding in the Weald', *Journal of the Ordnance Society*, Vol. 9 (1997), pp. 31–48

Hourcade, Juan Luis García, 'From Tartaglia to Lechuga. Artillery Engineers in the Renaissance', in Alicia Cámara Muñoz and Bernardo Revuelta Pol (eds), *Renaissance Engineers*, Fundacion Juanelo Turriano, 2013

Hrnčiřík, Pavel, 'The Emergence of Mobile Light Field Artillery during the Thirty Years' War', *Bohemiae Occidentalis Historica* (Pilsen, 2018)

Johnson, William, 'Ricochet of Non-Spinning Projectiles, Mainly from Water. Part I: Some Historical Contributions', *International Journal of Impact Engineering*, 9, Vol. 21, Nos. 1 2, (1998), pp. 15–24

Kahanov, T., Y. Me-Bar, D. Cvikel, A. Hillman (eds), 'Akko 1 shipwreck: the effect of cannon fire on the wooden hull', *Journal of Archaeological Science*, 39 (2012)

Keller, Alexander. 'Mathematics, Mechanics and the Origins of the Culture of Mechanical Invention', *Minerva*, vol. 23, no. 3, Springer (1985), pp. 348–61.

Konstam, Angus, 'A Gunner's Rule from the "Bronze Bell" Wreck', *Journal of the Ordnance Society*, Vol. 1 (1989), pp. 23–26

Leibnitz, Klaus, 'An Analysis of the Saltpetre Extraction and Purification Recipes in Biringuccio's Pirotechnia', *Journal of the Ordnance Society*, Vol. 16 (2004), pp. 83–92

Lewis, Michael, 'Armada Guns: Section I "Introduction"', *The Mariner's Mirror*, Vol. 28:1 (1942), pp. 41–72.

Lewis, Michael, 'Armada Guns: Section II "The Guns of the Queen's ships", 1569–1599', *The Mariner's Mirror*, Vol. 28:2 (1942), pp. 104–144

Lewis, Michael, 'Armada Guns: Section III "The Queen's Ships in July 1588"', *The Mariner's Mirror*, Vol. 28:3 (1942), pp. 231–245

Lewis, Michael., Armada Guns: Section IV 'The auxiliary English fleet', *The Mariner's Mirror*, Vol. 28:4, (1942), pp. 259–290

Lewis, Michael, 'Armada Guns: Section V "The Guns of the Spanish Fleet, 1588"', *The Mariner's Mirror*, Vol. 29:1 (1943), pp. 3–39

Lewis, Michael, 'Armada Guns: Section VI "Comparisons"', *The Mariner's Mirror*, Vol. 29:2, (1943), pp. 100–121

Lewis, Michael, 'Armada Guns: Section VII "The Lessons of the Armada Fight"', *The Mariner's Mirror*, Vol. 29:3, (1943), pp. 163–178

Lewis, Michael, 'Armada Guns: Appendices', *The Mariner's Mirror*, Vol. 29:4 (1943), pp. 203–231

López-Martin, Javier, 'Did Naval Artillery Really Exist during the Modern Period? A Brief Note on Cannon Design', in *Ships and Guns*, ed. Beltrame and Ridella, pp. 73–84

Martin, Colin, 'Stowed or Mounted: The Spanish Armada of 1588 and the Strategic Logistics of Guns at Sea', in *Ships and Guns*, ed. Beltrame and Ridella, pp. 85–97

McElvogue, D. M., 'A Description and Appraisal of Ordnance from Three Spanish Armada Transports *c.* 1588', *Journal of the Ordnance Society*, Vol. 14 (2002), pp. 31–50

Montojo, Vicente, 'Primitive Ballistics, 1586–1610', *Journal of the Society for Army Historical Research*, Vol. 12, No. 48 ('Winter' 1933–4), pp. 247–253

Morin, Marco, 'Morphology and Constructive Techniques of Venetian Artilleries in the 16th and 17th Centuries: Some Notes', in *Ships and Guns*, ed. Beltrame and Ridella, pp. 1–11

Muller H. G., *A Brief History of Powder Testers*, pp. 206–215 <http://www.reenactor.ru/ARH/PDF/Muller_H._G.pdf>

Murphy, Samuel, 'The Technology of Casting Cannon', *Journal of the Ordnance Society*, Vol. 13 (2001), pp. 73–96

Parker, Geoffrey, 'The *Dreadnought* revolution of Tudor England', *The Mariner's Mirror*, Vol. 82, No 3 (1996), pp. 269–300

Pepper, Simon, 'Castles and Cannons in the Naples campaign 1494–5', in David Abulafia, *The French Descent into Renaissance Italy, 1494–95: Antecedents and Effects* (London: Routledge, 2016)

Pepper, Simon, 'The Face of the Siege: Fortification, Tactics and Strategy in the Early Italian Wars', in Christine Shaw (ed.), *Italy and the European Powers: The Impact of War, 1500–1530* (Leiden: Brill, 2006)

Phillips, Gervase, 'Weapons Technology and Technology Transfer in Early Modern England', *Technology and Culture*, Vol. 40, No. 3 (July 1999), pp. 576–593

Puype, Jan Piet, 'Guns and their Handling at Sea in the 17th Century: A Dutch Point of View', *Journal of the Ordnance Society*, Vol. 2 (1990), pp. 11–23

Puype, Jan Piet, 'Three-Wheeled Gun Carriages: From the Late 16th-Century Shipwreck Scheurrak SO 1 near the Texel', *Royal Armouries Yearbook*, Vol. 5 (2000), pp. 106–116

Ridella, Renato Gianni and Antonina Palazzolo, 'A bronze falcon from Federico Musarra's workshop found off Genoa in 1959 and a brief survey of ordnance production in Sicily through the 16th century', *Journal of the Ordnance Society,* Vol. 21 (2009) pp. 5–22

Ridella, Renato G., 'Dorino II Gioardi: 'A 16th-century Genoese Gunfounder', *Journal of the Ordnance Society,* Vol. 16 (2004), pp. 27–42

Ridella, Renato G., 'Examining two bronze cannon recovered from the wreck 54 of the frigate Neustra Señora de la Mercedes', *Journal of the Ordnance Society*, Vol. 28 (2021), pp. 54–79

Ridella, Renato Gianni, 'Genoese Ordnance Aboard Galleys and Merchantmen in the 16th Century', in *Ships and Guns*, ed. Beltrame and Ridella, pp. 39–56

Ridella, Renato Gianni, 'Two 16th century Papal esmerils in the Cleveland Museum of Art, Ohio, and some notes on bronze pieces of ordnance with a polygonal section', *Journal of the Ordnance Society*, Vol. 19 (2017), pp. 5–38

Ritchie, Tessy S., Kathleen E. Riegner, Robert J. Seals, Clifford J. Rogers, and Dawn E. Riegner, *Evolution of Medieval Gunpowder: Thermodynamic and Combustion Analysis* (ACS Omega), <https://pubs.acs.org/doi/pdf/10.1021/acsomega.1c03380>

Rodger, N A M., 'The Development of Broadside Gunnery, 1450–1650', *The Mariner's Mirror*, Vol. 82, No. 3 (1996), pp. 301–324

Roland, Alex, 'Science, Technology, and War', *Technology and Culture*, Vol. 36, No. 2 (April 1995), pp. S83–S100

Rose, Paul-Lawrence, 'Galileo's Theory of Ballistics', *The British Journal for the History of Science*, Vol. 4, No. 2 (Dec. 1968), pp. 156–159

Roth, Rudi, 'The Renaissance Man (15/16th Century Master gunners)', *Journal of the Ordnance Society*, Vol. 6 (1994), pp. 1–8

Scaglia, Gustina, 'A Miscellany of Bronze Works and Texts in the "Zibaldone" of Buonaccorso Ghiberti', *Proceedings of the American Philosophical Society*, vol. 120, no. 6, 1976, pp. 485–513. *JSTOR*, < https://www.jstor.org/stable/986601>, accessed 22 July 2021.

Smith, Geoff, 'Medieval Gunpowder Chemistry: A Commentary on the Firework Book', *Icon*, Vol. 21 (2015), pp. 147–166

Smith, Geoff, 'Saltpetre in Medieval Gunpowder; Calcium or Potassium', *Journal of the Arms and Armour Society*, Vol. XIX, No.3, <https://www.medievalists.net/wp-content/uploads/2015/02/Calcium-gunpowder.pdf>, pp.1–4

Smith, Geoff, 'Serpentine Powder and the Guns that Used it', *Journal of the Ordnance Society*, Vol. 26 (2019), pp. 27–39

Smith, Geoff, 'Sulphur: the Trigger of Gunpowder', *Journal of the Ordnance Society*, Vol. 28 (2021), pp. 115–119

Smith, Robert D., 'All Manner of Peeces: Artillery in the Late Medieval Period', *Royal Armouries Yearbook*, Vol. 7 (2002), pp. 130–138

Smith, Robert D., 'A New Forme of Greate Artillerie: An early 17th Century Invention', *Journal of the Ordnance Society*, Vol. 15 (2003), pp. 89–92

Smith, Robert D., 'Casting Shot in the 16th Century', *Journal of the Ordnance Society*, Vol. 12 (2000), pp. 88–92

Smith, Robert D., 'Ships and Guns of the Tudor Navy 1495–1603', in *Ships and Guns*, ed. Beltrame and Ridella, pp. 109–112

Smith, Robert D., 'The Reconstruction and Firing Trials of a Replica of a 14th Century Cannon', *Royal Armouries Yearbook*, Vol. 4 (1999), pp. 86–94

Smith, Robert D., 'The technology of wrought–iron artillery', *Royal Armouries Yearbook*, Vol. 5 (2000), pp. 68–79

Stevenson, D., and D. Caldwell, 'Leather guns and other light artillery in mid-17th century Scotland', *Proceedings of the Society of Antiquaries of Scotland*, 108 (1977), pp. 300–317. Retrieved from <http://journals.socantscot.org/index.php/psas/article/view/8975>

Strzyž, Piotr, 'Characteristics of medieval Artillery in the light of written sources from Bohemia and Poland' (Fasciculi Archaeologiae Historicae Fasc.XV), *Ze starých latopisů českých*, ed. J. Porák, J. Kašpar, (Praha, 1980)

Thompson, I. A. A., 'Spanish Armada Guns', *The Mariner's Mirror*, Vol. 61 (1975), pp. 355–371

Tomkinson J., 'Gunpowder and the Introduction of Industrial Charcoal', *Journal of the Ordnance Society*, Vol. 28 (2021), pp. 97–114

Towes, R. M. and P. McCree, 'A Note on Drakes', *Journal of the Ordnance Society*, Vol. 6 (1994), pp. 39–46

Trollope, Charles J. N., 'The design and evolution of English cast-iron guns from 1543 to 1660', *Journal of the Ordnance Society*, Vol. 14 (2002), pp. 51–64

Trollope, Charles J. N., 'The Guns of the Queen's Ships during the Spanish Armada', *Journal of the Ordnance Society*, 2002, Vol. 6 (1994), pp. 22–38

Walton, Steven, 'Perceptions of the Performance of Cannon Shot Before 1700', *Journal of the Ordnance Society*, Vol. 20 (2008), pp. 69–80

Walton, Stephen, 'The Bishopsgate Artillery Garden and the First English Ordnance School', *Journal of the Ordnance Society*, Vol. 15 (2003), pp. 41–51

Waterhouse, Robert, 'Tudor Bronze Guns of the Channel Islands' (2015), pp. 82–109 <http://www.christopherlong.co.uk/aanna/papers/WaterhouseR/TudorBronzeGunsOfTheChannelIslands.pdf>

Webb, Henry J., 'Elizabethan Field Artillery', *Military Affairs*, Vol. 19, No. 4 (1955), pp. 197–202

Webb, Henry J., 'The Science of Gunnery in Elizabethan England', *Isis*, The History of Science Society, Vol. 45, No. 1 (May 1954), pp. 10–21

Pictorial Sources

Diebold Schilling, Amtliche Berner Chronik, vol. 3, 1478–1483 <https://www.e-codices.ch/en/searchresult/list/one/bbb/Mss-hh-I0003>

Philip Monch, *Kriegsbuch* (Heidelberg, 1496), Universitätsbibliothek Heidelberg, Cod. Pal. germ. 126-0105, <https://www.uni-heidelberg.de/en>

Feuerwerks- und Büchsenmeisterbuch. Rezeptsammlung – BSB Cgm 734, Image 1 of 472 | MDZ, <https://digitale-sammlungen.de/en/>

The 'Artilleriebuch' (BSB Cgm 909) by Walther Litzelmann <https://bildsuche.digitale-sammlungen.de/index.html?c=viewer&bandnummer=bsb00052357&pimage=293&v=100&nav=&l=en>

Zeugbuch Kaiser Maximilians I – BSB Cod.icon. 222, Innsbruck, um 1502 [BSB-Hss Cod.icon. 222] Digital Library – Munich Digitization Centre, <https://digitale-sammlungen.de/en/>

Website

Ulrich Bretscher's Black Powder Page <https://web.archive.org/web/20110520014651/http://musketeer.ch/blackpowder/bp_menu.html>

The Ordnance Society

The Ordnance Society was formed in June 1986 to promote, encourage and coordinate the study of all aspects of the history of ordnance and artillery. It is today an international society with members from more than 20 countries and all walks of life. Its primary aim is to provide a means of communication for individuals and organisations interested in any aspect of ordnance and artillery of all periods and to enable publication of their interests and researches.

The Society makes available a wide range of ordnance downloads, organises events, runs surveys of various ordnance items, and produces publications to further these aims:

- The annual journal of the Ordnance Society, produced to a high standard, includes well-illustrated articles and plans of historic cannons and their carriages, artillery, ammunition and explosives.
- The quarterly newsletter includes short articles, information, illustrations, queries and their replies, and news.
- Regular 'Members Day' events when individual members have the opportunity to present or display their research, interests or collections. These are often held jointly with various museums and ordnance related organisations.
- Visits to various military establishments, museums and other places of interest.

<https://ordnancesociety.org.uk/>
ordnance.society@btinternet.com

Linstock and Pledget

Linstock and Pledget is a small but dedicated re-enactment group representing a gun company in the Tudor period. The falconet referred to and illustrated in this book is theirs. If you would be interested in joining the group or hiring us for events, please contact us through our Facebook page.